INTRODUCTION TO ENGINEERING DESIGN

BOOK 4

HUMAN POWERED PUMPING SYSTEMS

2nd Edition

James W. Dally
University of Maryland, College Park

College House Enterprises, LLC
Knoxville, Tennessee

The manuscript was prepared using Word 2000 with 11 point Times New Roman font. Publishing and Printing Inc., Knoxville, TN printed the book from camera-ready copy.

College House Enterprises, LLC.
5713 Glen Cove Drive
Knoxville, TN 37919, U. S. A.
Phone   (865) 558 6111
FAX   (865) 584 1766
email   jdally@www.collegehousebooks.com
http://www.collegehousebooks.com

ISBN 0-9700675-0-X

ABOUT THE AUTHOR

James W. Dally obtained a Bachelor of Science and a Master of Science degree, both in Mechanical Engineering from the Carnegie Institute of Technology. He obtained a Doctoral degree in mechanics from the Illinois Institute of Technology. He has taught at Cornell University, Illinois Institute of Technology, and served as Dean of Engineering at the University of Rhode Island. He is currently a Glenn L. Martin Professor of Engineering at the University of Maryland, College Park.

Professor Dally has also held positions at the Mesta Machine Co., IIT Research Institute, and IBM, Federal Systems Division. He is a fellow of the American Society for Mechanical Engineers, Society for Experimental Mechanics, and the American Academy of Mechanics. He was appointed as an honorary member of the Society for Experimental Mechanics in 1983 and elected to the National Academy of Engineering in 1984. Professor Dally was selected by his peers in the College of Engineering to receive the Senior Faculty Outstanding Teaching Award. He was also a member of the University of Maryland team receiving the 1996 Outstanding Educator Award from the Boeing Co.

Dr. Dally has co-authored several textbooks books including: *Experimental Stress Analysis, Photoelastic Coatings, Instrumentation for Engineering Measurements, Packaging of Electronic Systems, Production Engineering and Manufacturing, Design, Analysis of Structural Elements, and Introduction to Engineering Design, Books 1, 2, 3, 4 and 5.* He has written over two hundred scientific papers and holds five patents.

THE CODE:

INSTRUCTOR AND STUDENT EXPECTATIONS

Learning and teaching require trust and mutual understanding between the students and the instructor. Trust and understanding lead to an enhanced learning environment. If we all recall a listing of expectations, a better relation can be established between the instructor and his or her students that improves the learning process. While many of these expectations may be self evident, we believe the list will remind everyone of his or her obligations for learning, increased understanding, communication, and respect. The students and instructors in the Introduction to Engineering Design course at the University of Maryland have developed the list shown below.

Students expect the instructor to:

- Respect all students.
- Be fair in grading.
- Provide leadership.
- Be committed to teaching and advising.
- Provide encouragement rather than discouragement.
- Clearly define course requirements and the grading algorithms.
- Balance course workload with credit hours.
- Schedule office hours and be available to help students.
- Provided candid and timely feedback on assignments.
- Arrive before the scheduled class time and prepare the classroom.

Instructors expect the student to:

- Show respect to everyone involved in the program.
- Be responsible for your own progress and learning.
- Be dedicated to understanding and learning.
- Stay current with materials and issues covered in class.
- Be a positive and creative team member.
- Be inquisitive and compete within the framework of a design team.
- Be interested in engineering and product design.
- Attend class or notify the instructor in advance if you intend to be absent.
- Arrive on time for class with a positive attitude.

PREFACE

This book is the second edition of the fourth book in a series of five dealing with Introduction to Engineering Design. Jim Dally prepares a new textbook each academic year for the first year engineering students of the University of Maryland at College Park. We are encouraged that other Colleges of Engineering are also adopting one or more of these books for their design classes.

PROCESS

The procedure we follow in offering a design experience to first year students is to:

- Teach the class in relatively small sections.
- Divide the class into product development teams with five or six members per team.
- Assign a major project entailing the development of a prototype that will require the entire semester.
- The student teams all develop the same product.
- In the product realization process, the students:
  - Design
  - Manufacture or procure parts for the prototype.
  - Assemble the prototype.
  - Test and evaluate the prototype.
- In developing the prototype, we have the opportunity to emphasize:
  - Communication skills.
  - Team building skills.
  - Engineering graphics.
  - Software applications including CAD.
  - Design methods and procedures.

The textbook is used to support the students during a semester long project. Some of the material may be covered in class or in a computer laboratory. Other material is covered with reading assignments. In other instances, the students use the text as a reference document for independent study. Exercises are provided at the end of each chapter that may be used for assignments when the demands of the project on the students' time are not excessive.

The book is organized into six parts to present various topics that the students need as they proceed through a significant portion of product realization process. We introduce the product realization process by considering the design of wind-powered water pump. By assigning a demanding overarching project, we employ a holistic approach that avoids compartmentalization of knowledge. Design of a product is treated by the instructor as an opportunity to integrate a spectrum of knowledge about many topics.

We have found that students serving on a design team developing a product are motivated. They learn much more on their own initiative than we could ever teach them in a more typical lecture course. The benefit of hands-on participation in the design, building and testing a product that they own significantly enhances the learning process.

CONTENT

The book is organized into six parts to present various topics that the students need as they proceed through a significant portion of product realization process. We introduce the product realization process by considering the design of a human powered water pump. By assigning a demanding overarching project, we employ a holistic approach that avoids compartmentalization of knowledge. Design of a product is treated by the instructor as an opportunity to integrate a spectrum of knowledge about many topics.

We have found that developing a product motivates the students. They learn much more on their own initiative than we could ever teach them in a more typical lecture course. The motivational benefit of hands-on participation in the design of their own product and building and testing the prototype significantly speeds the learning process.

Part I of the textbook covers material on the human powered water pump. We briefly introduce the product development process and the students' role in a development team producing a product in Chapter 1. The product is described in Chapter 2 together with some history about early methods of drawing water from streams and wells. Several concepts for the design of pumps are introduced briefly, and the idea of system design is described. In Chapter 3, work and human power as related to pumps are discussed. Mechanisms and components employed in drives for the pumping systems are described. Much more detail on the design of piston, diaphragm and centrifugal pumps is given in Chapter 4. The material presented in Chapter 5 is to support the students' efforts to measure flow rates generated by the pumping system. Several different methods for measuring flow velocity are described. We recognize that some of the analysis presented in Chapters 3, 4 and 5 may be beyond the scope of the ordinary first year student. However, we believe that many students will understand and lead others on their team to learn about design analysis and recognize its importance in the design process.

Part II presents three chapters on engineering graphics. We begin in Chapter 6 with a relatively complete treatment of three-view drawings. Pictorial drawings including isometric, oblique, and perspective are covered in Chapter 7. A treatment of Tables and Graphs is given in Chapter 8. The emphasis in Part II is on manual preparation of the graphics. CAD preparation of drawings and spreadsheet generation of graphs is covered in a separate textbook.

Part III describes two different software programs useful to the first year engineering student. A spreadsheet program, Microsoft Excel, is described in Chapter 9. The coverage is intended to provide the students with entry-level skills. We expect students to be able to perform calculations, to plot curves and prepare several types of charts using spreadsheets. In Chapter 10, we provide a brief description of a computer graphics presentation program—Microsoft PowerPoint. Our experience is that the student can learn to master this program in a few hours if they make use of the chart wizard incorporated in the program. The coverage of the capabilities of PowerPoint is related to preparing design briefings on a human powered pump.

Part IV is related to the product development process. We describe development teams in considerable detail. The idea of working on teams is new to most students who have been educated in the public school system in the U. S. Many of these students have been trained in K - 12 to act as individuals and to avoid cooperation in the learning process. In Chapter 11, we introduce a number of useful topics such as team member traits, positive and negative team behavior, and effective team meeting. Experience with several thousand students has shown that many of them initially have trouble adapting to the team concept. (Males appear to have more trouble than females.) However, over the semester they slowly learn how to work as effective team members and they appreciate the opportunity to do so. The social bonding that takes place on the team with the first semester students is interesting to observe. Chapter 12 deals with the product development process. We have tried to incorporate a very wide range of material in this chapter. For this reason, the coverage often is brief, and we refer the student to more thorough higher-level books on the topic. However, we cover that part of the product development process, which is important to the assigned project. We start with the product specification, cover material on design concept generation, and then concept selection. Methods are introduced that students may effectively employ to generate design concepts and to select the best concept though the use of systematic design trade-off analyses.

Part V treats the very important topic of communications. A chapter on technical reports describes many aspects of technical writing. The most important lesson here is that a technical report is different than a paper for the History or English Departments. An effective professional report is written for a predefined audience with specific objectives. We describe the technical writing process and give many suggestions to facilitate composing, revision, editing and proofreading. The final chapter in the book covers design briefings. We draw a distinction between speeches, presentations and group discussions. Then we focus on the technical presentation and indicate the importance of preparing excellent visual aids. We make extensive use of PowerPoint in illustrating visual aids to be employed in a design briefing. Finally, we include a discussion of the delivery of the presentation and the need for extensive rehearsing.

Part VI contains four chapters dealing with engineering and society. Chapter 15 provides a historical perspective on the role engineering played in developing civilization and on improving the lives of the masses. In this chapter, we move from the past into the present and indicate the current relationship between business, consumers and society. Safety, risk and performance are covered in Chapter 16. Here we discuss failure, and its implications on the safety and well being of those using our products. Theoretical methods are introduced to determine both component and system reliability. In Chapter 17, we discuss ethics, character and engineering. A large number of topics are covered so the instructor can select from among them. A description of the Challenger accident is also covered because it is an excellent case history covering safety related conflicts between management and engineers. Finally, in Chapter 18 we introduce sustainable engineering. This is a relatively new subject that is currently receiving attention in engineering colleges. We have described the problem of achieving a balance between economic growth and maintaining the quality of the environment. We have also described the important role technology and engineering will play as we design and produce products with manufacturability, remanufacturability and recycling as design objectives.

ABET AND EDUCATIONAL OUTCOMES

In the next few years, it is imperative that we think very seriously about ABET's new criteria for program accreditation. The process for Criteria 2000 is very different from the previous process for accreditation. ABET formerly accredited our engineering programs based on input and with many pages of instructions provided to guide in the assessment of this input. However, the new criteria are based primarily on educational outcomes. Very little information given in Criteria 2000 pertains to program input. The guide provided by ABET for assessment of educational outcomes is very terse. It appears that it will be a faculty responsibility to specify the intended educational outcomes for each course and each program, to define the metrics used to measure these outcomes, and to plan and place into effect methods of measurement of these metrics over extended periods of time.

We have considered Criteria 2000 in writing this book, and believe a first course in engineering design should have the expected educational outcomes listed below:

Communication Skills

1. Engineering Graphics
   - Understand the role of graphics in engineering design.
   - Understand orthographic projection in producing multi-view drawings.
   - Understand three-dimensional representation with pictorial drawings.
   - Understand dimensioning and section views.
   - Demonstrate capability of preparing drawings using both manual and computer methods.
   - Demonstrate understanding of engineering graphics by incorporating appropriate high-quality drawings in the design documentation.
2. Design Briefings:
   - Within a team format, present a design review for the class using appropriate visual aids.
   - Each team member demonstrates briefing skills.
3. Design Reports:
   - The team's design is documented in a professional style report incorporating time schedules, costs, parts list, drawings and an analysis.

Team Experience
- Develop an awareness of the challenges arising in teamwork.
- Demonstrate teamwork in the product realization process through a systematic design concept selection process involving participation of all team members.
- Demonstrate planning from conceptualization to the evaluation of the prototype.
- Understand and demonstrate sharing responsibility among team members.
- Demonstrate teamwork in preparing design reports and presenting design briefings.

Software Applications
- Demonstrate entry-level skills in using spreadsheets for calculations and data analysis.
- Show a capability to prepare graphs and charts with a spreadsheet.
- Show a capability to prepare professional quality visual aids.
- Understand entry-level skills in a feature-based solid modeling program.
- Demonstrate these computer skills in preparing appropriate materials for design briefings and design reports.

Design Project:
- The design project is the overarching theme of the course.
- Utilize all the skills listed above to assist in the product development process.
- Demonstrate competence in defining design objectives.
- Generate design concepts that meet the design objectives.
- Understand the basis for design for manufacturing, assembly and maintenance.
- Manage the team and the project effectively.

ACKNOWLEDGEMENTS

Acknowledgments are always necessary in preparing a textbook because so many people and organizations are involved. First, we thank the National Science Foundation for their support. Their funding was important, but more critical than money was our need for credibility. Without the Engineering Education Coalition Program, the need for curriculum reform would not have received adequate attention from the college administrators and our colleagues. The NSF basically called for a reform of the curriculum, and with generous funding gave it the required status.

Second, we need to recognize the contributions of many of our colleagues at the University of Maryland. To date over 60 different faculty members have taught this course on at least one occasion. They have all contributed to improving this textbook. In particular, I would like to thank, Dr. Guangming Zhang and Dr. Gary Pertmer for their continued commitment in providing the leadership necessary to operate a multi section offering and in providing support for the many instructors involved. We want to thank Ms. Jane Fines who continues to maintained contact with the students every semester. She provides us with very valuable feedback as we continue to modify the course. Thanks are also due to Ms. Janet Yowell from the University of Colorado, Dr. James Key, Dr. Sheryl Ehrman, and Dr. Kaye Brubaker from the University of Maryland for helpful editorial comments on several chapters that they reviewed. Thanks are also due to Mr. James Kramer and Ms. Lisa Wilson for their assistance with the material on sustainable engineering. Finally, the author appreciates the careful proofreading of Ms. Rosie Crowe.

As always, thanks are due to the administrators who encouraged and supported the development of this course. Dr. Thomas M. Regan, Associate Dean for Undergraduate Studies, lead the effort to scale up the course from pilot sections to an organization capable of instructing nearly 600 students each fall semester. Dr. George Dieter and Dr. William Destler, former Deans of Engineering, authorized the small class size essential for effectively teaching this course and committed to significant long-term expenditures necessary to support the faculty involved. They also publicly supported the efforts of the small group of instructors during the early years as we were institutionalizing the course. In fact, Dr. William Destler, first as a Department Chair and even as Dean, took time from his busy schedule to teach a section of this course.

Jim Dally
Knoxville, TN

ABOUT THE AUTHORS
CODE
PREFACE

CONTENTS

PART I HUMAN POWERED WATER PUMPS

CHAPTER 1 ENGINEERING DESIGN AND PRODUCT DEVELOPMENT

CHAPTER 2 HUMAN POWERED WATER PUMPS

CHAPTER 3 WORK, HUMAN POWER AND PUMPS

CHAPTER 4 RECIPROCATING AND ROTARY PUMPS

CHAPTER 5 FLOW MEASUREMENT

PART II ENGINEERING GRAPHICS

CHAPTER 6 THREE-VIEW DRAWINGS

CHAPTER 7 PICTORIAL DRAWING

CHAPTER 8 TABLES AND GRAPHS

PART III SOFTWARE APPLICATIONS

CHAPTER 9 MICROSOFT EXCEL–2000

CHAPTER 10 MICROSOFT PowerPoint 2000

PART IV PRODUCT DEVELOPMENT PROCESSES

CHAPTER 11 DEVELOPMENT TEAMS

CHAPTER 12 A PRODUCT DEVELOPMENT PROCESS

PART V COMMUNICATIONS

CHAPTER 13 TECHNICAL REPORTS

PART VI ENGINEERING AND SOCIETY

CHAPTER 15 ENGINEERING AND SOCIETY

CHAPTER 16 SAFETY, RISK AND PERFORMANCE

CHAPTER 17 ETHICS, CHARACTER AND ENGINEERING

CHAPTER 18 SUSTAINABLE ENGINEERING

PART I

HUMAN POWERED PUMPING SYSTEMS

CHAPTER 1

ENGINEERING DESIGN AND PRODUCT DEVELOPMENT

1.1 THE IMPORTANCE OF PRODUCT DEVELOPMENT

We use hundreds of products every day. Products surround us. This morning I prepared breakfast by toasting bread, frying eggs, and making coffee. How many products did I use in this simple task? The toaster (Black & Decker), the frying pan (T-FAL), the coffee maker (Krups) and the gas cooktop (Kitchen Aid) are all products that are designed, manufactured and sold to customers both in the U. S. and abroad. Some products are relatively simple, such as the frying pan with only a few parts, yet some are much more complex, such as your automobile with several thousand parts.

Corporations worldwide continuously develop their product lines with minor improvements introduced every year or two with more major improvements every four or five years. Product development involves many engineering disciplines and is a major responsibility of engineers entering the work place, as illustrated in Table 1.1. An examination of Table 1.1 shows that designing, manufacturing, and product sales represent more than 80% of the responsibilities of mechanical engineers in their initial position in industry. This distribution of activities is also typical of many of the other engineering disciplines.

Table 1.1
Responsibilities of Mechanical Engineers in Their First Position [1]

ASSIGNMENT	PERCENT of TIME
Design Engineering	40
Product Design	24
Systems Design	9
Equipment Design	7
Plant Engineering / Operations / Maintenance	13
Quality Control / Reliability / Standards	12
Production Engineering	12
Sales Engineering	5
Management	4
Engineering	3
Corporate	1
Computer Applications / Systems Analysis	4
Basic Research and Development	3
Other Activities	7

The financial well being of many corporations depends on the introduction of a steady stream of successful products to the marketplace. A successful product must satisfy the customer by providing robust and reliable service at a competitive price that completely meets the customer needs.

1.1.1 Development Teams

Interdisciplinary teams develop products with members from engineering, marketing, manufacturing, production, purchasing, etc. A typical organization chart for a team developing a relatively simple electro-mechanical component is shown in Fig. 1.1. The team leader coordinates and directs the activities of the designers (mechanical, electronic, industrial, etc.), and a marketing specialist, to ensure that the evolving product meets the needs of the customer, supports the team. Financial, sales and legal assistance is usually provided on an as-needed basis by corporate staff or outside contractors.

The team leader is also supported by a purchasing agent who establishes close relations with the suppliers who provide materials and component parts used in the final assembly of the product. In the past decade, the relationship between suppliers and end-product producers has changed significantly. Suppliers become part of the development team; they often provide a significant level of engineering support that is external to the internal (core) development team. In some developments, these external (supplier funded) development teams—when totaled—are larger than the internal development team. For example, the internal development team for the Boeing 777 airliner involved 6,800 employees, and the external teams were estimated to include about 10,000 employees [2].

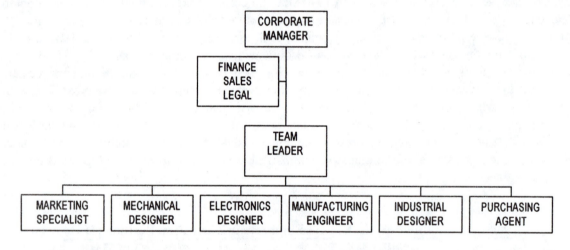

Fig. 1.1 Organization of a product development team for a relatively simple product.

1.2 DEVELOPING WINNING PRODUCTS

There are two primary aspects involved in developing successful products that win in the competitive marketplace. The first is the quality, performance and price of the product. The second is the time and the cost of the development, and also the manufacturing cost of the product in production.

1.2.1 Quality, Performance and Price

Let's discuss the product first. Is it attractive and easy to use? Is it durable and reliable? Is it effective and efficient? Does it meet the needs of the customer? Is it better than the products now available in the marketplace? If the answer to all of these questions is an unqualified **YES**, then the customer may want to buy the product if the price is right. Next, you need to understand what is implied by product cost and its relation to the price actually paid for the product. Cost and price are distinctly different quantities. Product cost clearly includes the cost of materials, components, manufacturing and assembly. The accountants also include less obvious costs such as the prorated costs of capital equipment (the plant and its machinery), the price of the tooling, the development cost, and even the expense of maintaining the inventory in establishing the total cost of producing the product.

Price is the amount of money that a customer pays to buy the product. The difference between the price and the **cost** to produce a product is profit, which is usually expressed on a per unit basis.

$$\text{Profit} = \text{Product Price} - \text{Product Cost}$$

This equation is the most important relation in engineering or in any business. If a corporation cannot make a profit, it soon is forced into bankruptcy, its employees lose their positions, and the stockholders lose their investment. It is this profit that everyone employed by a corporation seeks to maximize while maintaining the strength and vitality of the product lines. The same statement can be made for a business that provides services instead of products. If a business is to make a profit and prosper, the price paid by the customer for a specified service must be more than the cost to provide that service.

1.2.2 The Role of Time in the Development Process

Let's now discuss the role of the development process in producing a line of winning products. Developing a product involves many people with expertise in different disciplines. Also, it takes time, and it costs a lot of money. Let's first consider development time. Development time, as it is used in this context, is time to market, i.e., the time from the product development kickoff to the introduction of the product to the market. This is a very important target for a development team because of the many significant benefits that follow from being first to market. A corporation realizes many competitive advantages with a fast development capability. First, the product's market life is extended. For each month cut from the development schedule, a month is added to the life of the product in the marketplace with an additional month of sales revenue and profit. We show the benefits of being first to market on sales revenue in Fig. 1.2. The shaded area between the two curves to the left side of the graph is the enhanced revenue due to the longer sales.

A second benefit of early product release is increased market share. The first product to market has 100% of market share in the absence of a competing product. For products with periodic development of "new models," it is generally recognized that the earlier a product is introduced to compete with older models—without sacrificing quality and reliability—the better chance it has for acquiring and retaining a large share of the market. The effect of gaining a larger market share on sales revenue is also illustrated in Fig. 1.2. The shaded area between the two curves at the top of the graph shows the enhanced sales revenue due to increased market share.

A third advantage of a short development cycle is higher profit margins. If a new product is introduced prior to availability of competitive products, the corporation is able to command a higher price for the product, which enhances the profit. With time, competitive products will be introduced forcing price reductions. However, in many instances, relatively large profit margins can still be maintained because the company that is first to market has added time to reduce their manufacturing costs. They learn better methods for producing components and reduce the time needed to assemble the product. The advantage of being first to market, with a product where a manufacturing learning curve exists, is shown graphically in Fig. 1.3. The manufacturing learning curve reflects the reduced cost of manufacturing and assembling a product with time in production. These cost reductions are due to many innovations introduced by the workers after mass production begins. With time and manufacturing experience, it is possible to drive down production costs.

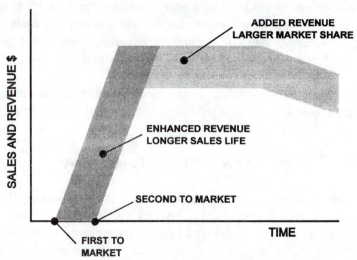

Fig. 1.2 Increased sales revenue due to extended market life and larger market share.

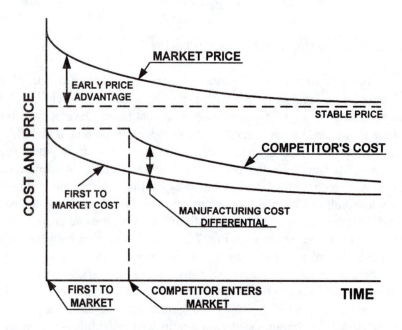

Fig. 1.3 The development team bringing the product to market first enjoys a higher initial price and subsequent cost advantages from manufacturing efficiencies.

Let's next consider development costs, which represent a very important investment for the companies involved. Development costs include the salaries of the members of the development team, money paid to subcontractors, costs of pre-production tooling, expense of supplies and materials, etc. These development costs can be significant, and most companies must limit the number of developments in which they invest. The size of the investment may be appreciated by noting that the development cost of a new automobile is estimated [2] at $1 billion, with an additional expenditure of $500 to $700 million for the new tooling required for high-volume production.

We have included this discussion on time and cost of product development to help you begin to appreciate some of the business aspects of developing winning products. Any company involved in the sale of products depends completely on their ability to continuously introduce winning products in the marketplace in a timely manner. To win, the development team must bring a quality product to market that meets the needs of the customer. The development costs must be minimized while maintaining a product development schedule that permits an early (preferably first-to- market) introduction of the product.

1.3 LEARNING ABOUT PRODUCT DEVELOPMENT

The best way to learn about product design is to work on a development team and develop a prototype. For this reason, we have selected a human powered water pump for your first development effort. The degree of complexity of the water pump will be defined in detail in Chapter 2. You are to work as a member in a team of students to develop a prototype of a human powered water pump. A prototype is the first working model of a product. In some instances, companies develop three or four prototypes before finalizing a particular product design. However, time available during this semester will limit each team to a single prototype of the pumping system.

We understand that the development of a pump system is a tremendous task particularly when the assignment is given so early in your engineering program. However, our experience with several thousand students beginning to study engineering indicates that you will gain immensely from your efforts. We know from previous classes that you are all very creative; you are cooperative; you work well within a team structure and you design and build successful prototypes. Of course, the product selected for your initial development project cannot be overly complex. Since beginning this course in 1990 at the University of Maryland at College Park, the students participating in this introductory course have developed:

- Playground equipment
- Windmills for the generation of electricity
- Furniture
- Human-powered water pumps
- Wind-powered vehicles
- Digital and analog bathroom scales
- A digital postal scale
- A solar-cooking unit
- A solar-desalination unit.

Extensive student surveys clearly indicate that you will spend many hours each week on the project, have fun in the process, learn a lot about engineering and appreciate the lessons learned in working as a member of a development team.

1.4 PROTOTYPE DEVELOPMENT

We often divide the prototype development process into three major phases that include:

1. Designing the prototype.
2. Preparing the assembly kit.
3. Assembling and evaluating the prototype.

We will begin the process by providing you with a typical description of methods to pump water, maximize the human power to the drive system, and to measure flow rate and/or the quantity of water delivered. Also, you will be provided with design requirements pertaining to the input:

- The availability of the water supply.
- The height to which the water is to be pumped.
- The details of the flow channel into which the water is delivered.

The description will also include the time allotted to your team for assembly of the system at the test site, the time available for pumping, and the time for disassembly and clearing the area.

We suggest that you make full use of the library to search for literature and patent files. There is absolutely nothing to be gained by reinventing the wheel (or the pump, drive and flow measuring system in this instance).

After you understand in general terms the design requirements for the human powered pumping system, we will introduce the product specification [3]. The product specification, written relatively early in the development process, is usually prepared in tabular format. It tersely states design requirements and design limits or constraints. Remember we will always design with constraints and/or limits. The specification also includes all of the performance criteria and provides design targets on weight, size, power, cost, etc.

We will place explicit design constraints, which limit your flexibility in designing the prototype. First, there is a maximum limit of $25.00 per team member for the cost of materials, supplies, and components purchased for the human powered pumping system and the displays. The idea is to keep your out-of-pocket expenses at an affordable level, while giving the team adequate funds to build a good prototype.

1.5 DESIGN CONCEPTS

After you and each member of your development team understand the design specification and the design limitations, you are ready to generate design concepts. (A much more complete description of the design requirements, constraints, etc. is given in Chapter 2 together with some initial design ideas for you to consider). Design concepts are simply ideas for performing each function involved in the operation of the pumping system, which will help you meet the product specifications. To illustrate design concepts, let's examine the functions (processes) involved in pumping water and measuring the flow rate, as depicted in the block diagrams presented in Fig. 1.4.

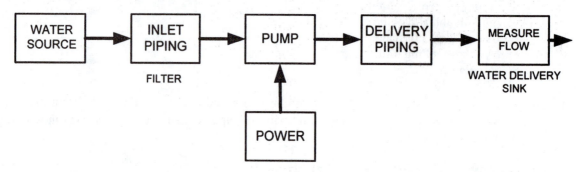

Fig. 1.4 Functions involved in pumping water and measuring the flow rate of the output.

1.5.1 The Design Concept—Design Function Relationship

In a typical pumping system, we have a source of water such as a stream or an underground source accessed by drilling a well. In this case, we will pump the water from a stream located on campus. Unless we want to work in the stream while operating the pump, it will be necessary to provide piping leading from the stream to the inlet of the pump.

No problem—a length of plastic pipe or rubber hose would serve to convey the water from the stream to the pump. Either is available at local retail outlets[1] in several diameters at reasonable cost.

---

[1] We suggest retail outlets such as Home Depot, Lowe's, Hechinger, or local hardware stores for procuring most of the components required for developing the prototype pumping system.

Of course, it will be necessary to select the most suitable diameter for the flow rate that you intend to develop with the pumping system.

Before turning our attention to the next function in the system depicted in Fig. 1.4, let's consider the quality of the water in the stream. Certainly, it is not pure enough to drink, but that is not a problem because the system is intended for irrigation. However, is the water clean enough to pump and to measure the flow rate? If the water in the stream contains debris, such as grass, twigs, mud, etc, the inlet conduit or the pump may foul and jam. If you manage to get the debris through the pump, it may foul the apparatus you have designed to measure the flow rate. Should we incorporate a filter on the end of the inlet pipe to screen out the debris that might interfere with either the pump or the flow measuring apparatus?

When we have a suitable inlet that is capable of providing a supply of clean water, we consider the pump. But what type of pump should we select? Our library search showed there were many different types of pumps. A search on the web site www.thomasregister.com indicated over 100,000 sources for pumps in the U. S. We appear to be trapped in a common modern dilemma— information overload. What type of pump should our team design and build?

Several different types of pumps are described in Chapter 2 together with some recommendations for your team to consider. The most important aspect is to keep the design of the pump simple. Simplicity is essential for three reasons:

1. The equipment available to your team is very limited; therefore pumps with complex designs will be very difficult to fabricate.
2. Leakage through the seals of pumps with complex design cause major problems in their operation.
3. The pump is intended for use in underdeveloped countries where facilities for repair are limited. Clearly, the availability of spare parts will also be limited.

In selecting the type of pump, we compare the qualities of each type of pump with a set of criteria for our pumping system. Indeed, we have already established a criterion—simplicity of the design. What are other criteria that your team will use in selecting the type of pump to design, build and test? Let's consider some qualities describing the pumping system that your team may wish to employ in the selection process.

Flow rate is an obvious criterion. The system is to deliver water for irrigation; the more water that is delivered the larger the area that can be cultivated. Many different types of pumps deliver high-flow rates including piston pumps, diaphragm pumps, rotary vane pumps, and centrifugal pumps.

Low-pressure discharge is another criterion. The pump is intended to lift water from a stream over a bank 6 feet high to a level field where it will flow through irrigation ditches to the plantings. Some pumps develop a high-pressure discharge that is important in fighting fires and/or delivering hydraulic fluid to force generating cylinders. However, these high-pressure pumping systems usually do not produce high flow rates. Pressure and flow rates trade off in pump design. Gear pumps are an example of a system that generates a high-pressure at an essentially fixed flow rate. Piston pumps, on the other hand, may generate either high or low-pressure discharge depending on the head against which they operate.

Ease of priming the pump is still another criterion. The pump is usually operated at a location some distance above the level of the stream. It draws[2] the water upward from the stream into its inlet, pressurizes the water, and then discharges it with both head and velocity. To draw the water into the inlet, the pump must develop a suction that permits atmospheric pressure to drive the water up through the inlet pipe. Priming a pump involves loading its chamber with water to effect seals (or valves) that enable the pump to draw a vacuum when it is operated.

---

[2] The suction developed by the pump is limited by the atmospheric pressure with a maximum equal to about 28 feet of water.

Easy to seal is an important criterion. With few exceptions[3], pumps contain one or more seals. Some seals are easy to design and to implement—for example; the rubber or plastic flapper valve in the closet of your toilet acts an inexpensive seal that is reliable for a few years of service. On the other hand, sealing either a rotary or reciprocating shaft is more difficult. In both cases the annular area between the shaft and the housing must be of uniform thickness around the shaft. This annular region is filled with a soft packing material that is squeezed against both the housing and the shaft. Unfortunately when the shaft moves, the packing material wears and must be tightened and/or replaced periodically. We illustrate the location of the seal relative to the shaft and housing of a hydraulic cylinder in Fig 1.5.

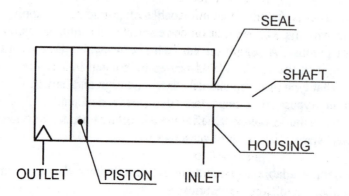

Fig. 1.5 A typical single-acting hydraulic cylinder showing the location of the shaft seal.

We would like the pumping system to be efficient. As we pump water, energy is expended—in this instance human work. This energy is consumed as we lift the water from the stream to the outlet of the delivery pipe. The efficiency $\mathcal{E}$ is defined as the ratio:

$$\mathcal{E} = Wk_{out} / Wk_{in} \qquad (1.1)$$

The work in, Wk_{in} is the force F we exert on the pump times the distance d through which this force moves.

$$Wk_{in} = F \times d \qquad (1.2)$$

The work output from the pump is the weight of the water W_t times the height h that it has been lifted.

$$Wk_{out} = W_t \times h \qquad (1.3)$$

We seek efficiencies of 100%, but in practice achieve much lower values. Friction that occurs when two surfaces in contact move relative to one another produces losses of energy. Turbulence in the water adds to the losses.

The pumping system must be compatible with the capabilities of the operator. The pump is human powered, and as such the force required to drive it must be matched with the force that can be sustained by the operator over a suitably long time interval. You may be able to bench-press 250

---

[3] Centrifugal pumps equipped with magnetic drives are designed without seals to avoid possible leakage of corrosive or toxic chemicals. Also the peristaltic pump that is used to transport blood in heart-lung machines does not contain seals.

pounds, but can you apply this magnitude of the force repeatedly at a high frequency for 15 to 20 minutes?

The pumping system will be manufactured in a few weeks after the initial design has been completed. The equipment, tools and support staff available to your team is very limited. The type of pump selected must be consistent with the manufacturing capability available to your team.

The pumping system is intended for use in underdeveloped countries where spare parts and skilled technicians are in short supply or non-existent. If the pump does fail, it should be easy to disassemble and to repair with parts that are available in these under developed countries.

The pumping system should be reliable. It should be capable of providing a significant quantity of clean water to the irrigation system for several years without the need for repair.

The system should be portable. We anticipate that the pumping system will be moved from one field to another to provide irrigation water on demand. Portability requires that the weight of the system should be limited, and that the system be easily assembled and disassembled repeatedly.

The cost of the pumping system should be minimized consistent with producing a reliable product. Funds available for procuring and distributing these pumping systems are limited. As such, the cost of each component employed in the system should be minimized.

We always seek a quality product. Quality means different things to different people. In this instance, we define a quality product as one that performs reliably, delivers the specified water flow, and is easy to use. Your team may wish to add to this definition.

Last, but certainly not least, the product should be safe to operate. Hazards associated with products are covered in detail in Chapter 16, Section 6 of this text. Before beginning the design of you system, we strongly recommend that you read and understand the many different hazards associated with the design and operation of products.

We will discuss the history of pumping water and measuring flow in Chapter 2. The principles upon which different types of pumps are based are also described in this chapter. The pumping system that is comprised of the pump, drive and measurement apparatus is discussed in Chapter 2.

Some initial ideas to consider in the design of the mechanical components for selected types of pumps are presented in Chapter 3. The drives, which interface with the pumps, are also described in Chapter 3. Human factors important in the design of the drive system are discussed. Chapter 4 provides information on flow measurement methods. Sensors used in recording the parameter related to the flow measurement are also described in this chapter.

Your team should develop many different ideas for each subsystem included in the design of the pumping system. Each concept should be described in considerable detail prior to developing preliminary design proposals incorporating each of the design concepts. The more completely you develop the design proposals, the better prepared you will be to conduct a design-trade-off analysis. In a design-trade-off analysis, you compare several design concepts and select the best one.

1.5.2 Design-trade-off Analysis

To perform a design-trade-off analysis, we consider each design proposal individually and list its strengths and weaknesses. Factors usually considered in this analysis include size, weight, cost, ease of manufacturing, performance, appearance, etc. After the strengths and weaknesses of each design proposal have been listed, we can compare and evaluate them and select the most suitable for our prototype development.

We have briefly discussed the design criteria to be employed in your design proposals and design trade-off analyses. The design concepts are developed for each function in the pumping system. You have several functions to consider in the design of the overall system:

1. Acquiring the water.
   - Inlet conduit
   - Filter or screen
   - Foot valve
2. Pressurizing the water
   - Pump
   - Seals and/or valves
   - Priming
3. Delivery of the water
   - Outlet conduit
4. Measuring the flow
   - Flow channel
   - Flow parameter
   - Sensor
   - Calibration and recording

The procedure in design is to generate ideas for performing each function, expanding these ideas with more detail until they can be treated as design proposals, and finally to perform a design trade-off analysis where the merits and faults of each proposal are evaluated.

When the design proposals are selected for each of the different functions involved in the weighing process, the winning concepts must then be integrated into a seamless system that provides a sufficient quantity of clean water suitable for irrigating fields in under developed countries.

We often refer to the part of a product that performs some function as a subsystem. The collection of all of the subsystems constitutes a complete system, which when manufactured and assembled, becomes the prototype. The integration of the various subsystems is often difficult, as one subsystem influences the design of the others. The manner in which the subsystems interact is defined as the interface between subsystems. In a mechanical application, the fit of a shaft in a bearing and/or the fit of a seal against a shaft are examples of interfaces. Frequently, interfaces between the subsystems are troublesome and difficult to manage. It is vitally important that you control the interfaces to permit seamless integration of all of the subsystems without loss of effectiveness and efficiency.

1.5.3 Preparing the Design Documentation

The final step in designing the prototype is the preparation of the design documentation package. The package, which is an extended engineering document, contains:

- Engineering sketches illustrating the design concepts.
- Engineering assembly drawings that describe how the various components fit together to form the complete system.
- Engineering drawings of all of the component parts in sufficient detail to permit the component to be manufactured by anyone capable of reading an engineering drawing.

We recognize that many of you may need instruction in graphics, and we have included three chapters in this text to help you learn how to prepare three-view and pictorial drawings and to make tables and graphs. In addition, a computer aided drafting program, that will be helpful in preparing the drawings for your design documentation package, will be introduced in the computer laboratory.

PARTS LIST FOR THE CHAIN DRIVE				
PART NO	NAME OF ITEM	DESCRIPTION OR DRAWING NO.	QUANTITY	PRICE
1	LARGE SPROCKET	CATALOG NO. 6236K27 McMASTER CARR	1	$ 23.48
2	PEDAL ASSEMBLY	LOCAL BYCYCLE SUPPLY HOUSE	1	$ 15.55
3	CHAIN ANSI NO. 35	CATALOG NO. 6261K531 McMASTER CARR	4 ft	$ 6.76
4	SMALL BRACKET	CATALOG NO. 6280K111 McMASTER CARR	1	$ 5.04
5	SPROCKET SHAFT	CATALOG NO. 88934K42 McMASTER CARR	1	$ 4.49
6	SLEEVE BEARINGS	CATALOG NO. 6391K214 McMASTER CARR	4	$ 3.76
7	FRAME SIDES	WOOD BARS-------DWG. NO. 100-01	2	$ 1.20
8	FRAME ENDS	WOOD BARS-------DWG. NO. 100-01	2	$ 0.60
9	SEAT	LOCAL BYCYCLE SUPPLY HOUSE	1	$ 8.50
10	SUPPORT	WOOD STRUT-------DWG. NO. 100-02	1	$ 0.60
11	SET SCREWS	1/4 - 20, 1/2 LONG, ALLEN HEAD	2	$ 0.85
12	WOOD SCREWS	NO. 8, 1-1/2 IN. LONG, FLAT HEAD, PHILLIPS	8	$ 0.80

Fig. 1.7 Example of a partially complete list of parts for a chain drive.

The design documentation package also contains list of parts that identifies every unique part employed in the assembly of the prototype. The quantity required of each component is also included on the line describing the component. For example, if you are going to use eight No. 8 wood screws to fasten together the joints of the support for the belt drive, you would:

1. Assign a part number for the screws to identify the need for these screws in the assembly of the prototype.
2. List the quantity required as "8" in the quantity column.
3. Describe the screws in sufficient detail so that another team member can go to a hardware store and procure exactly the type of screw that is needed.

The descriptions provided in the parts list are brief, but complete and precise. (i.e. No. 8 wood screw, flat headed, with a Phillips head, 1-1/2 in. long). In this description, **No. 8** gives the diameter; **wood** refers the type of application for the screw; **flat headed** describes its head; **Phillips** indicates that you will use a Phillips head screw driver in the installation; **steel**[4] is the material from which the screw is fabricated; **1-1/2 in**. is the length of the screw.

The parts list contains all of the items that are to be purchased for your prototype. It also contains the components that must be manufactured. If the component is to be manufactured, it is identified on the parts list by its name and a drawing number. We show an example of a parts list in Fig. 1.7.

An engineering report is also included as part of the design package. This report supports the design by describing the key features of your pumping system. Your report should treat each function involved in the pumping process and describe the design concepts that your team considered. A rationale for each design proposal that was adopted, based on systematic design trade-off studies, is an essential element. Additionally, the design report contains your theoretical analyses used to predict the performance of your prototype. The analysis is conducted before actual testing the prototype, and

---

[4] Unless otherwise stated in the parts list, steel is the default material for screws.

both the analysis and the test results are useful in assessing the merits of the product development. More information on the preparation of an engineering report is included in a chapter of this textbook on writing technical reports.

1.5.4 Final Design Presentation

The design phase of the product development process is concluded with a final design presentation. This is a formal review of the design of the pumping system that your team presents to the class (peer review), to the instructor and others in the audience. It is the team's responsibility to describe all of the unique features of the design and to predict the performance of the product. It is the responsibility of the class (peer group) and the instructors to question the feasibility of the design and the accuracy of the predicted performance. If, in your review, you note a shortcoming or a flaw in the design, identify the problem to the team presenting their work. As a peer reviewer of another team's design, be tactful in your critique. Criticism is always difficult to offer to another. Offer constructive suggestions in good faith and in good taste. If you are on the receiving end of criticism, do not be defensive. The individual offering the critique is not attacking your capabilities; he or she is actually trying to give you a suggestion that might help you achieve a better design.

The purpose of the design review is to locate deficiencies and errors. It is better to correct errors in the paper stage of the process, not later in the hardware phase, when it is much more difficult and costly to fix the problems.

1.6 PROTOTYPE ASSEMBLY AND EVALUATION

1.6.1 Preparing the Assembly Kit

The second major phase of the product development process is to prepare what is called an assembly kit. The assembly kit is a collection of all of the parts, in the appropriate quantities, which are required to completely build the prototype. The parts list, as described above, is the essential document that directs our procurement and manufacture of the components needed to build the prototype. The parts list identifies everything that we need and indicates the precise quantity of each item required to assemble a single unit.

An efficient method for collecting the required parts is to divide the list into three groups:

- Those parts to be purchased.
- The materials and supplies to be purchased.
- The components to be manufactured.

One or two team members handle purchasing the necessary components, materials and supplies while the other team members work in the workshop provided by the college to fabricate the parts according to the detailed drawings prepared in the design phase. It is important to recognize that an engineering drawing defines the part that is being manufactured. Always work from the engineering drawing and not from your memory or understanding of the part geometry.

It is suggested that one team member serve as the "inspector," checking the finished parts against the drawings for the parts that have been manufactured. If the parts have been purchased, the "inspector" should check the purchased items against the parts list to insure that the item exactly meets the description and that the correct quantity has been procured.

1.6.2 Prototype Assembly

When the assembly kit is complete and checked against the parts list, we can begin the final phase—prototype evaluation. The first part of the evaluation is to assemble or build the prototype. We often call this step the "first article build," because it is the first time that we have attempted to fit together all of the components needed to assemble the prototype. The "first article build" can go well if: all of the parts are available; they all fit together properly; the tolerances on each part and each feature are correct; and the surface finish on all of the parts is acceptable. Prototype design is often judged against the four Fs—form, fit, finish and function. The "first article build" permits us to assess how well the team performed with regard to the first three of the Fs—form, fit and finish.

If the parts do not fit, modifications of one or more parts are required. These modifications require design changes that are a dreaded and costly process in the real world. In fact, one of the most important criterion used by management to judge the quality of a development team is the number of design changes required both before and after the introduction of a product to the marketplace. Clearly, we want to design each part in a product correctly during our first effort and minimize the number of required design changes.

We anticipate that each team will make a few errors in the preparation of the detail design drawings and that changes will be required. The natural tendency is to take the offending part to the model shop and to correct (modify) it and move on with the "first article build." This behavior is acceptable only if the team revises the drawing of the offending part to reflect the modification made to correct the design deficiency. In the real world, the integrity of the drawing package is much more important than the prototype. The second, and all subsequent assemblies, will be fabricated from the details shown in your engineering drawings and not by examining the prototype. The prototype is often scrapped after it has been built and tested.

1.6.3 Prototype Evaluation

After the assembly of the prototype is complete, it should be carefully inspected to insure it is safe. Fortunately, a human powered pump is a relatively safe product since the parts do not undergo high-speed motion, and the hazards due to high pressures and temperatures should not exist. However, make sure that you have eliminated all sharp edges and points on the prototype, and that you have avoided introducing pinch points. (Pinch points are openings that close due to motion of one or more parts when a product is operated. Fingers, hair and/or clothing caught in pinch points may represent a hazard).

The final step in this development process is to test your prototype to determine if it is functional and accurate. This is a big day and indeed a big hour. Typically a one-hour time slot has been scheduled for evaluating each team's prototype in the testing facility available in a college laboratory. Your prototype will be judged based on the following criteria:

1. Performance: accuracy over the entire specified range and readability of the display.
2. Capability: ability to determine required postage for different sizes and types of letters and packages.
3. Cost: minimize.
4. Design innovation: novelty and creativeness.
5. Quality of parts and assembly: workman (women) ship.
6. Appearance: pleasant looking and in keeping with the decor in a typical office.

1.7 TEAMWORK

In this course, you are required—for three reasons—to participate on a development team. First, the project is too ambitious for an individual to complete in the time available. You will need the collective efforts of the entire team to develop a quality water pump during the semester. We plan on pressing the development teams to complete the project on a prescribed schedule. Second, we want you to begin to learn teamwork skills. Experience has shown us that most students entering the engineering program have not developed these skills. From K-12, the educational process has focused on teaching you to work as an individual often in a setting where you competed against other students in your class. We will demand that you begin functioning as a team member where cooperation, following and listening are more important than individual effort. Leadership is important in a team setting, but following is also a critical element for successful team performance. The final reason is that you will probably find yourself on a development team early in your career if you take a position in industry. A recent study [4] by the American Society for Mechanical Engineers (ASME), the results of which are shown in Table 2, ranked teamwork as the most important skill to develop in an engineering program. Teamwork was also the first skill, in a list of 20, considered important by managers in industry. We are hopeful that this course will be instrumental in exposing you to team working skills so necessary for a successful career.

As you work within your team in the development of a human powered water pump, you will be introduced to many of the 20 topics listed in Table 1.2. We hope that you will begin to develop an appreciation of the need for these skills and begin to enhance your level of understanding of the design process that requires you to utilize many of these skills.

TABLE 1.2
Skills Considered Important for New Mechanical Engineers
with Bachelor Degrees
Priority Ranking

1. Teams and Teamwork
2. Communication
3. Design for Manufacture
4. CAD Systems
5. Professional Ethics
6. Creative Thinking
7. Design for Performance
8. Design for Reliability
9. Design for Safety
10. Concurrent Engineering
11. Sketching and Drawing
12. Design for Cost
13. Application of Statistics
14. Reliability
15. Geometric Tolerancing
16. Value Engineering
17. Design Reviews
18. Manufacturing Processes
19. Systems Perspective
20. Design for Assembly

1.8 OTHER COURSE OBJECTIVES

While your experience in developing the human powered water pump is the primary objective of this course, there are several other related objectives. You will quickly recognize a critical need for graphics as you attempt to describe your design concepts to fellow team members. We have included three chapters on graphics to help you learn the basic skills required at this entry level. These chapters include materials on three-view and pictorial drawing, graphs and tables. Instructions for developing entry-level skills in a computer aided drawing program are provided in a computer laboratory.

You will also be required to become familiar with using three additional software application programs: word processing, spreadsheet, and graphics presentation. Our experience indicates that nearly all of you are already proficient with word processing; therefore, we will not cover this topic here. However, many of you are not familiar with how to use a spreadsheet. Accordingly, we have included a chapter describing Microsoft Excel to help you prepare your parts list, to perform calculations and to prepare graphs. Spreadsheets provide very powerful tools that you will find useful in many different ways for the remainder of your life. We hope that you will take this opportunity to learn how to use this important software. We have also included a description of Microsoft PowerPoint to aid you in the preparation of world-class slides for your design briefings.

As you proceed with the development of the pump and drive system, we frequently will require you to communicate both orally and in writing. Design reviews, before the class, provide you with the opportunity to learn presentation skills such as style, timing and the preparation of visual aids. The design report will give you experience in preparing complete, high-quality engineering drawings and in writing a technical report which contains text, figures, tables and graphs.

The final objective of the course is design analysis. We understand that your engineering analysis skills have yet to be developed. For this reason, we will present key equations that mathematically model the electrical and mechanical components used in the design of many of the subsystems that you employ in the design of the human powered pumping system. We do not expect you to completely understand all of the theoretical aspects of the relatively complicated subsystems involved. Most of you will take several courses later in the curriculum dealing with these subjects in great detail. However, at this stage of your career, we want you to begin to appreciate the relationship between analysis and design. To help you with the analysis, we have included two chapters on pumps and a chapter on work, human power and drive systems. Another chapter on flow measurements and some methods for measuring voltage is included. The coverage is very brief and introductory, but it should give you a direction and a start in understanding both mechanical and electronic systems.

REFERENCES

1. Valenti, M. "Teaching Tomorrow's Engineers," Special Report, Mechanical Engineering, Vol. 118, No. 7, July 1996.
2. Ulrich, K. T., S. D. Eppinger, Product Design and Development, McGraw-Hill, New York, NY, 1995, p. 6.
3. Cross, N., Engineering Design Methods, 2nd Edition, Wiley, New York, NY, 1994, p. 77-81.
4. Anon, "Integrating the Product Realization Process into the Undergraduate Curriculum," ASME Report to the National Science Foundation, 1995.
5. Macaulay, D., The Way Things Work, Houghton Mifflin Co., Boston, 1988.
6. Horowitz, P. and W. Hill, The Art of Electronics, 2nd Edition, Cambridge University Press, New York, 1989.
7. Dally, J. W., Riley, W. F. and K. G. McConnell, Instrumentation for Engineering Measurements, 2nd Edition, Wiley, New York, 1993.

EXERCISES

1.1   List ten products that you have used today and the companies that manufactured and marketed them.

1.2   Suppose that you do not want to work on product development, manufacturing or sales. What opportunities remain for you in engineering?

1.3   Write an engineering brief describing the characteristics of a winning product.

1.4   Write the most important equation in engineering or business.

1.5   Why is Ford Motor Company reluctant to develop a brand new model of an automobile?

1.6   What is a design concept? How many concepts will your team generate in developing the human powered pumping system this semester?

1.7   What are the three main subsystems in the development of the human powered pumping system?

1.8   Why is it important to prepare a parts list during a development program?

1.9   What is an assembly kit and why do we prepare one in the development of the first prototype?

1.10   Why do we invest scarce company funds in assembling a prototype?

1.11   What are the safety considerations with which you must be concerned in testing the human powered pumping system?

1.12   Why do you believe industry representatives ranked teams and teamwork as the most important skill for new engineers?

CHAPTER 2

HUMAN POWERED WATER PUMPS

2.1 IMPORTANCE OF WATER

Water is essential to life; we perish if deprived of it for only a few days. We must have about four to six pints of water daily to survive. Our bodies use water to process the food we consume, transport nutrients within our blood stream and to dissipate the excess heat we generate. Seventy percent of our body weight is comprised of water. If our bodies become dehydrated, we do not function well.

Water consumption for an American living in a typical city is about 1000 pints per day. We have grown accustomed to frequent changes of clean clothes, daily baths or showers, green lawns and sparkling cars. Also, a significant usage of water in the home is to transport sewage.

It appears that we are blessed with an abundant supply of water. Over 70% of the surface of the Earth is covered with seas or oceans that average about 13,000 feet in depth. With a world population of 5.8 billion, there is about 75 billion gallons of water per person [1]. Unfortunately, most of this supply is salt water and the available supply of fresh water is much less. Finally, the availability of clean (pure[1]) fresh water, suitable for drinking, is severely limited in many parts of the world.

The use of water for agriculture is equally important. On a worldwide scale, the amount of water consumed in agriculture is immense. Ten percent of the world's total land area is cultivated (about 250 million hectares) [2]. Rainfall and underground seepage provide water for most of this area; however, about one percent of this area is watered by irrigation. In many cases, the flow of the water from the source to the producing fields is achieved by building a dam, which raises the water level so that it is higher than the fields being irrigated. In other cases, water is pumped from streams or ditches into the fields. When wells are drilled, the water is pumped from underground to the producing fields.

Agriculture is not an efficient user of water because much of the available water is lost either by evaporation or transpiration. For example, it requires about 20,000 gallons of water to grow a bushel of wheat and 2300 gallons to grow a pound of beef [1]. In the U. S., we have made extensive use of dams and pumps to provide large irrigation systems with the capacity necessary for producing the huge quantities of water required for our crops. In under developed countries, either the funds and/or the skills required to construct irrigation dams are lacking. When water is available in natural streams it is necessary to pump it to the fields. Unfortunately electrical power is usually not available to drive the pumps needed for irrigation. In these cases the alternative is a human powered water pump [3]. Your design project for this semester is to design, construct and test a human powered

---

[1] Drinking pure water is not advisable because it lacks the dissolved minerals needed for good health. On the other hand, water with excessive quantities of dissolved minerals (about 0.1%) develops an unpleasant taste and upsets the digestive process.

water pump that would be useful for those living near a stream in rural regions of developing countries.

2.2 HISTORICAL DEVELOPMENT OF WATER PUMPS

We have been "raising" water since the beginning of recorded history. Most of the hydraulic machines used to acquire water for personal use or for agriculture were handed down to us from the Assyrians, Babylonians, Persians, and Egyptians. Some other devices came from the Romans, Greeks, Moors and Chinese. In a brief review of these ancient concepts, we attempt to show that product development is evolutionary. We rarely encounter an absolutely **new** product. Most products are slight modifications of products that have been around for many years or even centuries. Water pumps certainly are not new.

There are many different ways to pump water. Let's divide the pumping methods into four categories.

1. Lifting water in moveable vessels.
2. Raising water in tubes.
3. Piston pumps
4. Rotating pumps

2.2.1 Lifting Water in Movable Vessels

Ancient vessels for transporting water were taken from nature—the gourd or calabash. Later pottery was developed and vases or jars were used in transporting or storing water. The supply of clean water was under ground or in free flowing streams. In both cases, it was necessary to acquire and lift some quantity of water for personal use. The evolution of methods for lifting water in the simplest manner—with a container is depicted in the following figures that were taken from a reprint of the 1842 edition[2] of Thomas Ewbank's historical account of "Hydraulic and other Machines for Raising Water" [4].

The three plates reproduced in Fig. 2.1 show early progress in techniques for lifting small quantities of water. First, in Fig. 2.1 a, we note that a small vase is lowered into a well with the aid of a cord. If the well or cistern is shallow, the cord is replaced with a pole (see Fig. 2.1b) that allows for better control of the vessel in which the water is acquired. Finally, in Fig. 2.1 c, a pulley is added. The pulley permits one to pull at a convenient angle instead of lifting at an awkward angle. With a pulley larger quantities of water could be lifted.

Another approach to improve the process of lifting water was to reduce the force needed to lift it from wells. The swipe illustrated in Fig. 2.2 employs a counter weight to offset at least a portion of the weight of the water and its container. The windlass was another invention that was useful in reducing the force thereby permitting one to lift much larger quantities of water. Note the windlass incorporates a double crank, pulley and two drums to provide significant mechanical advantage.

When it was necessary to raise very large quantities of water, our forefathers became even more innovative. When animals were available they could be used to provide the force necessary to lift large containers of water as illustrated in Fig. 2.3. The pulley was essential of course to control the direction of the force. If an animal such as an oxen or horse was not available, human power was still required. To enhance the capabilities of men (or women) to lift load after load of water, a treadmill was introduced. The treadmill permits one to use the lower body muscles, which are stronger and more durable than the upper body muscles.

---

[2] We are indebted to Arno Press, for their efforts in reprinting a large number of classic historical treatments of many interesting topics in their series of books on Technology and Society.

Fig. 2.1 Evolution of product enhancements for lifting water in moveable containers.

Fig. 2.2 A swipe and windlass used to reduce the force needed to lift the water.

Fig. 2.3 Left—animals replace human operators in lifting very large quantities of water.
Right—treadmills introduced to permit the use of stronger more durable leg muscles.

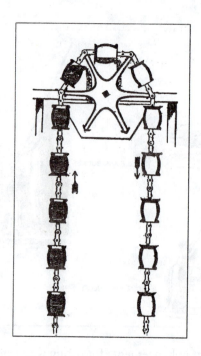

Fig. 2.4 An early move toward automation—the chain of pots.

One of the problems, with all of the schemes for raising water that have been described, is that they are limited to one container of water per cycle of operation. With a design utilizing a single container, it is necessary to lower it into the well, acquire the water, raise the container and finally dispose of the water. The cycle is repeated until a sufficient quantity of water is obtained. This process is slow and inefficient because of the time required raising and lowering the container. Our forefathers automated the process by developing the chain of pots illustrated in Fig 2.4. This invention incorporated a chain and a wheel with spokes. A large number of pots were fastened to the chain at closely spaced intervals. The chain was of sufficient length to form a loop that extended into the water near the bottom of the well. At the bottom loop of the chain, the pots rotate and fill with water. The full containers are lifted to the wheel at the top where they are emptied without spillage into a channel that carried the water away.

2.2.2 Raising Water in Tubes

Relatively early in history, methods for manufacturing tubes from leather were developed. These tubes were wrapped about shafts, which were then rotated in order to lift water from relatively shallow wells or streams with low banks. Three different devices incorporating closed tubes for raising water are illustrated in Fig. 2.5

The screw pump, illustrated in Fig. 2.5 a, is constructed with a tube wound in a helical spiral about a shaft. The shaft is inclined to the horizon; however, the angle of the helix must be inclined at an angle that traps the water when the shaft is rotated. Rotation of the shaft (in the correct direction) captures water in the tube and lifts it upward on each revolution. The water exits at the top of the helix and flows into a channel. The screw pumps are typically used with an inclination of about 45° although larger angles of inclination are possible if the helix angle is sufficiently small to prevent the water from escaping.

The spiral tube pump, illustrated in Fig. 2.5b, is like the screw pump since it traps the water as the shaft rotates. However, water is trapped only for a short interval during each revolution while the open end is under water. During part of each revolution, the open end of the spiral is in the air. As a consequence, slugs of both air and water are trapped in the tube in an alternating sequence. The weight of the water compresses the air and when the water exits the pump at its center it is pressurized. A spiral tube pump constructed in Florence, Italy in 1784 pumped a hogshead of water a minute to an elevation of 74 feet [3].

Fig. 2.5 Rotating devices that employ closed tubes to lift water.

There are many versions of what is known as a tympanum. The multiple spiral tube device, presented in Fig. 2.5c, is an advanced version. When the free end of each spiral tube sweeps through the stream, a quantity of water is trapped in the tube. As the shaft rotates the tube lifts the water to the center of the wheel. The water exits from the center of the wheel into a pipe, which also serves as the shaft.

The noria and the scoop wheel are similar to the tympanum because they both incorporate a wheel with a central shaft. The container to trap the water is different in each design, but the rotating wheel is used to elevate the water in all of the devices. In the noria, the water is elevated to the top of the wheel, but in the scoop, it is raised only to the center of the wheel.

2.2.3 Reciprocating Pumps

All of the pump designs discussed in the previous two sections were based on acquiring water by dipping a container below the water line to fill it. This dipping and filling action is a severe limitation because of the relative long time involved in the acquisition of the water. Water is also lost by spillage in transporting the containers to the surface. With reciprocating pumps, the filling time is reduced because the water is forced into the container by atmospheric pressure. Also, losses due to leakage are minimized.

2.2.3.1 Bag, Bellows and Diaphragm Pumps

There are two different classes of reciprocating pumps—the bag, bellows and diaphragm pumps and the piston pump. Examples of the bag, bellows and diaphragm pumps are presented in Fig. 2.6. The bellows and bag pumps are depicted in Figs. 2.6a and 2.6b respectively. Both the bag and bellows are flexible containers with internal volumes that can be varied from nearly zero to some maximum. When the bag is pulled up to increase its volume, as shown in Fig. 2.6b, suction is created. A flapper valve at the bottom of the bag opens and atmospheric pressure forces water to flow from the source into the bag. When the bag is full of water, the stroke is reversed and the top of the bag is forced downward. The flapper valve at the top of the bag opens and the water flows out of the bag into the chamber of the pump. The water drains from the chamber through a nozzle into a bucket.

The diaphragm pump is similar to the bag and bellows pumps because the container for the water has a variable volume. The container is the chamber of the pump with a diaphragm closing the top. The diaphragm flexes to permit the volume to increase or decrease as the pump is stroked. Two flapper valves each permitting flow in only one direction are incorporated in the design. When the diaphragm is pulled upward from its lower most position, suction is created, the flapper valve at the bottom opens, and a new charge of water is drawn into the chamber. At the same time, the flapper valve in the diaphragm plate closes and the water above the diaphragm is forced out of the pump. On the next stroke, the diaphragm is forced downward, the lower flapper valve closes, and the upper flapper valve opens. The water drawn into the chamber on the previous stroke is force into the chamber above the diagram. Note, in Fig. 2.6 c, that a lever is incorporated into the pump handle to decrease the force required to operate the pump.

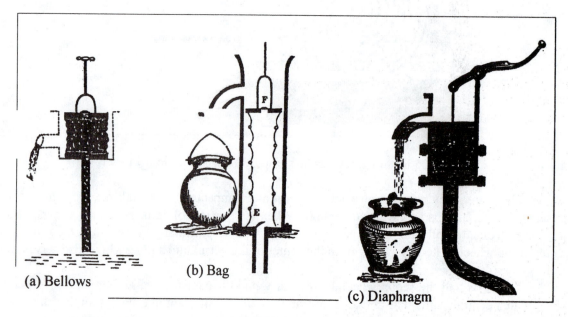

(a) Bellows   (b) Bag   (c) Diaphragm

Fig. 2.6 Examples of early bellows, bag and diaphragm pumps.

2.2.3.2 Piston Pumps

Piston pumps are similar to the bellows and diaphragm pumps because they both incorporate a reciprocating action together with one-way flapper values to control the flow direction. In bellows and diaphragm pumps, flexible containers are used to change the volume of the water container. However, in a piston pump the chamber is rigid—moving a piston inward and outward produces the volume variation.

A single acting piston pump is illustrated in Fig. 2.7a. When the piston is pulled upwards, the lower flapper valve opens and suction draws water into the expanding chamber. On the downward stroke, the bottom flapper valve closes, the upper flapper valve opens, and water is forced upward and outward by the piston.

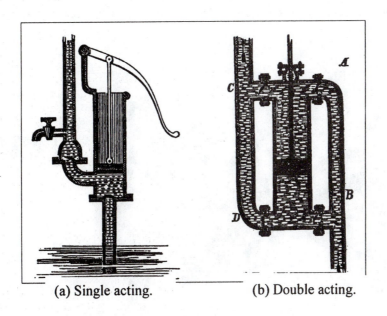

(a) Single acting.   (b) Double acting.

Fig. 2.7 Single and double acting piston pumps.

The double acting piston pump is illustrated in Fig. 2.7b. In this case, four flapper valves are used in the design to control flow into and out of the chamber above and below the piston. The concept is to provide increased flow rate by producing water on each stroke of the piston. Filling the chamber on one side of the piston at the same time as the water is forced out of the chamber on the other side provides increased flow. As with most improvements there is a price to pay. For the double acting pump, that price is the seal required on the plunger rod at the top of the chamber. We will discuss seals in more detail later in both Chapters 3 and 4.

The reciprocating pumps depend on one-way valves to control the flow of water into and out of the container. With these valves, it is possible to create a vacuum in the container and then fill it with flow generated by atmospheric pressure. The piston pumps are more complex than the bag or diaphragm pumps because of the seals required. Seals are needed to prevent leakage between the piston and the chamber wall, and on some designs around the plunger rod. The piston pump is also more difficult to manufacture since the piston and chamber must fit together with close tolerances.

2.2.4 Rotating Pumps

A rotating pump, as the name implies, utilize a rotating impeller to lift and pressurize a fluid. There are three different concepts used in the design of the rotary pumps as indicated in Fig. 2.8. A gear pump, depicted in Fig. 2.8a, employs two counter-rotating gears that sweep the water around the perimeter of the pump casing. The water is drawn into the pump by the suction that develops due to the rotation of the gears. The water exits through a pipe connected at the top of the pump.

The machine illustrated in Fig. 2.8b is similar to the modern vane pump. Protuberances on the central rotating element sweep around the perimeter of the casing. The water is entrained and forced out of the pipe located at the top of the pump. The operation of this pump is dependent on the rotating valve located at point B in Fig. 2.8b. This valve is activated by the protuberance, which causes it to rotate in passing. Like the gear pump, the vane pump draws the water into its chamber with suction developed by the rotating element.

A very early model of a centrifugal pump is presented in Fig. 2.8c. A belt located at the top drives the assembly, which consists of a T shaped pipe. The vertical stem of the pipe assembly is immersed in the water. As the assembly is rotated, suction is formed which draws water into the vertical pipe. The water is drawn up the vertical pipe by the centrifugal forces that develop due to rotation of the horizontal branches.

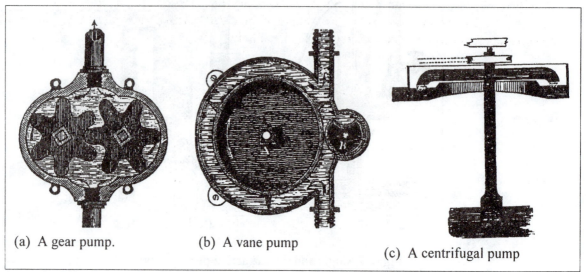

(a) A gear pump.   (b) A vane pump   (c) A centrifugal pump

Fig. 2.8 Early design features used in three rotating pumps.

Rotating pumps are the most difficult type to manufacture because the rotating element and the pump housing must fit together with tight tolerances. Also, the pumps require low friction bearings in addition to seals. Finally, the suction that draws the water into the inlet is achieved only if the pumps are operated at reasonably high revolutions per minute (RPM). As a consequence of these complications, rotary pumps were developed later in history.

These historical references to pump developments made many centuries ago are to indicate that you will probably not invent a new pumping system for this class project. In most cases, our forefathers developed the initial concepts. Our task is to improve on these concepts by making products smarter, more efficient, easier to use, more attractive, more reliable, etc.

2.3 MODERN PUMPING METHODS

Modern pumps resemble the machines designed two or three centuries ago. Of course, they are improved with electric motors or gasoline engines providing the driving power. They also incorporate high quality bearings and seals that improve their operation. Finally, they are made with better materials machined to close tolerances. Consequently, the size of the pumps has been reduced while increasing their output. Also, the efficiency of the pumps and the pressure at the exit has increased dramatically.

There are many different types of pumps that are sold in the U. S. today by a large number of companies. If you explore pump availability on the Web[3], the information available is overwhelming. Our initial attempt showed that 1140 companies were listed at the site maintained by the Thomas

---

[3] You may wish to examine the suppliers of pumps at the Web site http://www.thomasregister.com.

Register. Over a hundred of these companies had on-line catalogs. We recommend that you explore the Web to find information regarding pumps, drives and flow measurement systems.

To aid in your introduction to pumps in the next subsections, we will briefly describe the design concepts used in constructing seven different types of pumps.

2.3.1 Piston Pumps

Piston pumps are very common and they have been used for centuries to pump water. They are classified as positive displacement pumps because they displace a well-defined volume of water on each stroke. This volume is equal to the volume of the piston inserted into the pumping chamber. The essential components of a piston pump are depicted in Fig. 2.9.

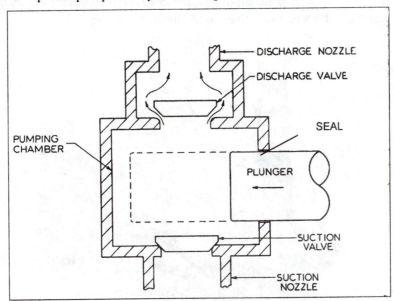

Fig. 2.9 Cross sectional view of a piston pump showing its essential components.

Suppose the pumping chamber is full of water, and the piston or plunger is forced inward. The water is compressed and its pressure increases until the discharge valve opens. A volume of water equal to the volume of the plunger inserted into the chamber is forced through the discharge valve into the discharge nozzle. When the plunger is withdrawn, the pressure within the chamber drops below atmospheric pressure forming a vacuum. The discharge value is pulled shut and the suction valve is pulled open. Water is drawn into the chamber by suction.

The piston or plunger is cycled in a reciprocating action. In the design illustrated in Fig. 2.9, water is discharged only during the inward stroke of the plunger. However, many piston pumps are designed to be double acting with discharge of water occurring on both the inward and outward strokes.

Piston pumps have several advantages:

- Relatively easy to manufacture from commercially[4] available components.
- Low cost.
- Positive displacement.
- Easy to prime.
- Simple to use with no minimum speed requirement.

---

[4] We define commercially available components as those items that can be procured from the local retail supply houses such as Lowe's, Home Depot, Home Base, and the better hardware stores.

- Easy to maintain and repair.

They also have a few disadvantages:

- Two one-way valves are required on single acting pumps.
- Four one-way valves are required on double acting pumps.
- At least one seal and as many as three seals may be necessary.
- The action is linear and not rotary.

The Mennonite Central Committee [3] has published a construction manual for the West African Rower Pump that is rated at 60 liters per minute. The pump is intended to supply water from wells for rural villages and to irrigate small gardens in developing countries. A drawing of the pump in operation, taken from the referenced manual, is illustrated in Fig. 2.10.

Fig. 2.10 The West African rower pump. Reproduced from reference [3].

2.3.2 Diaphragm Pumps

Diaphragm pumps are similar to piston pumps because they operate with a reciprocating action using inlet and outlet valves to control the flow of water in and out of the chamber. However, the piston is replaced with a diaphragm as shown in Fig. 2.11. The diaphragm divides the chamber into two parts with the upper portion used to contain the fluid. The lower portion is reserved for the actuating mechanism. Let's suppose the chamber above the diaphragm is full of fluid and a force on the plunger rod pushes it upwards. The diaphragm is fabricated from an elastomer (rubber) and is flexible. It stretches on the upward stroke and reduces the volume of the upper chamber. This action pressurizes the fluid and forces the outlet valve open permitting the fluid to flow out of the pump. On the downward stroke, the diaphragm moves into the lower chamber increasing the volume of the upper chamber. As the volume is increased, a vacuum is produced, the input valve is opened, and a fresh charge of fluid enters the chamber.

There are many different designs of diaphragm pumps some with springs and some without. Most designs incorporate a circular diaphragm because the stresses imposed on it during repeated cycling are minimized with this shape.

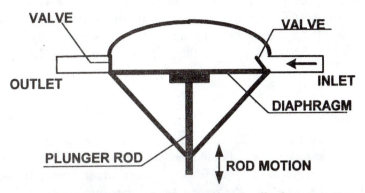

Fig. 2.11 Schematic illustration of a diaphragm pump.

A comparison of the piston and diaphragm pumps shows almost the same advantages and disadvantages. However, a significant advantage of the diaphragm pump is that seals are not required and leakage past worn seals is not a problem. Because leakage is controlled, diaphragm pumps are often used in automobiles for pumping gasoline from the tank to the fuel injector system, and often for pumping small quantities of corrosive chemicals.

The disadvantage of the diaphragm pump relative to a piston pump is that the length of the stroke is limited. Hence, the flow rate is relatively small.

2.3.3 Gear or Lobe Pumps

The gear or lobe pump, shown in Fig. 2.12, is a positive displacement pump because it traps a fixed volume of fluid with the rotating rotors. The rotors, which may be cams, lobes, or gears, mesh together and rotate in opposite directions. They fit closely to the pump casing and force the trapped fluid through and out of the pump. A specific amount of water is transported through the pump on each revolution depending on the geometry of the rotors. As such, this type of pump may be used to accurately measure the quantity of fluid pumped from one tank to another.

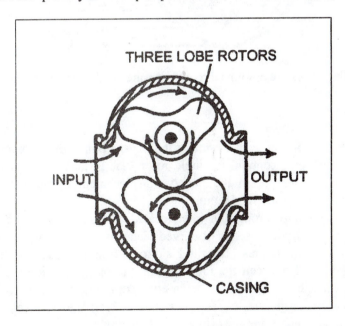

Fig. 2.12 A gear or lobe pump with counter-rotating impellers.

The gear pump has the advantage that it is self-priming and does not require valves. However, it is difficult to manufacture because the rotors must precisely fit the periphery of the casing as they rotate. Another disadvantage is the susceptibility of the rotors to wear due to sand or dirt entrapped in the water. As the rotors wear, their fit with each other and the pump casing deteriorates and leakage becomes appreciable.

2.3.4 Rotary Vane Pumps

The rotary vane pump, depicted in Fig. 2.13, is similar to the gear pump since it is a positive displacement pump. The main difference is in the manner in which the water is trapped as the shaft rotates. A vane pump contains a single rotor that fits loosely in a circular pump casing. Several vanes are fitted into slots cut into the rotor. As the rotor rotates, centrifugal forces act to slide the vanes against the side of the casing. As the vanes are self-adjusting, precision fit of the rotor within the casing is not required. Also, wear due to grit in the water is much less important.

The rotor is off center within the casing. As the rotor turns, the chambers between the vanes change volume. Near the inlet, the chambers increase in volume creating a vacuum that draws water into the pump. With further rotation, the chamber between two vanes decreases in volume pressurizing the fluid before it reaches the outlet conduit.

Since the vane pump is a positive displacement, it may be employed to measure the quantity of fluid pumped between two tanks. When filling the fuel tank on your automobile at a service station, you will probably operate a vane pump.

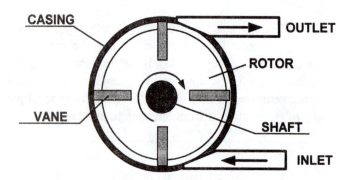

Fig. 2.13 Sliding vanes are employed to entrap water with rotation of the rotor.

2.3.5 Centrifugal Pump

The centrifugal pump, presented in Fig. 2.14, is similar to the vane pump, but it operates on a different principle. It is similar because it contains a single rotor that is located within a pump casing. However, the centrifugal pump is not in the positive displacement category because the water displaced per revolution is not a well-defined volume.

The water enters the pump through an inlet located at the center of the rotating element (impeller). The impeller is fitted with a series of curved blades, which act on the fluid causing it to flow along the vanes as the impeller rotates. The fluid is thrown from the blades with a relatively high velocity into the channel formed between the impeller and the pump housing. The housing for a centrifugal pump is called a volute. The volute is carefully designed to reduce the velocity of the water before it reaches the discharge nozzle. The idea is to convert the flow velocity into pressure before it exits the pump thereby increasing the efficiency of the pump.

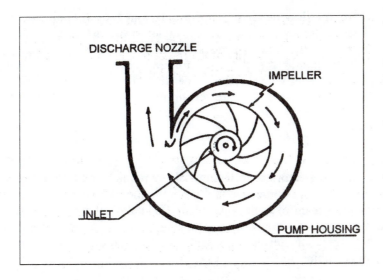

Fig. 2.14 A centrifugal pump incorporates an impeller that energizes the water.

There are several different types of centrifugal pumps depending on the impeller design and the location of the inlet. The impeller with an open radial design and a central inlet, depicted in Fig. 2.14, is the easiest to manufacture.

Centrifugal pumps are used for a very large number of applications. They operate at relatively high efficiency and are easy to manufacture and repair. They are usually driven at high revolutions per minute (RPM) by electric motors or gasoline engines. They are so insensitive to the presence of debris in the fluid that they are often used to pump sewage.

2.3.6   Peristaltic (Tubing) Pumps

Peristaltic pumps are very simple devices as indicated by the drawing presented in Fig. 2.15. The primary element in the pump is a length of tubing, into which the water enters and exits. The tubing is looped through a casing with a circular shape. A shaft drives a rotor fitted with a number of rollers. As the rotor rotates the rollers squeeze the tubing and force the fluid to flow through the tube.

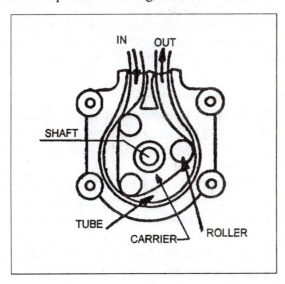

Fig 2.15 A peristaltic pump consists of a tube that is progressively squeezed by rollers.

Peristaltic pumps are low-flow devices. They are used in special applications such as blood transfer. The mechanical action of most pumps is violent and destroys the blood cells. With relatively low speed rotation of the rollers, the tube is squeezed gently and the blood is transferred without damaging the red cells. Tube pumps are also used for transferring small quantities of corrosive chemicals. Nothing comes in contact with the fluid other than the tube itself. There are no seals or gaskets to leak and contaminate. These pumps are self-priming[5] and are resistive to wear from small abrasive particles.

2.3.7 Siphon Pumps

Siphon pumps are also known as ejectors or jet pumps. They require a source of pressurized fluid to operate—usually steam is used. As indicated in Fig. 2.16, the siphon pump is very simple and contains no moving parts. A stream of high-pressure fluid (likely steam) enters the small diameter pipe and is ejected into the throat of a converging diverging nozzle. The jet creates a vacuum at the throat of the nozzle in the larger diameter pipe and water is drawn into it from the left side. The two flows mix at the nozzle and combine and exit the nozzle at the right side.

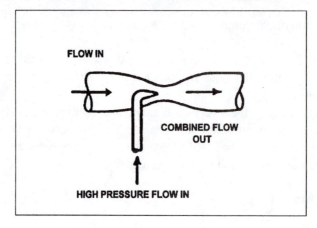

Fig. 2.16 A siphon or jet pump.

2.4 PUMPING SYSTEMS

The discussion in Section 2.3 concentrated on pumps, but a pumping system consists of much more than a pump. The pump is only one part of a complex system—albeit the most important part. Let's consider the complete pumping system, and develop a block diagram, as illustrated in Fig. 2.17, to represent the entire system.

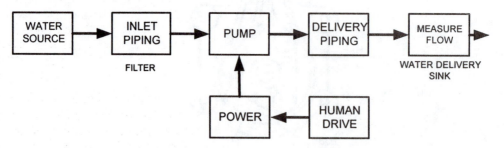

Fig. 2.17 A block diagram for the pumping system.

---

[5] Many types of pumps must be primed (filled with water) before they are capable of drawing a vacuum that draws water into the pumping chamber. If a pump is capable of drawing water into the pumping chamber without prior filling it is call self-priming.

The block diagram has several purposes:

- It helps to define all of the components in the system.
- It permits us to divide the system into subsystems.
- It defines the interfaces between the subsystems.
- It guides the design team in the division of labor.
- It provides a basis for checking the progress in developing each subsystem.
- It provides an opportunity for the assignment of tasks and responsibilities.

2.4.1 The Water Source

In constructing the block diagram, we divided the system into parts beginning with the water source. The water source is a stream from which the water is to be drawn as shown in Fig. 2.18.

Fig. 2.18 Paint Branch Creek is a representative stream providing a source of water.

Inspect the stream to determine the details that will affect the design of the water inlet piping. You should note the gravel bed to the lower right of the photo and the abrupt and overgrown banks on both sides of the stream. The stream is shallow near the gravel bed, but it is a few feet deep near the bank on the left side of the stream. The stream is an interface with the inlet piping; therefore, the inlet piping must be compatible with the stream. For example, the inlet pipe must be long enough to extend beyond the shallow region near the gravel bank. If the inlet to the pipe is too near the bottom of the stream, a filter may be required to prevent loose gravel from being sucked into the inlet pump.

2.4.2 Inlet Piping

The inlet piping is a subsystem. Its purpose is to provide a conduit to transport water from the stream to the pump. One of your team members will design this subsystem. What does this mean? What is entailed in the design task? Clearly the designer will select a conduit. There are many choices:

- Metal pipe.

- Plastic pipe.
- Hose (but be certain it does not collapse on the intake stroke).

After you have selected the type of conduit and justified the choice based on sound reasons, it is necessary to specify the dimensions. What dimensions (diameter and length) are required? Clearly, the length must be sufficient to extend from the pump to the location of the deep water in the stream. The diameter is not so obvious. There is an interface between the inlet piping subsystem and the pump subsystem. The diameter must be compatible with the pump. If the diameter is too large, cavitation of the flow into the pump will interfere with the efficient operation of the pump. If the diameter is too small, the flow will be constrained with losses in output. While the design of both subsystems should be concurrent, the design of both subsystems must be compatible.

2.4.3 The Pump

The pump is the most critical of the subsystems defined in the block diagram of Fig. 2.17. There are many types of pumps available for your selection. Which type should you choose? Why is this the best of the available choices? The best approach to making decisions regarding the type is to carefully consider each pump in a team discussion. Some types can be eliminated quickly for very apparent reasons. For example, the siphon pump is impossible because a source of steam or some other high-pressure fluid is not available at the pumping site. After eliminating those pumps that are clearly inappropriate, several types will remain as possible choices.

List the criteria upon which to base your choice and evaluate each type of pump against these criteria. There are many criteria, but four are paramount:

- System safety.
- Quantity of water pumped.
- Ability to fabricate the required components in the limited manufacturing facilities available to your team.
- The cost of the system.

A systematic procedure for evaluating each design concept is described in Chapter 12. We suggest you read this chapter early in the design process.

Suppose your team decides to design and construct a piston pump or a centrifugal pump. Both of these pumps are reasonable choices. What do you do to design the pumps? We suggest that you prepare a drawing of the entire assembly, identify all of the components required, and design the individual parts. It is necessary to size each component, and specifying the size requires an understanding of the behavior of the water under pressure and the mechanical actions of the pump. We have provided information in Chapter 3 that will help you in the detailed design of many of the components.

2.4.4 Power

The pump is powered by a human operator and must be designed to be compatible with several human characteristics. Human factors are important because they will markedly influence the quantity of water that you will be able to pump. The performance of your system is dependent on the strength and range of motion of the human operator. Should you expect the operator to be able to provide a reciprocal motion exerting a 50-lb. force with a stroke of ± three feet for 15 minutes? Who is the operator? Do you know? Should we drive the pumps with muscles from the lower body, upper body or both combined? We have raised many questions, and will provide some helpful information for your consideration in Section 3.6.

The study of compatibility among man (woman), machine, process and environment is known as ergonomics. It is an important subject to engineers responsible for people working with machinery in production.

2.4.5 Delivery Piping

The delivery piping carries the water from the pump to the irrigation ditches. As with the inlet piping, it must be sized in diameter and length to transport the water to the specified location without significant pressure loss or leakage. The pipe or hose must connect with the pump in a manner that does not impede its operation. Also, the exit must be designed so that the output flows into the channel provided at the test site without significant loses.

2.4.6 Flow Measurement

The flow from the pump will be measured using two different techniques. First, containers will be provided to capture the water delivered by your pumping system. These containers will be calibrated, like a measuring cup, so that the volume, Q, of water pumped is determined by measuring the height h of water in the container. The average flow rate q_{ave} from the pump is then determined from:

$$q_{ave} = Q/t \tag{2.1}$$

where t is the time required to fill the container to a height h.

The instantaneous flow rate q_{inst} is also to be measured. The flow meter to measure the instantaneous flow rate is an integral part of the design of the pumping system. An open channel constructed from wood will be provided to accommodate the flow from the delivery piping. This channel may be employed to house the instruments designed by your team to measure the instantaneous flow rate.

Upon the completion of the pumping interval the accuracy of the flow meter will be determined from:

$$Q = \int_{0}^{t^*} q_{inst} \, dt \tag{2.2}$$

where t* is the time of the pumping interval.

2.5 THE PRODUCT SPECIFICATION

Your team is to design, assemble and evaluate a complete pumping system as described in Section 2.4. The water will be drawn from a stream located on campus and pumped to an open channel located on the adjacent bank. The head from the stream to the open channel is approximately eight feet. The water is to be returned to the stream without disturbing the environment.

The stream identified in Fig. 2.17 will provide an abundant source of water. An open channel, eight inches wide and 6 inches high with a length of six feet, will be available to receive the flow from the pump. The water from this open channel will be collected in calibrated containers. Measuring the height of the water collected in these containers during the pumping interval will indicate the volume of water pumped.

The time available for evaluating your system is limited. You will be allotted 15 minutes for pumping. Criteria that will be used to judge the merits of your team's development effort include:

1. Performance
   - Total quantity of water delivered to the containers.
   - Maximum flow rate.
   - Minimum leakage.
2. Measurement method
   - Accuracy of the measurement of the instantaneous flow rate.
   - Resolution of the measurement.
   - Ease of recording measurement.
3. Cost
   - Minimize the cost of materials and components used in the prototype.
   - Minimize cost consistent with maintaining a quality product.
4. Design innovation
   - Simplicity.
   - Novel concepts.
   - Ease in manufacturing and assembly.
   - Ease of use by operator.
   - Minimize the number of components required.
5. Quality
   - Reliability of the flow measurements.
   - Fit of parts.
   - Minimum friction.
   - Superior workman(women)ship.
   - Alignment.
   - Durability
6. Appearance
   - Pleasing to the eye.
7. Package size
   - See extended description below.
8. Safety
   - Avoid all hazards[6].

The criterion—package size—requires more explanation because it is more involved than the other criteria used in determining the adequacy of the pumping system. Consider package size as a design constraint. The space available for construction and storage is limited. We strongly suggest you make every effort to keep each part of the system as small as possible, and reduce the overall volume of storage space required during the build and evaluation period.

You will also be required to assemble your prototypes using simple hand tools such as a drill, screwdrivers, hammer, pliers, soldering iron, etc. You will be allotted 15 minutes immediately prior to the class period scheduled for the evaluation test to assemble the prototype. You will also be required to disassemble the pump and remove all of the parts from the test site in the 10-minute period immediately following the evaluation test. Unless you receive instructions to the contrary from your instructor, you will have only one opportunity to evaluate "officially" the performance of your pump. However, the calibrated containers and the open-channel will be available at the test site six weeks prior to the scheduled evaluation test. You are encouraged to check your design concepts well in advance of your final class. When you have completed the evaluation test, dispose of the pumps and the piping in an environmentally sensitive manner. If the system is functional, perhaps you will use it or give it to a charitable organization interested in providing aid to developing countries.

The measuring system may incorporate sensors that require electrical power. The usual supply of 120 V ac power is not available at the test site. Also, we believe that it represents a safety

---

[6] See Chapter 16 on Safety, Risk and Performance.

hazard if used in your early prototype development of a flow rate measurement instrument. Accordingly, <u>the use of a 120-volt ac power supply is strictly prohibited</u>. We recognize that an electrical supply may be essential for the sensors and/or instruments used in measuring the instantaneous flow rate. However, we anticipate that you will employ batteries to provide the necessary power. <u>Do not employ more than one nine-volt battery without permission in writing from your instructor.</u>

REFERENCES

1. Overman, M., <u>Water: Solutions to a Problem of Supply and Demand</u>, Double Day, Garden City, NY, 1969.
2. Hunt, C. A. and R. M. Garrels, <u>Water: The Web of Life</u>, Norton & Co., New York, NY, 1972.
3. Anon, "MMC West African Rower Pump", Mennonite Central Committee, Akron, PA, 1992.
4. Ewbank, T., <u>Hydraulic and other Machines for Raising Water</u>, Tilt and Bogue, London, 1842.
5. Linsley, R. K. and J. B. Franzini, <u>Water-Resources Engineering</u>, 2nd Edition, McGraw Hill, 1972.
6. Macaulay, D., <u>The Way Things Work</u>, Houghton Mifflin, Boston, MA, 1988.

LABORATORY EXERCISES

2.1 Find an old air pump used to inflate tires or balls. Disassemble this pump, and catalog all of the parts. Describe each part, cite its function and identify the material from which it was made. Identify the critical components and provide the reasons for classifying them as important. Reassemble the pump and check its operation to ascertain that it is functional.

2.2 A centrifugal pump has been disassembled and is available for your inspection in the laboratory. Examine the individual parts and identify the components. Does the pump utilize sleeve bearings or roller bearings? Classify the type of seals used. Inspect the impeller. Do you believe that your team is capable of manufacturing the impeller? Comment on the importance of alignment in the assembly process.

2.3 From an auto repair shop, obtain a gasoline fuel pump that has failed in service. Disassemble this pump, and catalog all of the parts. Describe each part, cite its function and identify the material from which it was made. Identify the critical components and provide the reasons for classifying them as important. Determine the reason for the failure of the pump. Recommend an improvement that would extend its service life. If possible, replace or repair the component that has failed. Reassemble the pump and check its operation to ascertain that it is functional.

EXERCISES

2.1 What quantity of water is stored in the flush tanks on toilets manufactured in recent years? How does this compare with older models?

2.2 Write an engineering brief for the local newspaper describing the status of the water system serving your municipality.

2.3 Estimate the quantity of water that you could draw from a well 10 m deep in 20 minutes if the method used was a rope and bucket. A cylinder 200-mm in diameter and 300 mm high represents the bucket. State all of the assumptions made in your analysis.

2.4 Estimate the quantity of water that you could draw from a well 20 m deep in 20 minutes if the method used was a rope and bucket. A cylinder 250-mm in diameter and 400 mm high represents the bucket. State all of the assumptions made in your analysis.

2.5 Estimate the quantity of water that you could draw from a well 15 m deep in 60 minutes if the method used was a rope and bucket. A cylinder 150-mm in diameter and 300 mm high represents the bucket. State all of the assumptions made in your analysis.

2.6   Determine the density of water at 68° F in units of:

- lb/gallon
- lb/in^3
- grams/mm^3

2.7   Suppose you design a chain of buckets like the one shown in Fig. 2.4. If 26 buckets each with a capacity of 1.5 gallons are deployed on the chain, determine the weight of the water being raised. Is the weight of the chain and the buckets important in the operation of this pump? State why or why not.

2.8   What is the primary advantage of the treadmill type of pump depicted in Fig. 2.3? What change would you recommend to make this concept much more productive?

2.9   Consider the tympanum pump shown in Fig. 2.5. If the wheel is six feet in diameter and each of the four tubes are four inches in diameter, determine the volume of water raised to the center of the wheel on each revolution. Assume that the wheel is immersed in the water source to a depth of 1.5 feet. State any other assumptions made in your solution.

2.10   Compare the characteristics of the bellows and piston pumps.

2.11   What is the primary advantage of the double acting piston pump?

2.12   Prepare a sketch of a piston pump showing all of the critical components.

2.13   Prepare a sketch of a piston suitable for a piston pump.

2.14   Write an engineering brief describing the methods and materials that you would use in fabricating the piston described in the sketch prepared for Exercise 13.

2.15   Prepare a sketch of a diaphragm pump showing all of the critical components.

2.16   Write an engineering brief comparing the merits and deficiencies of the gear and rotary vane pumps.

2.17   Prepare a sketch of a centrifugal pump showing all of the critical components.

2.18   Prepare a sketch of an impeller suitable for a centrifugal pump.

2.19   Write an engineering brief describing the methods and materials that you would use in fabricating the impeller described in the sketch prepared for Exercise 18.

2.20   Prepare a discussion of your views of the suitability of the peristaltic pump for this project.

2.21   Each subsystem defined in the block diagram of Fig. 2.17 may be expanded based on their detailed functions. Prepare a block diagram showing the functions of each of the following subsystems:

- Inlet piping.
- Pump.
- Power drive.
- Flow measuring device.

2.22   Prepare a written statement representing a consensus of the team specifying the following items:

- The volume of water to be pumped in 15 minutes.
- The accuracy of the flow measurement device.
- The cost of the pumping system.
- The importance of safety in building and operating the prototype pump.

CHAPTER 3

WORK, HUMAN POWER AND PUMPS

3.1 WORK AND ENERGY

To pump water from a stream up and over to an open irrigation channel requires work. Since a human being operating the pump provides this work, it is essential that we appreciate the amount of work involved in pumping a specified quantity of water. Let's suppose that we have ten pounds of water (1.2 gallons) and lift it through a distance of 10 feet as shown in Fig. 3.1.

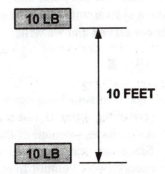

Fig. 3.1 Lifting a weight of 10 lb through a distance of 10 feet.

To lift a weight, we oppose the gravitational force field and must perform work. The amount of work, Wk required to lift the weight, W through a height, h is given by:

$$Wk = W \times h \qquad\qquad (3.1)$$

For the example presented in Fig. 3.1, the work W is given by:

$$Wk = 10 \text{ lb} \times 10 \text{ ft} = 100 \text{ ft-lb}$$

The unit for work is ft-lb in the U. S. Customary system and N-m or Joule in the SI system.

It is important to recognize in Eq. (3.1) that the distance h is in the vertical direction. If we transport a ten-pound weight upwards by ten feet and horizontally through a distance of 20 feet as shown in Fig. 3.2, the work required is the same 100 ft-lb. Work is not required to move the weight in the horizontal direction because it was not necessary to apply a horizontal force to accomplish this movement. In making this statement, we consider that the horizontal force required, to accelerate the body and overcome air resistance, is negligible.

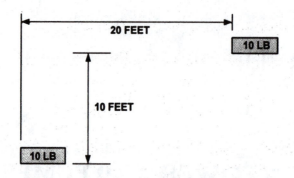

Fig. 3.2 Transporting a weight both horizontally and vertically.

An equivalent method for determining work Wk is to equate it to force, F and distance, d as:

$$Wk = F \times d \qquad (3.2)$$

The pumping force, F = W the weight of the water. The distance d = h the lift height.

We often discuss work and energy as if they were the same, and in a perfect system they are identical. Let's again consider that ten-pound bucket of water illustrated in Fig. 3.1. When the bucket is at sea level the system[1] is without energy. However, when the water is lifted to a height of ten feet, work is performed on the system (the bucket of water) and energy is added. In this new state (the elevated position of the bucket), we have a system with potential energy, E. If there are no loses of work performed on the system, energy is **conserved** and we write:

$$Wk = E \qquad (3.3)$$

There will be many instances in later courses in engineering when we will consider systems where energy is **conserved**. However, pumping water is **not** a process where all of the work is converted into a useful form of energy. A significant amount of the work is converted to heat, which is a much less useful form of energy. Since the heat generated in a human powered pump is not recovered, it is lost to the system. The energy (work) output from the system[2] is less than the work performed on the system. When dealing with processes involving the generation of heat that cannot be recovered, the "usable" or "extracted energy is given by:

$$E = Wk - Wk_L \qquad (3.4)$$

Where W_L is the work lost to the system.

When the work lost in a process, such as lifting, is zero the efficiency of the process is 100%. However, when a loss occurs, the efficiency decreases as indicated below.

$$\mathcal{E} = (Wk - Wk_L)/ Wk \qquad (3.5)$$

Let's consider an example to illustrate the relationship among work, work lost and efficiency.

---

[1] The system is the bucket of water located on the surface of the Earth that is subjected to the Earth's gravitational force field.
[2] We define the human powered water pump as the system in this paragraph. The heat generated when we pump water is not a useful form of energy because it cannot be economically recovered. In this sense the heat represents work lost by the pump. The heat is transferred to the atmosphere and increases the temperature of our environment.

EXAMPLE 3.1

A centrifugal pump is operating at an efficiency of 38%. Determine the work lost for each 100,000 ft-lb of work added to the system.

Solution:

From Eq. (3.5), we write:

$$\mathcal{E}\,Wk = Wk - Wk_L$$

Solving this relation for Wk_L yields:

$$Wk_L = Wk\,(1 - \mathcal{E}) = 100,000\,(1 - 0.38) = 62,000 \text{ ft-lb}$$

Clearly, this pump is not performing well because its losses are larger than the energy gained by the system, which is only 38,000 ft-lb.

3.1.1 Power

Power P is the amount of work performed per unit time. The watt is the measure of power in the SI system and in the U. S. Customary system power is expressed in horsepower. Both quantities are defined as indicated below:

$$P = \text{Work/Time} = (F \times d)/t \qquad (3.6)$$

$$1 \text{ Watt} = N \times m/\,s = \text{Joule/s} \qquad (3.7)$$

$$1 \text{ Horsepower} = 33,000 \text{ ft-lb/min} = 550 \text{ ft-lb/s} \qquad (3.8)$$

This listing of equations may appear to be abstract to you since they represent definitions and conversions from one system of units to another. To demonstrate the use of these relations in an analysis of power expended in performing a task, we present the example shown below:

EXAMPLE 3.2

A crew of eight team members row to propel a shell through the water at high speed. Each crewmember exerts a force of 45 lb while pulling an oar through a stroke of 2.8 ft. If a count of 20 strokes per minute is maintained, determine the horsepower provided by the crew.

Solution:

Recall Eqs. (3.6) and (3.8) and write:

$$P = (F \times d)\,n/t = 45 \times 2.8 \times 8 \times 20 = 20,160 \text{ ft-lb/min}$$

$$P = 20,160 \text{ ft-lb/s} \times 1 \text{ HP}/33,000 \text{ ft-lb/s} = 0.611 \text{ HP}[3]$$

---

[3] The symbol HP represents horsepower. James Watt conducted experiments in 1782 to determine that a "brewery horse" was able to produce 32,400 ft-lb per minute. He standardized the number at 33,000 ft-lb per minute to classify the Boulton and Watt steam engines being produced at that time.

Let's interpret the results. A crew of eight is only delivering 0.611 HP or 0.0763 HP/person. This result appears to indicate that the crew is not in a race situation. What changes would you suggest in their operation to increase the power input to the racing shell?

Let's consider another example showing an analysis of power to demonstrate use of the relationships in evaluating pumps.

EXAMPLE 3.3

You have designed a pump intended to deliver 20 gallons per minute (GPM) of water with a head of eight feet. If the efficiency of the system is 50%, determine the power required to operate the pump.

Solution:

Let's begin by determining the work required in lifting 20 gallons of water through a distance of eight feet. Recall Eq. (3.2), note that a gallon of water weighs 8.34 lb[4], and write:

$$Wk_{out} = 20 \times 8.34 \times 8 = 1334 \text{ ft-lb} \qquad \text{(a)}$$

Since the efficiency of the system is only 50%, we require an input of:

$$Wk_{in} = Wk_{out}/\mathcal{E} = 1334/0.50 = 2668 \text{ ft-lb} \qquad \text{(b)}$$

Using Eq. (3.6) we determine the power needed to operate the pump as:

$$P = Wk_{in}/t = 2668/1 = 2668 \text{ ft-lb/min} = 0.0809 \text{ HP} \qquad \text{(c)}$$

This example indicates the amount of power needed to pump 20 GPM of water with a head of eight feet and a system efficiency of 50%. These numbers are consistent with the development of the pumping system described in Chapter 2 for the class project. In beginning the design of the pumping system, consider using the results from this example and other examples that will be described later in this Chapter as a guide in determining the capacity of your pump. However, two questions are in order:

1. Is 0.09 HP the most appropriate level of power for an operator to deliver for an extended period of time?
2. How can we assure that the efficiency of the pumping system will be at least 50%?

The information presented in Sections 3.2 and 3.3 of this chapter is to aid you in answering the first of these two questions. The material provided in Chapter 4 is to facilitate your understanding of the influence of the mechanical details of pump design on efficiency.

3.2 HUMAN POWER

The pump is to be employed for supplying water to villages and for irrigation in rural areas where electricity is not available. These are primitive localities and even gasoline engines are not feasible because they are too expensive to operate and too difficult to maintain. The pumps are operated with human power. This fact should not surprise you because many pumps in the U. S. are operated by hand. Lever pumps are often found in State and National Parks to provide water for hikers and

---

[4] The density of water depends on its temperature and the amount of dissolved minerals it carries in solution. At 68 °F the density of fresh water is 62.4 lb/ft^3 and that of seawater is 64.4 lb/ft^3.

campers. Small hand operated pumps, with a crank or lever, are often used on drums to transfer liquids at rates up to about 28 GPM[5].

Why are cranks and levers employed in the design of these commercially available pumps? Why are cranks used for one type of pump and levers used for another type? To answer these questions, let's consider the lever first.

3.2.1 Levers

Levers are used with reciprocating pumps because the movement of the lever is consistent with the cyclic in—out action of the pumps. The lever is also a force amplifier as indicated in Fig. 3.3.

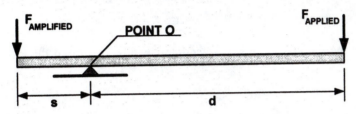

Fig. 3.3 A lever of length L = s + d with a fulcrum positioned at point O.

To determine the relationship between the applied and amplified forces, we consider the equilibrium of moments about the fulcrum. The clockwise moment about point O is:

$$M_{CLOCKWISE} = F_{APPLIED} \times d \qquad (a)$$

The counterclockwise moment about point O is:

$$M_{COUNTERCLOCKWISE} = F_{AMPLIFIED} \times s \qquad (b)$$

For the lever to remain in equilibrium under the action of these two applied forces, the two moments must be equal. Hence, we write:

$$M_{CLOCKWISE} = M_{COUNTERCLOCKWISE} \qquad (c)$$

Substituting Eqs. (a) and (b) into Eq. (c) yields:

$$F_{AMPLIFIED} = (d/s)\, F_{APPLIED} \qquad (3.9)$$

It is clear from Eq. (3.9) that the amplifying factor for a lever is the ratio (d/s). By increasing d and decreasing s, we can generate large forces on the short end of the lever with relatively small ones on the long end. This sounds good—but do we amplify the work performed on a system with a lever? Unfortunately the answer is NO. Let's show that the work-in and work-out of the system are the same by considering the motion of the lever in action as shown in Fig. 3.4.

The work applied to the system (lever) is given by Eq. (3.2) as:

$$Wk_{IN} = F_{APPLIED} \times d \times \theta \qquad (a)$$

The work removed from the system is:

$$Wk_{OUT} = F_{AMPLIFIED} \times s \times \theta \qquad (b)$$

---

[5] See the McMaster-Carr Supply Company catalog which describes piston, diaphragm and rotary hand pumps for use in transferring or dispensing liquids from drums.

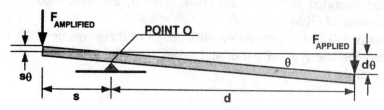

Fig. 3.4 Illustration of lever motion.

Substituting Eq. (3.9) into Eq. (b) yields:

$$Wk_{OUT} = F_{APPLIED} \times (d/s) \times s \times \theta = F_{APPLIED} \times d \times \theta = Wk_{IN} \qquad (c)$$

From Eq. (c) it is clear that there is no gain in work accomplished by the use of the lever.

EXAMPLE 3.4

You are to analyze a large-diameter, short-stroke, piston pump that is to deliver water at a pressure of 7 psi[6]. If the diameter of the piston is 12 inches and the stroke of the pump is 5 inches, determine the force applied to a lever adapted to this pump if it is designed with a ratio of d/s = 4. Also compute the distance through which the long end of the lever is moved.

Solution:

The force, F that must be exerted on the piston to develop the specified pressure, p is given by:

$$F = pA_p = p \times \pi r^2 = 7 \times \pi (6)^2 = 792 \text{ lb} \qquad (a)$$

where A_p is the area of the piston.

Clearly, this force is much too large for anyone to apply directly to the piston. If we consider a lever with a ratio of (d/s) = 4, the applied force is given by Eq. (3.9) as:

$$F_{APPLIED} = (s/d) \, F_{AMPLIFIED} = \frac{1}{4} \times 792 = 197.9 \text{ lb} \qquad (b)$$

This force is much lower than that directly exerted on the piston, but it is still much to large to consider for a human powered pump. How many times a minute could you push on a lever with a force of about 200-lb?

From the dimensions for the lever, it is evident that the stroke, u of its long end is given by:

$$u = d/s \, v = 4 \times 5 = 20 \text{ in.} \qquad (c)$$

where v is the stroke of the short end of the lever, which is the same as the stroke of the pump.

---

[6] The notation psi is the designation for pressure in pounds per square inch in the U. S. Customary system of units. In the SI system, pressure is designated with Pascal (Pa = N/m² or MPa = N/mm²).

What design information has this example provided for us? Is our large-diameter, short-stroke piston pump a good idea? How can we refine the design concepts? Should we conclude that the piston pump is not feasible? It is extremely important that you interpret an analysis to provide guidance for design decisions.

3.2.2 Cranks

The impeller of rotary pumps must be rotated to draw the water into the chamber and then force it from the discharge nozzle. We usually provide this rotary motion by using a crank. The crank, illustrated in Fig. 3.5, is operated by turning it about its center of rotation with a force F. A torque, T developed about the center of rotation is given by:

$$T = F \times r \tag{3.10}$$

The force F is perpendicular to the crank radius r.

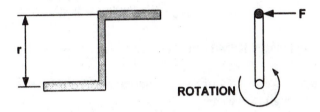

Fig. 3.5 A crank used to produce rotary motion.

Perhaps the most common example of a crank mechanism is found on the bicycle as shown in Fig. 3.6.

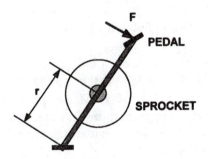

Fig. 3.6 Illustration of dual cranks used to power a bicycle with a sprocket and chain drive.

The relation describing the work performed with rotary motion is similar to Eq. (3.2), which applies to linear motion:

$$\mathit{Wk} = T \times \theta \tag{3.11}$$

T is the applied torque and θ is the angle through which the torque is applied. The angle θ is expressed in radians. The units for work performed with rotary motion are the same as those for linear motion—ft-lb in the U. S. customary system and N-m in the SI system.

The power P is given by:

$$P = T \times \omega \tag{3.12}$$

The symbol ω is the angular velocity given in units of radians/s or radians/min.

EXAMPLE 3.5

You are to analyze the power available to a rotary pump if it is driven with a dual crank mechanism taken from a bicycle. The radius of the crank is 6 inches and the force applied to the active pedal is 70 lbs. If the crank mechanism is operated at 45 revolutions per minute (RPM), determine the horsepower available to drive the pump. Assume that the losses in the chain drive are 5%.

Solution:

We will assume that the force on the active pedal is perpendicular to the arm of the crank; therefore, the torque, T is given by Eq. (3.10) as:

$$T = F \times r = 70 \times (6/12) = 35 \text{ ft-lb} \qquad (a)$$

The power into the sprocket of the chain drive is determined from Eq. (3.12) as:

$$P_{IN} = T \times \omega = 35 \times 45 \times 2\pi = 9896 \text{ ft-lb/min} \qquad (b)$$

Note we have converted RPM to the number of radians per minute by multiplying by 2π.

Since there are power losses of 5% in the chain drive, the power delivered to the pump is reduced. Accordingly, we write:

$$P_{OUT} = \mathcal{E} \, P_{IN} = (1 - 0.05) \times 9896 = 9401 \text{ ft-lb/min} \qquad (c)$$

Finally, we convert the power from the chain drive available at the pump to horsepower by using the definition in Eq. (3.8).

$$P = 9401/33,000 = 0.285 \text{ HP}$$

This example shows that a chain drive on a bicycle driven by the strong leg muscles of a cyclist generates slightly more than ¼ HP. Is this a reasonable example? A dedicated cyclist attempts to maintain a cadence of 60 to 80 RPM, and professional racers pump at 80 to 100 RPM; hence, the rate of 45 RPM appears to be feasible. The force of 70 lb depends on the weight and muscle structure of the cyclist, but sustaining a force of about ½ of one's body weight appears to be reasonable.

3.2.3 Human Factors

In the previous example, we questioned the capabilities of the human body to deliver forces at a sustainable cyclic rate. Clearly, if we are to effectively design a pumping system, knowledge of the physical capabilities of a typical man or woman must be understood. Fortunately, a significant body of knowledge has been developed over the past 50 years on the interface between man (woman) and machine. Ergonomics is the study of human characteristics for the appropriate design of the living and working environment. Its objective is that all tools, devices, machines and environment should advance the safety, welfare and performance of humans.

One of the essential elements of ergonomics is designing to match the size and posture of the human body. Tasks are accomplished in one of three major body positions: lying, sitting or standing. It is interesting to think about working while in a lying position, but it is a difficult position. Our muscle structure is conditioned to function more effectively if we are standing or sitting. In the design of the pumping system, the operator may stand or sit while providing power to the drive mechanism; however, sitting is preferred to standing if the operation is to be continued for long periods of time.

The work should be adapted to the operator[7], which implies that you should consider the following list of factors as your team designs the drive for the pump:

- Strength requirements shall be within reasonable limits.
- Body movements should follow natural rhythms.
- Posture, strength and movement should be harmonized.
- Appropriate support should be provided.
- The chain of force vectors through the body should be short and simple.
- Sufficient space for body movements should be allowed.
- Unnecessary strain should be avoided.

3.2.3.1 Pushing and Pulling with the Upper Body

Pump mechanisms operate with either reciprocating or rotary motion, both of which may be matched to the natural rhythms of the human body. The inward and outward motions of the reciprocating pumps may be applied directly by pushing and pulling with the upper body muscle structure. Capabilities of male soldiers exerting horizontal push and pull forces for short periods of time are presented in Table 3.1.

Table 3.1
Horizontal Push Pull Forces
Capabilities of Male Soldiers[8]

Horizontal Force N	Force Applied with	Traction Condition Friction coefficient μ at the floor
100	Both hands, one shoulder or back	Low traction — $0.2 < \mu < 0.3$
200	Both hands, one shoulder or back	Medium traction $\mu \cong 0.6$
300	Both hands, one shoulder or back	High traction $\mu > 0.9$
500	Both hands, one shoulder or back	Anchoring feet against foot stop

Inspection of Table 3.1 indicates that male soldiers can exert a relatively high force for short periods of time. With proper positioning and stops to provide adequate traction for the feet, it is reasonable to expect an average male to pull or push with a force of 500 N (112 lb). Data collected by NASA indicates significant variation in the capabilities of the male population. This variation expressed in terms of the standard deviation σ was about 120 N.

A normal distribution curve showing the frequency of the population as a function of the push or pull force capabilities is presented in Fig 3.7. If the drive mechanism is operated with the mean value of the force less one standard deviation, then:

$$F_{DESIGN} = F_{MEAN} - \sigma = 500 - 120 = 380 \text{ N} \qquad \text{(a)}$$

Most of the male population (84.1%) will be physically capable of operating the pump. It appears then that a push or pull force of 380 N (85 lb) is an appropriate design parameter if a typical male operates the pump.

We have presented a statistical approach for determining the ability of a population of males to apply forces with the upper body muscles. However, women may operate the pumping system. Do

---

[7] ISO standard 6385 addresses factors to consider in adapting the task to be performed to the operator.
[8] See MIL STD 1472 that is used in specifying equipment designed for the Department of Defense.

you expect a 110-pound woman to match the physical capabilities of a 160-pound male? Data on the physical capacity of women is not readily available. You may consider reducing the force capacities for the other gender by the ratio of the body weights. (For example, multiply by a factor of 110/160 ≅ 0.70 for a woman weighing 110 pounds).

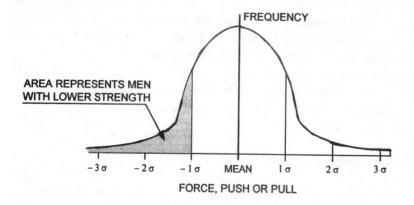

Fig. 3.7 The normal distribution is a statistical function used in describing the characteristics of a population.

3.2.3.2 Pushing with Legs

It is possible to generate higher forces for longer periods of time with the lower body muscles. We stand, walk and run using our leg muscles; consequently, they develop into larger and stronger muscles than those in our arms. The maximum force that can be developed by pushing with the legs depends on one's body position. If one is standing while pushing, the maximum force is limited to the body weight. However, if one is in a sitting position with back support, the maximum force is significantly larger.

Studies have shown that the maximum force developed depends on the angle of the legs in the sitting position, and the location of the pedals relative to the sitting plane. We show a sketch of a male in the sitting position in Fig. 3.8. The angle of rotation at the hip and knee are defined by α and β respectively. The back of the operator is supported and the pedal against which the force is imposed is located slightly above the sitting plane.

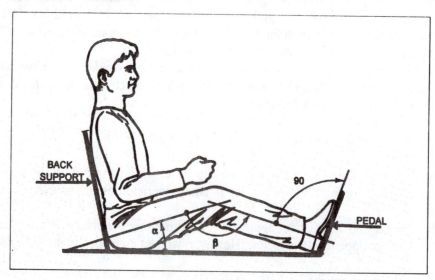

Fig 3.8 Geometric parameters affecting maximum force delivered with a leg.

The best performance is achieved with a hip angle α from 15 to 20°. The maximum force developed on a pedal with the action of a single leg is a function of the angle of knees, β as illustrated in Fig. 3.9. Inspection of this figure indicates that a typical male can develop a maximum force of about 2000 N (450 lb) with a knee angle of about 160°. However, this force decreases rapidly as the legs are straightened beyond this point. The force is imposed on a pedal and the position of that pedal influences the force that is developed. The highest forces are developed if the pedal is positioned at the sitting plane or above it by less than 300 mm. If the pedals are positioned below the sitting plane, the maximum force decreases significantly with the distance below the plane.

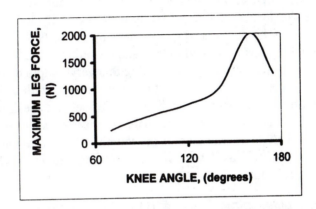

Fig. 3.9 Force developed by a typical male operator as a function of the knee angle β.

Again, the data available for the forces developed with legs are for males. If you are designing a pump for operation by females, you may wish to reduce these forces by the ratio of body weight as discussed previously. Alternatively, you may visit a weight room at a well-equipped health organization and measure the ability of several people to repetitively generate forces using different sets of muscles.

3.3 DRIVE MECHANISMS

In the previous section, we described both the lever and the crank as devices to facilitate the human interface with a machine. The lever is used to amplify a person's force capability by trading off additional stroke for more force. Work is not added to the system, but the lever makes it possible for a small person to operate a machine requiring large forces.

The crank is another device that is commonly employed to convert human motions into a smooth rotation. The human body is made up of many straight members connected with joints. Our fingers, arms and legs are essentially rods with joints at the knuckles, wrists, elbows, knees, etc. While we are very mobile, the motions created are rarely perfectly circular. The crank aids us in creating rotary motions required to drive wheels on vehicles and impellers on pumps. Like the lever, the crank also permits us to amplify the torque that can be produced by a specified force. The length of the handle on a crank is similar to the length of a lever.

While the significant benefits of both the lever and the crank are inherently apparent, they must be adapted to each engineering application. The simple lever shown in Fig. 3.3 is adequate to describe the principle of force amplification, but it is deficient in several respects if it is to be applied to power a reciprocal pump. Similarly, the crank illustrated in Fig. 3.5 describes the concept of converting force into torque, but it is not suitable to drive a pump. For both rotary and reciprocating pumps, the drives will include one or the other of these basic devices. However, the drives incorporate several other mechanical components that are essential for a successful application of human power to the operation of a pump.

3.3.1 Oscillating Drives

3.3.1.1 Direct Oscillating Drives

Reciprocal pumps with their in-out action are operated with oscillating drives. The drive may be direct or it may contain linkages that amplify the force available to operate the pump. Let's consider a direct drive for a diaphragm pump as illustrated in Fig. 3.10. The diaphragm in the pump is moved to the left and right by the application of the forces applied to the handle. A rod connects the handle to the diaphragm. A stop is incorporated on both sides of the diaphragm to prevent it from being cut by impact with either of the two heads. A sleeve bearing is inserted between the rod and head to reduce friction and losses of work. The pump is held stationary by supports that fix it to some immovable base. Without the supports the forces applied to the handle would tend to move the entire pump.

There are many decisions to make in designing this very simple drive. Consider the handle. Suppose it is fabricated from a circular rod so that it can be grasped with both hands in a power grip. The diameter of the handle must be large enough so that it does not break under the action of large forces from both hands. On the other hand, the diameter must be small enough for the operator's fingers to encircle it with a slight overlapping of the thumbs to generate the maximum force. If the pump is to be operated for long periods of time, rubber grips contoured to fit the hand may be added to prevent severe tissue compression.

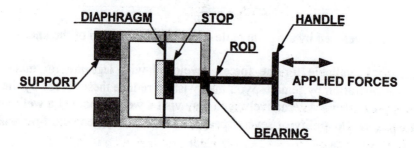

Fig. 3.10 Schematic illustration of a direct drive for an oscillating diaphragm pump.

3.3.1.2 Oscillating Drives with Force Amplification

If the pump is designed with a large diameter, the force required to generate the specified pressure may be higher than a human operator is capable of providing for sustained periods of time. In these applications, an oscillating drive with force amplification similar to that shown in Fig. 3.11 is needed.

An examination of Fig. 3.11 indicates that several components are required for an effective oscillating drive that provides an amplified force. The lever is the central element with the amplification determined by the ratio d/s. Two pins replace the fulcrum on the simple lever, illustrated in Fig. 3.3. The first pin connects the lever to a base with a clevis. The second pin connects the lever to the piston rod through a link. It is not possible to connect the lever directly to the piston rod. The pin at position s moves along an arc instead of a straight line. The purpose of the link is to accommodate the rotation of the pin at one end while providing linear motion to the piston rod at the other.

The entire system is connected to a base plate that holds the cylinder in a fixed position relative to the lever. A cross handle is provided at the end of the lever to permit the operator to provide alternating forces with both hands.

Energy (work) is lost in the drive because of friction at each of the three pins. As the lever is moved back and forth, the pins rotate in the holes and frictional forces develop that resist the motion. The losses may be mitigated by incorporating a self lubricating sleeve bearing in the holes provided for the pins as shown in Fig. 3.12.

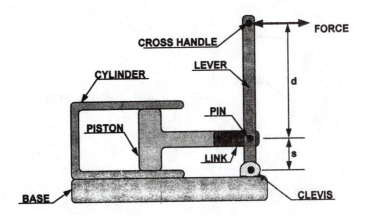

Fig. 3.11 An oscillating drive mechanism incorporating a lever for force amplification.

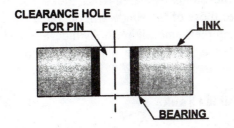

Fig. 3.12 Section view of a sleeve bearing in the connecting link.

Sleeve bearings are usually fabricated from porous brass or bronze with a lubricant impregnated in the pores. The lubricant serves to reduce the friction between the pin and the clearance hole in the link. In some applications with lower loads on the pins, the sleeve bearings are fabricated from plastics such as nylon or teflon compounds that exhibit low coefficients of friction. Frictional losses are reduced by a factor of about three when sleeve bearings are used together with very smooth and hardened steel pins.

3.3.2 Rotary Drives

Rotary drives are employed with centrifugal pumps to apply torque to the impeller at relatively high speed. In many applications where electric power is available, a motor is connected directly to the shaft of the pump to provide the required torque at a suitable angular velocity (often at 1750 RPM). However, if electrical power is not available and the pump is operated with human power a suitable rotary drive must be provided. We will describe a simple rotary drive based on a crank mechanism, and then illustrate more complex rotary drives where the speed and/or torque may be amplified.

3.3.2.1 Simple Rotary Drive with a Single Crank

Consider the simple rotary drive utilizing a crank mechanism shown in Fig. 3.13. The crank is attached to the shaft extending from the rotary pump with a setscrew or key. At the other end, a handle is attached with a pin through the crank. The pin passes through a clearance hole in the handle so that the handle will rotate when the crank is turned. Rotation of the handle is important to avoid rubbing blisters on the operator's hands. A support is attached to the pump to keep the entire assembly fixed as the crank is rotated.

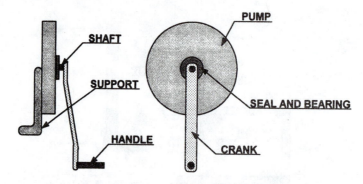

Fig. 3.13 A simple rotary drive incorporating a crank to generate torque.

The rotary drive presented in Fig. 3.13 is driven with the upper body muscles; consequently, the torque that can be developed is less than a double crank mechanism found on a typical bicycle. The double crank used on a bicycle is operated with the lower body muscles and the torque capabilities are two to three times those of the single arm operated cranks. Of course, the mechanism for the double crank is more complex since it must include a seat for the operator and a spatial arrangement that avoids interference between the pump and the drive mechanism. We describe a double crank rotary drive in the subsection below.

3.3.2.2 Rotary Drive with a Double Crank

A typical bicycle employs a double crank with a pedal on each side of the sprocket—one for each foot. This arrangement permits the lower body muscles to apply forces first with one foot and then the other. By alternating the action of the legs as the double crank rotates, the application of high forces is nearly continuous and significant power is transmitted.

A double crank drive similar to that used on a bicycle is illustrated in Fig. 3.14. We have adapted the double crank and the sprocket from a bicycle to provide a comfortable arrangement for a human operator. The entire assembly is mounted on a base that provides the required support for all of the components in the pumping system. The forces on the pedals are transmitted to the large sprocket and then to a chain. The chain transmits the forces to a smaller sprocket, which in turn drives a rotary pump. The operator, not shown in Fig. 3.14, sits on the base with a back support and uses his or her legs to apply forces to the pedals.

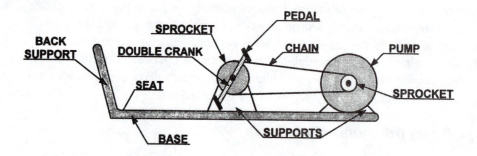

Fig. 3.14 Rotary drive incorporating a double crank and sprockets with chain.

This double crank chain drive performs five different functions as listed below:

1. Forces applied to the pedals are converted to a torque about the center of rotation of the large sprocket.
2. The torque on the large sprocket is converted to a force that is applied to the chain.

3. The chain transmits this force from the large sprocket to the smaller sprocket.
4. The small sprocket converts the chain force to a torque at the center of rotation of the shaft driving the pump.
5. In transmitting forces from one center of rotation to another, a change in angular velocity occurs.

Let's examine each of these functions and describe the relations necessary for your team to determine the torque and angular velocity (power) applied to the shaft of the pump.

Step 1: Forces applied to the pedals are converted to a torque about the center of rotation of the large sprocket. From Eq. (3.10) it is clear that this torque, T_{sp} is given by:

$$T_{sp} = F_p \times r_{cr} \tag{3.13}$$

Where F_p is the component of force on the pedal that is perpendicular to the arm of the crank, and r_{cr} is the radius of the crank.

Step 2: The torque, T_{sp} on the large sprocket is converted to a force, F_{ch} that is applied to the chain. Let's draw a free body diagram of the large sprocket showing the torque and the forces as shown in Fig. 3.15.

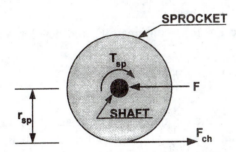

Fig. 3.15 Free body diagram showing the forces and torque acting on the large sprocket.

Again, Eq. (3.10) applies and we write:

$$T_{sp} = F_{ch} \times r_{sp} \tag{a}$$

Substituting Eq. (a) into Eq. (3.13) and eliminating T_{sp} yields:

$$F_{ch} = (r_{cr}/r_{sp}) F_p \tag{3.14}$$

From Eq. (3.14), it is evident that the force on the chain may be amplified by the ratio (r_{cr}/r_{sp}). The crank and the sprocket act like a lever providing the opportunity for force amplification by appropriate specification of the radii of the crank and sprocket.

Step 3: The chain transmits this force from the large sprocket to the smaller sprocket. As the large sprocket rotates, a tension force is developed on the lower chain while the upper segment of the chain is slack. The tension force, F_{ch} in the lower segment of the chain is transmitted to the smaller sprocket as shown in Fig. 3.16.

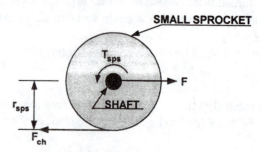

Fig. 3.16 Free body diagram of the small sprocket showing the chain force F_{ch} and torque T_{sps}.

Step 4: The small sprocket converts the chain force to a torque about the center of rotation of the shaft driving the pump. Let's again employ Eq. (3.10) with the free body diagram shown in Fig 3.16 to determine the torque T_{sps} acting on the shaft of the small sprocket.

$$T_{sps} = F_{ch} \times r_{sps} \qquad \text{(b)}$$

Substituting Eq. (3.14) into (b), we eliminate F_{ch} and obtain T_{sps} in terms of the pedal force, F_p.

$$T_{sps} = (r_{cr}/r_{sp})r_{sps} \times F_p \qquad \text{(c)}$$

Finally, we substitute Eq. (3.13) into Eq. (c) to eliminate F_p and develop a relationship between the torques applied to the two sprockets.

$$T_{sps} = (r_{sps}/r_{sp})T_{sp} \qquad \text{(3.15)}$$

Examination of Eq. (3.15) indicates that the torque acting on the small sprocket is less than the torque acting on the large sprocket because the ratio (r_{sps}/r_{sp}) is less than one. By using two different sizes of sprockets, we have changed the amount of torque delivered to the pump. With the arrangement of the sprockets—large to small—we have reduced the torque.

Step 5: In transmitting forces from one center of rotation to another, a change in angular velocity occurs. In examining Eq. (3.15), we noted a decrease in the torque about the small sprocket when compared to the torque about the large sprocket. In this case we traded torque for speed (angular velocity). A chain provides a positive type of drive. Teeth on the sprocket engage holes in the chain; consequently, no slippage occurs as the sprockets rotate. When the large sprocket revolves once, the chain moves a distance S.

$$S = \pi D_{sp} \qquad \text{(d)}$$

The distance the chain moves on the small sprocket is the same; hence, we may write:

$$S = k\pi D_{sps} \qquad \text{(e)}$$

Where D_{sp} and D_{sps} are the diameters of the large and small sprockets respectively, and k is the number of revolutions of the small sprocket for each revolution of the large sprocket.

It is evident from these equations that:

$$k = D_{sp}/D_{sps} = r_{sp}/r_{sps} \qquad \text{(f)}$$

From this relation, it is clear that the angular velocity n_{sps} of the small sprocket is given by:

$$n_{sps} = (r_{sp}/r_{sps})n_{sp} \qquad (3.16)$$

The sprocket set changes the speed of the shaft driving the pump. If a human operator is pumping at 60 RPM and we employ a small sprocket that is half the diameter of the large sprocket the speed of the shaft driving the pump is 120 RPM.

EXAMPLE 3.6

A dual crank chain drive similar to that shown in Fig 3.14 is to be constructed with the following design parameters:

$$r_{sp} = 5 \text{ in.}, \ r_{sps} = 3 \text{ in.}, \ r_{cr} = 7 \text{ in.}$$

The operator is a physically fit male weighing 190 lb capable of exerting an average pedal force of 160 lb at 50 RPM for an extended period of time. The component of the average pedal force, perpendicular to the crank rod for an entire rotation of the sprocket, is 140 lb. You are to perform a design analysis to determine the feasibility of the proposed drive.

Solution:

Let's first determine the torque applied to the large sprocket by using Eq. (3.13).

$$T_{sp} = F_{cr} \times r_{cr} = 140 \times 7 = 980 \text{ in.-lb} = 81.67 \text{ ft-lb} \qquad (a)$$

The force on the roller chain is given by Eq. (3.14) as:

$$F_{ch} = (r_{cr}/r_{sp}) \, F_p = (7/5)(140) = 196 \text{ lb} \qquad (b)$$

Let's examine this result for the chain force. Is 392 lb too large a force for the roller chain to support? To ascertain if the roller chain is safe we compare the applied force to the rated breaking load for the chain. Reference to a chain catalog[9] indicates that ANSI No. 35 steel, single-strand, roller chain, with a pitch of 3/8 in., has a rated breaking load of 2400 lb. It is clear that the force applied to the roller chain is not excessive and that the chain will not fail by breaking.

Next, let's determine the torque acting on the small sprocket from Eq. (3.15)

$$T_{sps} = (r_{sps}/r_{sp})T_{sp} = (3/5)(81.67) = 49.00 \text{ ft-lb} \qquad (c)$$

This is the torque that is delivered to the pump.

Since the angular velocity of a centrifugal pump is extremely important, we determine this quantity from Eq. 3.16.

$$n_{sps} = (r_{sp}/r_{sps})n_{sp} = (5/3)(50) = 83.33 \text{ RPM} \qquad (d)$$

---

[9] McMaster-Carr Supply Co. catalog lists more than 155,000 products with specifications provided for most of them.

We should question if the rotational speed of about 80-RPM is adequate for the operation of the centrifugal pump; however, the answer cannot be provided at this stage in the analysis. More information on pumps to be provided in Chapter 4 is required.

We continue the analysis by determining the power available to the pump from Eq. (3.12).

$$P = T \times \omega = 49.00 \times 83.33 \times 2\pi = 25,660 \text{ ft-lb/min} \qquad (e)$$

Converting this result to horsepower using the conversion factor given in Eq. (3.8) yields:

$$P = 25,660/33,000 = 0.777 \text{ HP} \qquad (f)$$

This result indicates a very large power output for a human—even a well-fit male weighing 190 pounds. You may visit a well-equipped weight room to determine if the data provided in this example are realistic.

In the previous example, we did not account for the efficiency of the chain drive. The power to the pump will be reduced by any losses in the drive mechanism. These losses will depend on the alignment of the two sprockets and the bearings employed to support the shafts of the large and small sprockets. If the two sprockets are properly aligned and ball bearings are used to support the shafts, the losses in a well-lubricated roller chain drive are relatively low. Properly designed roller chain drives exhibit efficiencies of 90 to 95%.

3.3.2.3 Belt Drives

Belt drives are similar to roller chain drives. Belts are used instead of the roller chain and sheaves are used instead of sprockets. Today V-belts are the most common of the various types of belts employed in the design of drives. As the name implies, the cross section of a V-belt has slanted sides like the letter V. We show a typical cross section of a V-belt in Fig. 3.17.

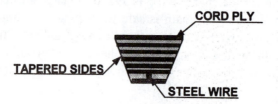

Fig. 3.17 Cross-section of a V-belt showing a steel reinforcing wire and cord plies.

The sheave or V-belt pulley has a V shaped groove about its perimeter. The V-belt fits into the groove on the pulley, and as the belt moves, the pulley is rotated about its center. The driving force on the pulley is due to friction acting on the sides of the V-belt. The bottom flat of the V-belt does not contact the base of the groove in the pulley. A V-belt drive may be examined by opening the hood of your auto and finding the alternator. The alternator is driven by a V-belt that transmits power from the engine.

Belt drives have several advantages when compared to roller chain drives. They cost less, are quieter and are more tolerant of poor alignment. The disadvantages are lower efficiency and lower capacity for transmitting power. They also tend to slip and cannot be consider positive drives.

3.3.2.4 Gear Drives

Gear drives are also used to transmit power from one shaft to another. Spur gears illustrated in Fig. 3.18 are the most common although many other types of gearing are often used. The two gears are in mesh with teeth from one gear inserted in the grove between two teeth from the other gear. The profile (involute) of the teeth is formed to minimize slipping between the mating surfaces as the gears rotate. The smaller of the two gears is called the pinion and the larger is simply the gear or if it is very large the bull gear.

The spur gears are keyed to two parallel shafts that rotate in opposite directions. Spur gears provide a positive drive since slippage cannot occur without breaking teeth. The pitch velocity v at the contact point where the two gears interact is:

$$v_1 = r_1 \times \omega_1 \tag{a}$$

The subscript one identifies the larger gear.
The pitch velocity for the pinion gear is:

$$v_2 = r_2 \times \omega_2 \tag{b}$$

Since the engagement of the teeth provides a positive drive:

$$v_1 = v_2 \tag{c}$$

Substituting Eqs. (a) and (b) into Eq. (c) yields:

$$\omega_2 = (r_1/r_2)\,\omega_1 = (D_1/D_2)\,\omega_1 = (N_1/N_2)\,\omega_1 \tag{3.17}$$

Where r is the pitch radius of the gears, D is the pitch diameter and N is the number of teeth on each gear. Examination of Eq. (3.17) indicates that it is possible to change the angular velocity of a shaft by using gears of different diameters. We transmit power from a large diameter to a small diameter to increase angular velocity and from a small diameter to a large diameter to decrease angular velocity. Also, note that the directions of rotation of the input and output shafts are different.

Fig. 3.18 Illustration of spur gears.

Now that we may determine the angular velocity of one shaft or the other from Eq. (3.17), let's next consider the change in torque produced by a pair of spur gears. We will assume that no power is lost in the gear drive[10]. With this assumption, we write:

$$P_2 = \omega_2 \times T_2 = P_1 = \omega_1 \times T_1 \qquad (3.18)$$

Solve Eq. (3.19) for the output torque yields:
$$T_2 = (\omega_1/\omega_2)T_1 \qquad (3.19)$$

This result shows that the output torque T_2 decreases if the angular velocity increases as the power is transmitted through the gear drive.

Gear drives are very efficient and are capable of transmitting large quantities of power. If you shift gears in your automobile, you are making use of a gear drive to adapt the torque from the engine to the requirements of the road. With good design, the size and weight of the drive can be minimized.

Gear drives are the highest performance of the various types of drives; however, they suffer from several disadvantages. They are the most costly of the drives considered here, and alignment and spacing of the shafts must be precise to avoid interference. Spur gears are also noisy at high speed.

3.3.3 Bearings and Alignment

Chain, belt and gear drives are similar in that they all incorporate two or more shafts as well as rotating components. The shafts must be supported with bearings and the rotating elements must be aligned with respect to each other if the drives are to function effectively. Let's first consider bearings that support the shafts employed in the drive and the pump.

3.3.3.1 Sleeve Bearings

There are two different classes of bearings—sliding and roller. The sliding bearings are the simpler of the two classes because they consist of a cylindrical sleeve or a sleeve with an attached flange as illustrated in Fig. 3.19.

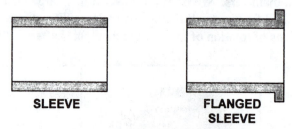

SLEEVE   FLANGED
                SLEEVE

Fig. 3.19 Section views of a sleeve and flanged sleeve bearing.

The sleeve bearing is inserted into a support block with a light interference fit. The support block is fastened to a stationary structure. The journal of the shaft that contacts the bearing is fitted to the sleeve. The inside diameter of the sleeve is slightly larger than the outside diameter of the shaft journal so that the shaft is free to rotate within the bearing. Since the sleeve is fixed to the support block, it serves to constrain the shaft in the radial direction.

A simple sleeve bearing cannot constrain the shaft in the axial direction. For this reason, a flanged sleeve bearing is employed on one end of the shaft as illustrated in Fig. 3.20. The flanged

---

[10] This is an excellent assumption since the power lost in a properly aligned spur gear drive is usually less than one or two percent.

sleeve bearing is inserted into the supporting block with a light interference fit. The journal end of the shaft, where contact with the bearing occurs, is fitted into the flanged sleeve bearing with the shaft journal contacting the sleeve and shoulder on the shaft contacting the flange. The radial forces acting on the shaft are resisted by the sleeve portion of the bearing and the flange resists the axial forces. The usual practice is to fit one end of the shaft with a simple sleeve bearing and the other end with a flanged sleeve bearing.

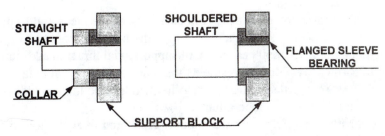

Fig. 3.20 Two methods for using flanged sleeve bearings to achieve axial constraint.

In many maintenance free applications, the sleeve bearings are impregnated with solid lubricants such as graphite or MoS_2 to provide lubrication over the life of the product.

3.3.3.2 Roller Bearings

The roller bearing is superior to the sleeve bearing because it provides the necessary constraint for the shaft with much lower friction losses. However, roller bearings are more expensive than sleeve bearings and the importance of reduced friction must be traded off with the increased price of the bearings. A cut away view of a typical roller bearing is presented in Fig. 3.21.

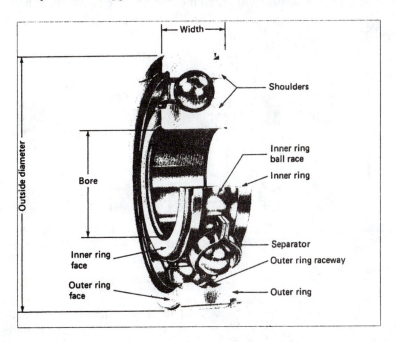

Fig. 3.21 A cut away view of a ball bearing showing its essential components.

Roller bearings transfer load from the shaft to the support block through roller elements rather than sliding elements. The rolling elements can be balls or rollers (straight, spherical, needle or tapered). For relatively light loads, as we expect to encounter in the design of a pump, ball bearings will be adequate.

The bearing has four different parts including:

- The outer race
- The inner race
- The balls
- The separator.

The separator is used to space the balls about the periphery of the races and to prevent the balls from contacting each other thereby reducing friction. Since the balls are contained in raceways (grooves) cut into the inner and outer races, ball bearings will support axial thrust in addition to the normal radial loads. Many ball bearings are fitted with shields that enclose the two annular openings between the inner and outer races. These sealed bearings are filled with grease that provides adequate lubrication for several years of maintenance free operation.

The fit of a roller bearing within the support block and over the journal of the shaft is very important because the bearing is manufactured to extremely small tolerances. The clearance between the two races and the balls is usually less than 0.001 inch. To avoid seizure, the ball bearing has to fit on the shaft with essentially no clearance or a very slight interference. Similarly the outer race must fit into the bore of the support block with essentially no clearance or an extremely small interference. Often screws, collars and flanges are employed to exert pressure on the faces of the races to insure that the shaft does not rotate within the bore of the inner race or that the outer race does not rotate within the bore of the support block.

3.3.3.3 Alignment

Drives for rotary pumps will contain two or more shafts and two or more rotating components such as sprockets or sheaves. For the drive to function without difficulty, the shafts must be parallel and the rotating components must lie in the same plane. The illustration presented in Fig. 3.22 represents three different chain drive assemblies.

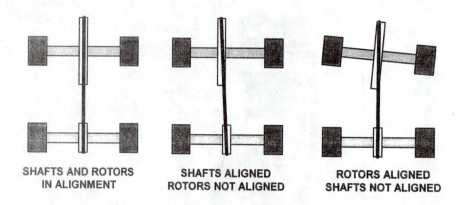

SHAFTS AND ROTORS   SHAFTS ALIGNED   ROTORS ALIGNED
IN ALIGNMENT   ROTORS NOT ALIGNED   SHAFTS NOT ALIGNED

Fig. 3.22 Examples of alignment and misalignment.

In Fig. 3.22 (left) the shafts and rotors are both in alignment; the chain lies in the plane of the rotors and it will move freely without binding. If one of the sprockets is displaced relative to the other as in Fig. 3.22 (center), even if the shafts are in alignment, the chain will bind and friction losses increase dramatically. Finally, if one shaft is not parallel to the other (as in Fig. 3.22 right), the rotor is forced out of the plane of the chain and again binding occurs. Alignment is a difficult problem that requires careful attention during the assembly of your system. We recommend that you carefully plan the assembly operation to obtain the best possible alignment.

3.4 SUMMARY

We have described work and energy and showed that they are equivalent in mechanical systems in which energy is conserved. However, when energy is not conserved, losses of work occur and the efficiency of a system is less than 100%. Equations (3.1) through (3.5) have been introduced to show you methods for determining, work, losses, energy and efficiency.

Power P is the amount of work performed per unit time. In the U. S. Customary system power is expressed in terms of horsepower and in the SI system it is in terms of watts. Examples related to the project are provided to illustrate the techniques used in determining work, work losses, efficiency and power.

Since the pump is to be operated with human power, we have discussed two simple devices used to enhance the capability of an operator. The first is a lever, which amplifies the force, and the second is a crank that permits one to impart a circular motion using either upper or lower body muscles. Relations are developed that enable you to design a lever or crank system to develop a specified output (force or torque) for a given applied force. Again, examples related to the design of the pumping system are provided to guide you in the application of these equations.

We briefly introduced the subject of ergonomics, which is the study of human characteristics for the appropriate design of living and working environments. Some data is provided to indicate the capabilities of men in developing forces with both the upper and lower body muscles. Data on women is sparse, and we recommend that the force capabilities of men be scaled downward based on the ratio of body weight to arrive at an estimate of the capability of women to generate forces. Alternatively, a trip to a well-equipped weight room in a health club will provide even better data.

Since a drive mechanism is required to power the pump, we have provided schematic illustrations of the components involved in oscillating and rotary drives. For oscillating drives, the direct and lever systems are illustrated. For rotary drives, we have described systems based on chain drives, V-belts, and gears. Relations have been introduced that will enable you to perform design analyses of these drives.

Methods for design analysis have been developed in conjunction with the description of the various drives. Examples have been provided to illustrate method of analysis. We have also shown an interpretation of the results of these analyses to modify and improve the design. We stress the importance of the interpretation of the analysis to give insight pertaining to the design. If an analysis fails to provide insight pertaining to the design it is of limited value.

Finally, we introduce the important topic of bearings and alignment. The selection of the bearing influences, to a significant degree, the amount of friction losses in the system. Self-lubricated sleeve bearings have much large frictional losses than the more expensive roller bearings. In either case, the alignment of the bearings to produce parallel shafts in the drive system is of paramount importance. The concept of alignment of the bearings and the rotating elements has been introduced and illustrations of perfect alignment and misalignment provided.

REFERENCES

1. Hibbeler, R. C., Engineering Mechanics: Statics, 8th Edition, Macmillian Publishing Co. New York, 1998.
2. Kroemer, K. H. E., H. B. Kroemer, and K. E. Kroemer-Elbert, Ergonomics: How to Design for Ease and Efficiency, Prentice Hall, Englewood Cliffs, NJ, 1994.
3. Kroemer, K. H. E., "Human Strength: Terminology, Measurement and Interpretation of Data," *Human Factors*, (12), p. 279-313, 1970.
4. Kroemer, K. H. E., "Pedal Operation by a Seated Operator," SAE Paper 72004, Society of Automotive Engineers, 1972.

5. NASA, "Man Systems Integration Standards (Revision A). NASA-STD 3000, L. B. J. Space Center, Houston TX, SP 34-89-230, 1989.

6. Juvinall, R. C. and K. M Marshek, Fundamentals of Machine Component Design, 2nd Edition, John Wiley, 1991.

EXERCISES

3.1 Determine the work required carrying a bucket containing two gallons of water from a stream to a house located at the top of a hill 1000 feet away from the stream. The hill has an average grade of 7%.

3.2 Susan pumps weights. In one of her routines, she lifts a 110 lb barbell from the floor to an outstretch position above her head. The lift is a distance of 6 ft-9 in. If she performs this lift 16 times, determine the work performed. Does she perform work in lowering the weight to the floor during each cycle?

3.3 A crew of 10 men is rowing to propel a shell through water at high speed. If the crew is expected to deliver two horsepower, while exerting a force of 90 lb through a stroke of 3.1 ft, determine the count that they must maintain.

3.4 Write an engineering brief providing your estimate of the maximum horsepower that an Olympic rowing team can develop during competition.

3.5 Both horsepower and watt are units of power. Derive the conversion factors from the basic definitions of these two quantities to enable you to convert from one to the other.

3.6 Suppose your team develops a drive that is 80% efficient and a pump that is 43% efficient. What is the efficiency of the pumping system?

3.7 Your team establishes a design specification for the pumping system of 40 GPM against a head of 10 ft. Prepare a graph showing the required power to the pumping system as a function of the total system efficiency.

3.8 Sketch a design of an oscillating drive for a large diameter piston pump. Specify the diameter of the piston and the length of the stroke. Then perform a design analysis to determine the flow rate from the system. State all of the assumptions made in the analysis. Finally, interpret the results in terms of design modifications.

3.9 A cyclist is travelling up a long hill with an 8% grade at a velocity of 10 MPH. He is maintaining a constant velocity with a count of 70 RPM. If the radius of the crank on the bicycle is 170-mm and the cyclist and bicycle weigh 200 lb, determine the average force applied by the cyclist to the pedal. Is this realistic? What assumptions did you make in the design analysis?

3.10 Conduct an experiment to verify at least one of the results presented in Table 3.1.

3.11 Suppose you find an old DOD report describing the mean upper body strength of seasoned marines as 135 lb with a standard deviation of 25 lb. If you use this data to design your pumping system, specify the force applied by the human operator if you expect that most of the male population will be able to operate the pump. Write a brief statement describing your rationale in establishing this specification.

3.12 Sketch a design of a seat and pedal mechanism for a chain drive. Cite reasons for the placement of the backrest and the pedal height. Have you incorporated features to permit the seat to be used by both large and small operators?

3.13 Sketch an oscillating drive mechanism with a pump incorporating a lever for force magnification. Specify the dimensions of each component and suggest materials to be used in their fabrication.

3.14 For the oscillating drive mechanism of Fig. 3.11, perform a design analysis that indicates the force requirements and the flow rates. State all of the assumptions required performing the analysis. Interpret the solution and recommend design changes to improve performance of the pumping system.

3.15 For the simple rotary drive shown in Fig. 3.13, the torque on the shaft at the pump is given by T = F × r. You are to design the crank. How long do you make it and why? What are the tradeoffs?

3.16 A dual crank chain drive similar to that shown in Fig. 3.14 is proposed for a pumping system.

The design parameters are r_{sp} = 7 in, r_{sps} = 3 in. and r_{cr} = 8 in.

The operator is a fine specimen—weighing 220 lb capable of exerting an average pedal force of 165 lb at 65 RPM for an extended period of time. Perform a design analysis and interpret your solution as a basis for design modifications. Make any reasonable assumption required to pursue the analysis.

3.17 Prepare an engineering brief describing the relative advantages and disadvantages of the three types of rotary drives (roller chain, V-belt and spur gear) discussed in Section 3.3.2.

3.18 Locate a typical consumer product like a power drill, or a leaf blower, car vacuum, etc. Disassemble the product with care and identify the types of bearings used. Note the method of lubrication. How was alignment achieved? Reassemble the product.

3.19 Suppose that your team is to design a centrifugal pump requiring an angular velocity higher than that, which can be achieved with a single stage chain drive. Prepare a sketch showing the design features of a multistage chain drive. Then show the relations necessary to perform a design analysis of this drive.

CHAPTER 4

RECIPROCATING AND ROTARY PUMPS

4.1 PUMP SELECTION

The selection of the type of pump to employ is the most important design decision facing the team. The historical review of the development of pumps showed that there are many different types of pumps that your team can design and build. To help your team make this decision, we will provide more detail pertaining to three different types of pumps commonly employed today. This pre-selection on the part of the author should not discourage your team from selecting another type of pump. If your team should choose an Archimedes screw or a chain of pots for your design project, that is your prerogative. However, the choice should be made for valid reasons after considerable team discussion.

Three different pump types are reviewed in detail—the piston pump, the diaphragm pump and the centrifugal pump. In each case, we describe the mechanical details of the pump, discuss the forces required during discharge, the work performed, and efficiency in operating the pump. Finally, we provide an overall assessment of a particular type of pump.

4.2 PISTON PUMPS

4.2.1 Mechanical Details

The piston pump, shown in Fig. 4.1, operates with a reciprocating action. On the intake stroke, water is drawn into the chamber, and on the output stroke this charge of water is forced from the chamber with sufficient pressure to transport it through the delivery piping to the open channel. The body of the pump is cylindrical—often fabricated from a section of tubing. The ends of the pump are fitted with two heads. One head is blank, but the other has a central hole to accommodate the piston rod. We have not shown the seals that must be incorporated to prevent leakage. We will describe the seals in more detail later in this Section.

The piston is usually attached to the piston rod with a threaded connection. The outside diameter of the piston is specified so that it fits with the inside diameter of the cylinder while accommodating the thickness of the seal.

Two nozzles are incorporated in the design. They may be placed in the head opposite the piston or in the sides of the cylinder. One nozzle connects to the inlet piping and the other to the delivery piping. Inlet and outlet valves are placed in these nozzles to control the flow of water into and out of the cylinder.

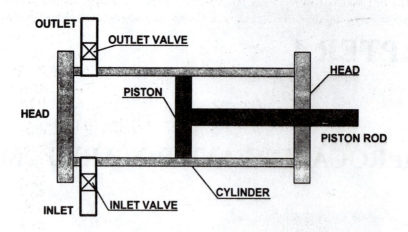

Fig. 4.1 Section view of a single acting piston pump.

4.2.1.1 Seals

O-rings are perhaps the most commonly used type of seal. O-rings are simply a rubber torus that is fitted into a groove cut into one of the two mating parts. The action of the O-ring in sealing a joint is illustrated in Fig. 4.2. O-rings are effective for both static and dynamic applications. In static applications, both surfaces are stationary, but in the dynamic case, one surface slides relative to another.

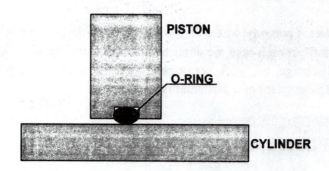

Fig. 4.2 An O-ring is fitted into a groove and then squeezed into the joint at the interface between two components to effect a seal.

A cup seal, an alternative to the O-ring, is commonly used to prevent leakage between the cylinder wall and the piston. As the name implies, the seal is cup shaped so that it fits over the end of the piston. A back up plate fastened to the piston with a bolt serves to hold the seal in place as the piston is stroked. The cup seal is fabricated from either rubber or leather. The thickness of the material is slightly larger than the gap between the piston and the cylinder so that the walls of the cup are squeezed together. If leather is used, it is often lubricated with oil or grease to reduce friction.

Rope packing is also used to seal around cylindrical objects such as a piston rod as shown in Fig. 4.4. The rope packing is wound about the shaft and forced into an annular groove cut into the head of the cylinder. A plug that fits into the groove is forced down on the packing to squeeze it against the piston rod. To reduce friction between the packing material and the piston rod, the packing is impregnated with graphite, grease or teflon. With time of operation, the rope wears and the plug must be adjusted to force additional packing material into the gap between the piston rod and the cylinder head.

There are additional methods for sealing against leakage. Reference to a typical machine design textbook will provide a more complete review of commercially available seals.

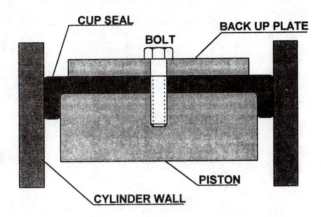

Fig. 4.3 Mechanical details of a cup seal preventing leakage between a cylinder and a piston.

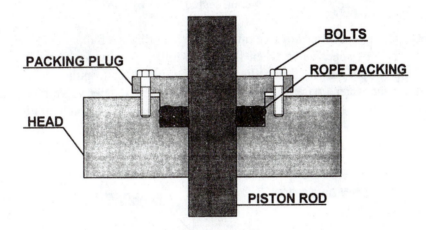

Fig. 4.4 Forming a seal with rope packing wound around the piston rod.

4.2.1.2 Valves

Several different types of pumps require valves to control flow into and out of the pump chamber. These are one-way valves, often called check valves, permitting flow in one direction but not the other.

You are familiar with valves; in fact, you open a valve when filling a glass with water or washing your hands. The valve on a water faucet is manually operated because you usually turn a screw to remove an obstruction that had previously blocked the flow. In addition to manual valves, there are those operated with motors, air pressure or solenoids. However, check valves are automatic because they do not require intervention by an operator to function. The differential pressure of the water across the active element of the valve opens them. They are closed either with a force generated by a spring or gravity as shown in Fig. 4.5.

On the intake stoke of a piston pump, a vacuum is created in the pump chamber and at the entrance to both the inlet and discharge pipes. This suction pulls the check valve on the discharge line closed and opens the check valve on the inlet line. For the check valve to operate without leakage, the fit of the valve cone against the valve seat must be perfect. The use of soft rubber on the face of the valve cone often alleviates problems associated with uneven contact between the cone and the seat.

The active element in this type of check valve is the valve cone. It moves to open and close the valve. Other geometries, such as the ball or disk, are often used instead of the cone; however, the principle of operation is the same in all cases.

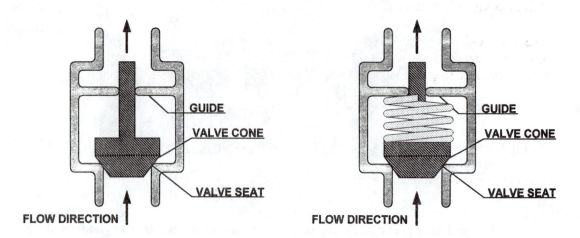

Fig. 4.5 Check valve with a valve cone seated either by a gravitational or spring force.

Another directional control valve often employed utilizes a flapper as the active element. You have probably used a flapper valve when flushing a toilet. The water stored in the water closet is contained with a flapper valve. When flushing you turn a lever to pull a small chain that lifts the flapper to release the water. After the closet has emptied, the flapper valve falls back into place and seals the tank. We show a typical flapper used as a directional control valve in Fig. 4.6.

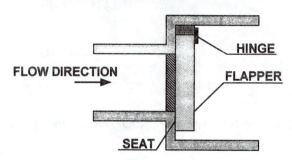

Fig. 4.6 A schematic illustration of a flapper type directional flow control valve.

The flapper valve shown in Fig. 4.6 opens due to differential pressure[1] across the valve. The suction pulls the flapper open and atmospheric pressure forces the water through the valve in the flow direction. On the inward stroke of the piston pump, the pressure developed acts against the flapper to close and press it against the seat. The rotation of the flapper as it opens and closes is controlled with a hinge. The hinge maintains the position of the flapper relative to the seat of the valve.

In addition to valves for the inlet and the discharge piping near the pump, a foot valve will be necessary at the mouth of the inlet line. For a piston pump to operate, it must be primed so that a vacuum is developed on the intake stroke. Before we begin pumping for the first time, the inlet pipe must be filled with water (primed). However, this charge of water will drain from the pump into the stream (water source) unless a one-way valve prevents the flow. We maintain water in the inlet line

---

[1] The differential pressure across the valve is because the pressure on the flow side of the valve is greater than the pressure behind the flapper.

with a foot valve. The foot valve is simply another check valve placed at the mouth of the inlet line. It permits water to flow into the line, but prevents it from draining back into the source.

4.2.2 Force, Work and Power for a Piston Pump

The work and power required to pump water from the source to the open channel located eight feet above the stream depends on the volume of flow. Let's begin by determining the flow rate for a typical single acting piston pump. The volume of water, Q drawn into the pump chamber is given by:

$$Q = \pi r^2 d \tag{4.1}$$

Where r is the radius of the chamber and d is the length of the stroke.

The flow rate, q_v in terms of volume per unit time is given by:

$$q_v = Q \times n = \pi r^2 \times d \times n \tag{4.2}$$

where n is the number of cycles per unit time.

The flow rate, q_w in terms of the weight of water pumped per unit time is given by:

$$q_w = q_v/\gamma = \pi r^2 \times d \times n/\gamma \tag{4.3}$$

Where γ is the density of water (62.4 lb/ft^3 or 0.0361 lb/in^3).

EXAMPLE 4.1

Determine the flow rate in terms of both volume and pounds per minute for a single acting piston pump with the following design parameters: diameter = 6 in., stroke 18 in. and cyclic rate = 30 cycles per minute.

Solution:

From Eq. (4.2), we write:

$$q_v = \pi r^2 \times d \times n = \pi \times 3^2 \times 18 \times 30 = 15{,}268 \text{ in}^3/\text{min} \tag{a}$$

$$q_v = 15{,}268 \text{ in}^3/\text{min} \times \text{gallon}/231 \text{ in}^3 = 66.1 \text{ GPM}[2] \tag{b}$$

$$q_w = q_v/\gamma = 15{,}268 \text{ in}^3/\text{min} \times 0.0361 \text{ lb/in}^3 = 551.8 \text{ lb/min} \tag{c}$$

Examine the results shown in Eqs. (a), (b) and (c). Do they seem reasonable? They appear to be feasible, but it is necessary to learn more regarding the force and power requirements before committing to these design parameters.

---

[2] The volume conversion is accomplished by noting that one-gallon contains 231 in^3.

Let's next explore the force required to drive the piston pump and the work involved in lifting the water to the height of the open channel. A relation exists between the head[3] delivered by a pump and the pressure, p developed in its chamber.

$$p = \gamma h \tag{4.4}$$

The force, F required to develop this pressure is given by:

$$F = pA_p = p \times \pi r^2 \tag{4.5}$$

The work, Wk performed during the pressure (discharge) stroke of the piston is:

$$Wk = F \times d = p \times \pi r^2 \times d \tag{4.6}$$

The power required is obtained from Eq. (3.6) and (4.6) as:

$$P = Wk/t = p \times \pi r^2 \times d \times n \tag{4.7}$$

EXAMPLE 4.2

For the design parameters presented in Exercise 3.6, determine the following quantities:

1. The pressure developed in the pump chamber.
2. The force exerted on the inward stroke.
3. The work performed on a single discharge stroke.
4. The power required if the pump is operating continuously.

Recall that the pump is operating against a head of eight feet.

Solution:

Let's first determine the pressure required to lift the water a vertical distance of eight feet by using Eq. (4.4):

$$p = \gamma h = 62.4 \text{ lb/ft}^3 \times 8 \times \text{ft}^2/144 \text{ in}^2. = 3.47 \text{ psi} \tag{a}$$

The force, F applied to the piston to produce this pressure is given by Eq. (4.5) as:

$$F = pA = p \times \pi r^2 = 3.47 \times \pi \times 3^2 = 98.0 \text{ lb} \tag{b}$$

The work, Wk performed during the pressure (discharge) stroke of the piston is determined from Eq. (4.6) as:

$$Wk = F \times d = 98.0 \times 18 = 1,764 \text{ in-lb} = 147.0 \text{ ft-lb} \tag{c}$$

The power required is obtained from Eq. (4.7) as:

$$P = Wk/t = F \times d \times n = 147.0 \times 30 = 4410 \text{ ft-lb/min} \tag{d}$$

We use the conversion factor given in Eq. (3.8) to determine the horsepower.

---
[3] The head is the height to which the water is lifted by the pump.

$$P = 4,410/33,000 = 0.134 \text{ HP} \qquad\qquad (e)$$

Again, examine the results. Do you believe the analysis? Are the numbers accurate? Can you apply 98 pounds on the power stroke? Is a 1/8 horsepower requirement reasonable for all of your team members to produce during the prototype evaluation? Are the design parameters consistent with the physical capabilities of the operators?

4.2.3 Efficiency of a Piston Pump

The results shown above for Example 4.2 assumed a pump efficiency of 100%. We did not take into consideration many different losses that occur in operating the pump. We did not consider the work performed on the intake stroke. We did not account for friction between the piston and the cylinder or between the piston rod and the head. More losses result from turbulence in the flow and from friction of the water flowing over the inside walls of the pipes.

It is dangerous to quote typical efficiencies of piston pumps because the efficiency depends on the quality of the construction and the fit of the moving parts. To a large degree, the skills that all team members exercise in fabricating each part of the pump control the efficiency.

Having announced the dangers of quoting efficiencies, we present typical results for commercially produced pumps in Fig. 4.7.

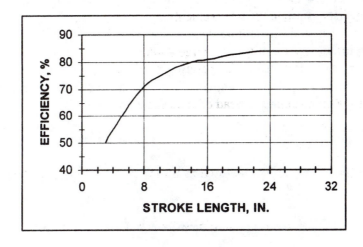

Fig. 4.7 Mechanical efficiency of piston pumps depends on stroke length.

The efficiency of a piston pump depends strongly on the length of the stroke. The efficiency is poor for short strokes, but as the stroke increases to 18 to 24 inches, the efficiency approaches a maximum of about 84%. The efficiency also depends on the pump speed with higher efficiencies associated with the slow speeds.

4.2.4 Assessment

Piston pumps have several advantages. They are very efficient particularly at the lower speeds. Since plastic pipe and fittings for the pipe are available in many different diameters at local retail outlets, the chamber, piston and piston rod are not difficult to fabricate. Joining these components with adhesive is feasible, and in certain designs, adhesive joints eliminate the need for some of the seals.

Piston pumps are easily adapted to drives operated by either the upper body or the lower body muscles. The lever mechanism is particularly well suited to the in—out action of the reciprocating pumps.

We have discussed the single acting pump that is usually operated by a single person. If two people operate the pump, there are significant advantages to a double acting piston pump that draws water on one side of the piston while discharging water simultaneously from the other side. Of course, more valves and seals are required with the double acting pump, but the output is nearly doubled with two operators.

The most significant disadvantage of the piston pump is the need for a minimum of three direction flow control valves. These valves may be difficult to manufacture with the workshop facilities available to you.

4.3 DIAPHRAGM PUMPS

The diaphragm pump is similar in many respects to the piston pump. It operates with a reciprocating action drawing water into a chamber on the intake stroke and expelling it on the exhaust stroke. Its chamber is usually cylindrical to minimize the strain imposed on the diaphragm when it is stretched in operating the pump. It requires the same directional control valves as the piston pump. The significant difference is the replacement of the piston with a diaphragm. The advantage of the diaphragm is the elimination of the seal required to prevent leakage past the piston. The disadvantage is the limited stroke possible with a diaphragm. The deflection of the diaphragm must be limited to avoid damaging it by overstressing.

4.3.1 Mechanical Details

A section view of a typical diaphragm pump is illustrated in Fig. 4.8. The key element is the diaphragm that divides the pump casing into nearly two equal volumes.

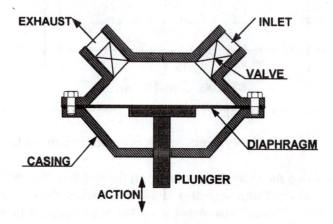

Fig. 4.8 Illustration of components in a diaphragm pump.

The top half of the casing serves as the water chamber. On the downward stroke, water is drawn into this chamber; on the upward stroke, water is forced through the exhaust port. One-way valves control the direction of the flow of water into and out of the chamber. The bottom half of the casing is never exposed to water. It provides housing for the diaphragm as it is pulled downward by the plunger. Because the diaphragm contains the water, a seal between the lower casing and the plunger is not necessary. The diaphragm also serves as a gasket aiding in the seal required at the joint between the two halves of the casing.

The diaphragm is attached to a flat circular plate, which in turn is connected to the plunger. The attachment technique is not illustrated in Fig. 4.8; however, the diaphragm may be sandwiched between two circular plates that are clamped together with screws. Adhesives may also be employed, but care must be exercised to insure that the adhesive is compatible with the rubber of the diaphragm and the material used for the circular plate.

The diaphragm is fabricated from an elastomer—usually neoprene rubber. At the center position the diaphragm is tight, but not strained. However, when the pump is stroked in either direction the diaphragm is stretched and subjected to stress. Since the life of the diaphragm is controlled by this stress, the stroke of pump is limited to avoid over stretching the rubber. We calculate the strain, ε from a very simple relation:

$$\varepsilon = \delta L/L \tag{4.8}$$

Where δL is the change in length of a line segment in the direction the diaphragm is stretched.

Let's analyze a typical diaphragm to show the application of Eq. (4.8) in determining the allowable stroke for a diaphragm pump.

EXAMPLE 4.3

Determine the strain induced in the diaphragm shown in Fig. 4.9 if we select $L_o = 4$ in., and $d = 2$ in.

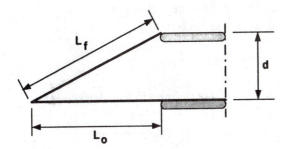

Fig. 4.9 As the pump is stroked upward through the distance d, the diaphragm is stretched from a length L_o to a length L_f.

Solution:

Note that the change in length, δL of a radial line on the diaphragm is given by:

$$\delta L = L_f - L_o$$

The stretched length, L_f is given by:

$$L_f = \sqrt{L_o^2 + d^2} \tag{a}$$

Substituting this relation into Eq. (4.8) yields:

$$\varepsilon = \frac{\sqrt{L_o^2 + d^2} - L_o}{L_o} = \sqrt{1 + \left(\frac{d}{L_o}\right)^2} - 1 \tag{b}$$

Substituting the design parameters L_o = 4in. and d = 2 in. into Eq. (b) gives the strain in the radial direction on the diaphragm as:

$$\varepsilon = 0.118 = 11.8\%$$

This value of the strain must be compared with the allowable strain that may be imposed on the neoprene rubber sheeting without causing failure. To determine this material property, cut a strip of the rubber that you intend to use for the diaphragm and subject it to repeated stretching that produces a strain of 15%. After 10 cycles, describe the appearance of the rubber. Try 100 cycles. If the rubber returns to its original length and shows no signs of distress, you may be able to increase the amount of strain. Continue these fatigue experiments to determine the safe limit of the strain that can be imposed on the rubber material used in constructing the diaphragm.

4.3.2 Force, Work and Power for a Diaphragm Pump

The analysis of the forces, work and power required to deliver water with a diaphragm pump are essentially the same as that presented previously for the piston pump. The only significant difference is the relation used to determine the volume of the pump's chamber. The top half of the pump casing is represented with a two dimensional view of a truncated cone shown in Fig. 4.10. The volume of the top chamber of the pump casing is the part of the cone that is not shaded.

The volume, Q of a truncated cone is given by:

$$Q = (\pi/3)(r_1^2\, h_1 - r_2^2\, h_2) \tag{4.9}$$

Where the dimensions r_1, r_2, h_1, and h_2 are defined in Fig. 4.10.

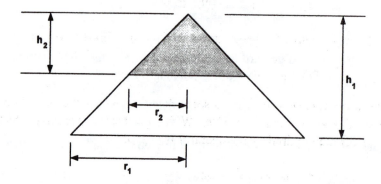

Fig. 4.10 Dimensional parameters defining the pump chamber.

The volume of the water drawn into a diaphragm pump equal to the volume of both the top and bottom chambers because the diaphragm is pulled into the lower half of the casing on the downward stroke. For this reason, the total volume of the water drawn into the pump is:

$$Q = (2\pi/3)(r_1^2\, h_1 - r_2^2\, h_2) \tag{4.10}$$

Let's consider an example to illustrate a design analysis to determine the forces, work and power required in operating a diaphragm pump.

EXAMPLE 4.4

Perform a design analysis of a diaphragm pump with a shape as illustrated in Fig. 4.8. The dimensions of both chambers are identical with dimensions as defined in Fig 4.10 given by:

$$r_1 = 6 \text{ in., } h_1 = 6 \text{ in., } r_2 = 3 \text{ in. and } h_2 = 3 \text{ in.}$$

The stroke utilized is $d = 2(h_1 - h_2) = 2(6 - 3) = 6$ in. The pump is employed with a lever mechanism that provides a mechanical advantage of three to one, and is operated at 50 cycles per minute.

Solution:

The flow rate, q_v in terms of volume per unit time is given by combining Eqs. (4.2) and (4.10) to give:

$$q_v = Q \times n = (2\pi/3)(r_1^2 \, h_1 - r_2^2 \, h_2) \times n \qquad (4.11)$$

$$q_v = (2\pi/3)(6^2 \times 6 - 3^2 \times 3) \times 50 = 19,790 \text{ in}^3/\text{min} \qquad (a)$$

$$q_v = 19,790 \text{ in}^3/\text{min} \times \text{gallon}/231 \text{ in}^3 = 85.7 \text{ GP} \qquad (b)$$

The flow rate, q_w in terms of the weight of water pumped per unit time is given by:

$$q_w = q_v/\gamma = 19,790 \text{ in}^3/\text{min} \times 0.0361 \text{ lb/in}^3 = 714.5 \text{ lb/min} \qquad (c)$$

Next, let's explore the force required to drive the diaphragm pump and the work involved in lifting the water to the height of the open channel. From Eq. (4.4), we convert the head to the pressure, p developed in the pump's chamber.

$$p = \gamma h = 62.4 \times 8 = 499.2 \text{ lb/ft}^2 = 3.47 \text{ psi}$$

The force, F that is required on the plunger to develop this pressure is given by Eq. (4.5) as:

$$F = pA = p \times \pi r^2 = 3.47 \times \pi \times 6^2 = 392.4 \text{ lb} \qquad (d)$$

Clearly, this force is excessive without some device to reduce the amount of effort required by the operator. The lever is such a device. With a mechanical advantage, MA = 3 as defined in Eq. (4.12), the force required by the operator is reduced to:

$$MA = F_{AMPLIFIED}/F_{APPLIED} \qquad (4.12)$$

$$F_{APPLIED} = 392.4/3 = 130.8 \text{ lb} \qquad (e)$$

The applied force at the long end of the lever is still large, but not impossible if the muscles in the lower body can operate the lever. The stroke of the lever is equal to the stroke of the pump times the mechanical advantage or $6 \times 3 = 18$ in. This result appears consistent with the rowing action of the upper body or the flexing action of the legs.

The work, **Wk** performed during the pressure (discharge) stroke of the piston is determined from Eq. (4.6) as:

$$\textbf{Wk} = F \times d = 392.4 \times 6 = 2354 \text{ in-lb} = 196.2 \text{ ft-lb} \qquad (f)$$

The power required is obtained from Eq. (4.7) as:

$$P = Wk/t = F \times d \times n = 196.2 \times 50 = 9,810 \text{ ft-lb/min} \qquad \text{(g)}$$

We use the conversion factor given in Eq. (3.8) to determine the horsepower.

$$P = 9,810/33,000 = 0.297 \text{ HP} \qquad \text{(h)}$$

These results assume 100% efficiency and accordingly they are optimistic. The efficiency of diaphragm and piston pumps is similar. Using the diaphragm instead of the piston eliminates the losses due to friction between the piston and the cylinder wall. The work that is required to stretch the diaphragm is largely returned to the system. Losses occur when the diaphragm bulges at the end of the exhaust stroke and the pressurized water is not expelled from the chamber. To avoid these losses the diaphragm should be sufficiently thick relative to its diameter.

4.3.3 Assessment

The diaphragm pump is a feasible concept for the proposed irrigation project. Because its stroke is limited, high flow rate must be achieved with a relatively large diameter chamber. This in turn increases the force that must be applied to develop the necessary head. Fortunately, in this project, the head is low, and as a consequence, the forces required to produce high flow rates remain within reasonable bounds if a simple lever system is employed.

Sealing the diaphragm pump is relatively easy because the soft rubber diaphragm serves as an effective gasket. The gasket fills any gap occurring at the joint between the two halves of the casing, thereby aiding in the prevention of leakage. It is important to place the bolts that clamp the two casing halves together on relatively close centers to provide a uniform pressure on the joint around the diameter of the casing.

Manufacturing the one-way valves with the limited shop facilities may be difficult. In addition, the construction of the casing for the diaphragm pump will require considerable imagination. You may wish to consider two different manufacturing processes in fabricating the conical casing for the diaphragm pump. The first is a process involving fiberglass cloth and a liquid plastic such as polyester or epoxy. A mold of the part is made from wood or clay that is easy to form from these materials. Then pieces of the fiberglass cloth are saturated with liquid plastic and draped over the mold. When the plastic cures and hardens, it forms a rigid fiberglass shell that is the same shape as the mold.

Another manufacturing technique involves the use of rigid sheets of thermo-plastic. Thermo-plastics soften, but do not melt, when heated above a critical temperature. In this softened state, the plastic sheet is pressed into a cavity mold to form the shape of a complex shell. When cooled the plastic hardens and forms a rigid, thin-walled shell with the shape of the mold.

4.4 CENTRIFUGAL PUMPS

The centrifugal pump is widely used throughout the world to transport a variety of fluids at either high or low pressure. Some centrifugal pumps are huge and can deliver up to 300,000 GPM. Other pumps are small with capacities of less than one GPM. Like the piston and diaphragm pumps, the power required to drive the centrifugal pump depends on the flow rate and the head.

The centrifugal pump operates with a rotary motion. A circular impeller rotates within the pump's housing. The vanes on the impeller engage the incoming water—imparting a radial velocity that carries the water out of the impeller into the discharge nozzle. Because of the centrifugal force produced by the rotation of the impeller, a vacuum is formed at the center of rotation. This vacuum draws the water from the source into the pump. Since the motion is rotary and not reciprocating, there is no requirement for one-way valves at the inlet and outlet nozzles for the pump to operate. However, a foot valve on the inlet line is necessary to maintain the charge of water in the pump in the event that the rotor stops.

The piston and diaphragm pumps are positive displacement pumps. They capture a specified volume of water on each intake stroke and expel that volume on each exhaust stroke. The centrifugal pump is not a positive displacement type. If a valve on the delivery line is closed, the pump will continue to rotate without water flowing in or out of the system. The water in the housing is mixed by the action of the impeller and the work input to the pump is dissipated in the form of heat.

4.4.1 Mechanical Details

The centrifugal pump incorporates an impeller, housing, shaft, bearing and seals. We will provide information on the design of the impeller, housing and shaft. Information on bearings and seals was provided in a previous section in this Chapter.

4.4.1.1 The Impeller, Shaft, Bearing and Seal

The impeller is a critical component in the design of a centrifugal pump. An illustration of a simple six-vane impeller is presented in Fig. 4.11. An inlet is placed in the center of the impeller plate permitting an axial flow of water into the impeller. The number of vanes may vary from three to ten; however, six or seven vanes usually give the best performance.

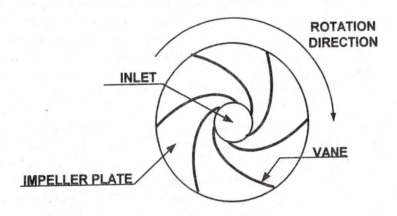

Fig. 4.11 Mechanical characteristics of an impeller for a centrifugal pump.

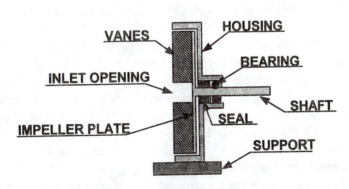

Fig 4.12 The impeller is supported by a shaft and bearing.

In small pumps, such as the one considered in this project, the impellers are supported on only one side by a shaft and bearing as indicated in Fig. 4.12. The shaft is fastened to the center of one side of the impeller plate, and extends through the housing. The center region of the housing accommodates a seal and a bearing. The seal is to prevent leakage past the shaft. The bearing is to enable the impeller to rotate freely within the housing. The housing is attached to a support to fix the pump so that it does not move as power is applied to the shaft.

Not shown in Fig. 4.12 is the drive mechanism for providing the power to the shaft. The drives have been discussed previously in Chapter 3, and the description of suitable rotary drives for operating the pump is not repeated. If the rotary drive and the pump are designed as separate items, a coupling is used to join the shaft from the pump with the shaft from the drive.

The housing shown in Fig. 4.12 for the centrifugal pump is not clearly illustrated. We have not closed the housing completely about the impeller to more clearly show the impeller, the inlet opening, and the vanes. We will describe the housing in more detail in a subsequent section.

4.4.1.2 The Housing or Volute and the Diffuser

A side view of a centrifugal pump, illustrated in Fig 4.13, shows the housing (volute), the impeller and the diffuser. In this illustration the impeller is rotating counterclockwise.

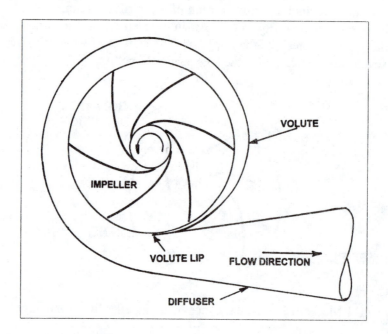

Fig. 4.13 An assembly view showing the volute, impeller and diffuser.

The water is drawn through an inlet located at the center of the impeller. The water interacts with the curved vanes of the impeller and is forced radially outward through the channels formed between adjacent vanes. The water is thrown from the impeller into the annular opening between the impeller and the volute. This opening is a minimum at the volute lip and increases with angular position about the volute. The volute lip serves to cut the fluid from the impeller and direct it into the diffuser. The diffuser is simply a conduit with increasing area. It serves to reduce the velocity of the flow and increase its pressure. After increasing to a suitable diameter, the diffuser is connected to the exhaust piping which transports the water to the delivery site.

The water intake passes through the housing into the inlet at the center of the impeller. The complete housing (volute) is shown in Fig. 4.14, with water input on the left, and power input on the right. The water output through the diffuser is not shown in this view.

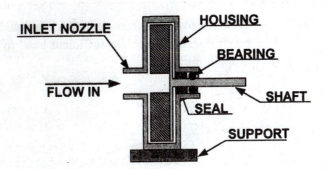

Fig. 4.14 Sketch showing water input and power input to the centrifugal pump.

4.4.2 Characteristics of Centrifugal Pumps

Centrifugal pumps are usually characterized as a function of two parameters—the flow rate and the head or pressure. Other characteristics such as efficiency and power requirements are strongly dependent on the flow rate q_v and head h, and consequently, these characteristics are often shown on the same q_v – h graphs.

4.4.2.1 System Requirements

Let's begin with the system requirements. In this case, the amount of water that we plan to pump and the head that must be developed. In general, the system requirements may be represented with a q_v – h curve as shown in Fig. 4.15.

Inspection of Fig. 4.15 shows that the head required from the pump is comprised of two parts—the static head and head losses due to friction. In this project, the static head is the distance from the stream to the delivery point, about 8 feet. The head losses due to friction are due to losses in the pump, wall friction on both input and output lines, and losses of excess head at the exhaust. These friction losses are dependent on the flow rate. In fact, the head losses increase approximately as the cube of the flow rate.

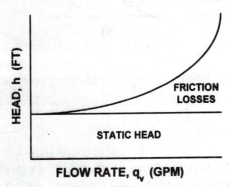

Fig. 4.15 System requirements include static head and friction losses.

4.4.2.2 Characteristic Curve for Centrifugal Pumps

The q_v – h characteristic curve for a centrifugal pump is depicted in Fig. 4.16. At low flow rates the head produced by a centrifugal pump is relatively high; however, with increased flow the head produced decreases. The operating point depends on both system requirements and pump characteristics. The maximum operating point is established by superimposing the curves of Figs. 4.15 and 4.16 as shown in Fig. 4.17.

Centrifugal pumps may be operated at flow rates from zero to a maximum as determined by the operating point on Fig. 4.17. The pump does not generate sufficient head to provide flow rates above the maximum corresponding to the operating point.

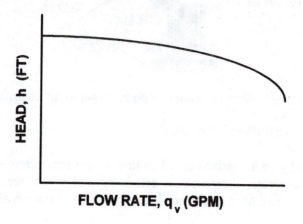

Fig. 4.16 Characteristic flow rate – head curve for a centrifugal pump.

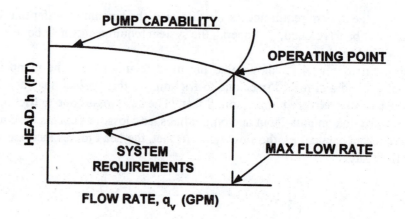

Fig. 4.17 The operating point for a pumping system is the crossing point of the two characteristic curves.

4.4.2.3 Efficiency

A centrifugal pump is not a positive displacement type of device. If a valve on the output blocks the flow, the pump will continue to rotate churning the water trapped inside. In the zero flow state, the efficiency of the pump is also zero. As the flow increases, the efficiency increases to a maximum before dropping off near the maximum flow rate expected from the pump. We show the efficiency on a q_v – h curve in Fig. 4.18.

It is clear that the pump is operated at high flow rates to achieve the higher efficiencies. Efficiency of the centrifugal pump is not as high as either the piston pump or the diaphragm pump even when it is operating under favorable flow rates. We will discuss design of the pump for efficiency in more detail in the next Section.

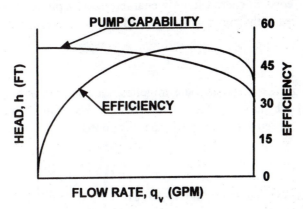

Fig. 4.18 Efficiency of a centrifugal pump increases with flow rate.

4.4.3 Design Parameters for Centrifugal Pumps

When designing a centrifugal pump several dimensions must be specified to adequately describe the pump. The design parameters include:

- Outside diameter of impeller
- Diameter of the eye of the impeller
- The number of vanes
- The discharge angle of the vanes
- The width of the impeller channel
- The area of the volute at the exit
- The volute width

The procedure for selecting the correct dimensions for a centrifugal pump is based on many empirical formulas. We will make no attempt to outline the approach here. Instead, the reference by Lobanoff and Ross [1] provides the design curves and equations.

4.4.3.1 Pump and Suction Specific Speeds

In spite of the empiricism of the design procedure, there are some generalities that are important to keep in mind as you specify the various design parameters. Two important numbers that markedly affect the design of a centrifugal pump are the pump specific speed N_s and the suction specific speed N_{ss}.

The pump specific speed is defined as:

$$N_s = (n \times q_v^{0.5})/h^{0.75} \qquad (4.13)$$

And the suction specific speed is defined as:

$$N_{ss} = (n \times q_v^{0.5})/h_s^{0.75} \qquad (4.14)$$

Where n is the speed of the pump in RPM.
q_v is the capacity in GPM
h is the head in feet.
h_s is the suction head at the pump inlet.

The specific speeds are reference numbers useful in characterizing a pump of any type. Let's consider an example to illustrate the use of the pump specific speed.

EXAMPLE 4.5

A centrifugal pump is designed with a six-vane impeller. It operates at a speed of 600 RPM and produces 100 GPM against a head of 10 feet. The distance from the stream to the inlet is 4 feet. Determine the pump specific speed and the suction specific speed.

Solution:

The pump specific speed is given by Eq. (4.13) as:

$$N_s = (n \times q_v^{0.5})/h^{0.75} = [600 \times 100^{1/2}]/10^{0.75} = 1067$$

The suction specific speed is computed for Eq. (4.14) as:

$$N_{ss} = (n \times q_v^{0.5})/h_s^{0.75} = [600 \times 100^{1/2}]/4^{0.75} = 2121$$

Typical numbers for the pump specific speed vary from about 400 to 3500. The value of $N_s = 1067$ is somewhat lower than the typical design, but it is reasonable for a pump of low capacity with a small flow rate.

4.4.3.2 Efficiency

The efficiency of a centrifugal pump depends on the flow rate and the specific speed. Typical results of an experimental program where efficiencies were measured are shown in Fig. 4.19.

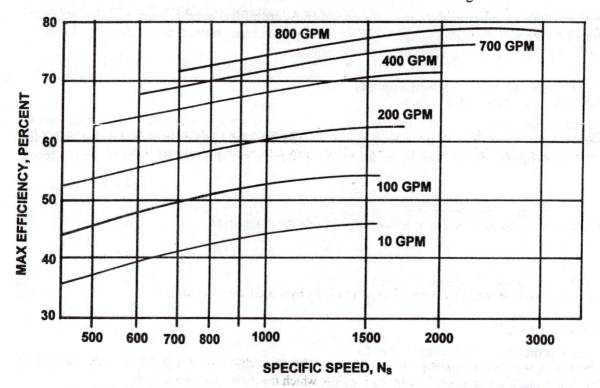

Fig. 4.19 Efficiency of a single-stage centrifugal pump with specific speed. After [1].

It is evident by an examination of Fig. 4.19 that the efficiency increases monotonically with the specific speed. It is also clear that the efficiency of centrifugal pumps is relatively poor for low capacity pumps. For example, a pump with a capacity of 100 GPM only achieves an efficiency of 54% when the specific speed is 1500. Efficiencies above 80% are achieved only in high capacity pumps (1000 GPM and above).

EXAMPLE 4.6

Determine the angular velocity required in driving a centrifugal pump with a capacity of 100 GPM if an efficiency of 50% is specified. The head against which the pump operates is 10 ft.

Solution:

From Fig. 4.14, it is clear that the pump must be operated with a specific speed $N_s = 1000$. Then, using Eq. 4.13 we write:

$$n = N_s \times h^{0.75}/q_v^{0.5} = 1000 \times 10^{0.75}/100^{0.5} = 562 \text{ RPM}$$

Let's interpret the result. Is 562 RPM a reasonable number? Suppose you are planning to use a single stage chain drive to provide the power for the pump. Will you be able to achieve the required angular velocity for the pump if your team members can pedal at a rate of only 80-RPM? Is it possible to increase the angular velocity by a factor of seven in a single stage drive? Probably not and if your team is planning on a centrifugal pump, plan also on a double stage drive that permits very high speeds to be achieved from human operators.

4.4.4 Torque and Power for a Centrifugal Pump

The flow rate from a centrifugal pump provides a means of calculating the power required for delivering water to a site with an elevation h.

$$P_{out} = q_w \times h \tag{4.15}$$

Where q_w is the flow rate in lb/min.

However, we recognize that the centrifugal pump is operating with significant losses, which leads to lower efficiencies. We account for the efficiency $\mathcal{E}$ by determining the power input as:

$$P_{in} = P_{out}/\mathcal{E} \tag{4.16}$$

Finally, we determine the torque required for driving the pump from:

$$T = P_{in}/\omega \tag{4.17}$$

Let's demonstrate the application of these relations by considering an example.

EXAMPLE 4.7

Determine the torque required in driving a centrifugal pump with a capacity of 100 GPM if an efficiency of 50% is specified. The head against which the pump operates is 10 ft.

Solution:

In Example 4.6, we have already determined that the speed needed to achieve the specified efficiency of 50% was 562 RPM. Next, let's determine the power output from the specified flow rate using Eq. 4.15.

$$P_{out} = q_w \times h = 100 \text{ gallon/min} \times 8.34 \text{ lb/gallon} \times 10 \text{ ft} = 8340 \text{ ft-lb/min} = 0.2527 \text{ HP} \quad \text{(a)}$$

Since the efficiency is only 50%, the power input required is given by Eq. 4.16 as:

$$P_{in} = P_{out}/\mathcal{E} = 8340/0.5 = 16,680 \text{ ft-lb/min} = 0.5054 \text{ HP} \quad \text{(b)}$$

Finally, the torque required at the shaft to the pump is given by Eq. 4.17 as:

$$T = P_{in}/\omega = 16,680/(562 \times 2\pi) = 4.72 \text{ ft-lb} \quad \text{(c)}$$

These results indicate that it should be possible to design a centrifugal pump with a flow rate of 100 GPM. However, the torque T determined in Eq. (c) is at the shaft of the pump. The torque to be provided by the operator is much higher. Recall that a double stage drive that increases the rotary speed from 80 RPM to 562 RPN (a factor of 7) will be required. This implies that the torque to the first shaft in the double stage drive system will be $7 \times 4.72 = 33$ ft-lb. A human operator can achieve this torque and sustain it for some time but it is challenging.

4.4.5 Assessment

Centrifugal pumps are widely employed for many different applications because they are very easy to control. They can be operated at any flow rate from zero to the maximum capacity of the pump. A control valve on the output is all that is required to adjust the flow to meet the momentary demand. Piston and diaphragm pumps are more difficult to control because they are positive displacement pumps. Adjusting the cyclic rate at which the pump is operated controls the flow rate. Alternately if the speed is held constant, a pressure relief valve is necessary to drop the excess flow to the sump (i. e. the source of water).

A disadvantage of the centrifugal pump for a human operator is the requirement of high speed to achieve reasonable efficiencies[4]. This high speed can be achieved with a two-stage drive mechanism—either belt or chain. However, the drive will add to the weight and cost of the pumping system.

An advantage of the centrifugal pump when compared to the piston or diaphragm pumps is the fact that valves are not required for controlling the direction of flow. The only valve needed to operate a centrifugal pump is the foot valve on the inlet line. Directional control of the flow is achieved with the centrifugal force generated by the impeller.

Manufacturing the impeller and the volute (housing) for the centrifugal pump will be difficult. In most commercial centrifugal pumps, a casting process is used in producing the impellers. This process enables the production of shapes more complex than those shown in Fig. 4.11. However, with the limited production processes available for this course, casting is not a viable option. It is possible to individually produce the vanes and the impeller plates from a sheet of thermoplastic plastic. The vanes may be cut as flat rectangular blanks from the plastic sheet. They are then heated until the

---

[4] When electric power is available these high-speed requirements are not a disadvantage. Electric motors operating at 1780 or 3560 RPM are well suited for powering centrifugal pumps.

plastic is soft. The soft blanks are pressed against a mold to form the specified curvature. The vanes are then fastened to the impeller plates by adhesive bonding or by screws.

4.5 SUMMARY

Three different types of pumps—the piston, diaphragm and centrifugal—have been reviewed in considerable detail. In the discussion of the piston pump, we showed the important components in a section view in Fig. 4.1. Then detailed descriptions of seals and check valves were provided. Relations were developed to determine the force, work and power required for operating a piston pump. Examples were given to demonstrate the use of these equations in a design analysis of the piston pump. Finally, we showed that the efficiency obtained in operating a piston pump depended strongly on its stroke with high efficiencies (84%) achieved with strokes in excess of 18 to 24 in.

The mechanical details of the diaphragm pump are illustrated in Fig. 4.8. In operation, the diaphragm pump is similar to the piston pump. It operates with a reciprocating action and has the same one-way valve requirements as the piston pump. However, the use of the diaphragm alleviates the sealing requirements. However, the presence of a diaphragm limits the stroke because of the danger of overstressing it. However, high volume flow rates can be achieved with a diaphragm pump if the diameter of the diaphragm is increased. Again, relations were developed to determine the force, work and power required for operating a diaphragm pump. Examples were provided to demonstrate the use of these equations in a design analysis. Efficiencies are similar to those achieved with the piston pump even though the stroke is much shorter.

The centrifugal pump is more complex than either the piston or the diaphragm pumps. The pump components include an impeller, housing (volute), bearing and seal. The mechanical details of this pump are described in Figs. 4.11, 4.13 and 4.14. We have described the general characteristics of centrifugal pumps and have shown that the higher efficiencies are achieved at higher flow rates. The design of the impeller and the housing for a centrifugal pump is based on a series of empirical relations. A reference is provided that outlines this empirical approach should your team attempt to design these components. A method of analysis is described and demonstrated based on determining the efficiency of the pump from experimentally derived curves for single stage centrifugal pumps. An example is presented that shows the feasibility of producing flow rates of 100 GPM; however, a two-stage rotary drive is necessary to achieve the high rotary speeds used in driving the pump.

Assessments are provided that indicate the advantage and disadvantages associated with each type of pumping system. We summarize these assessments below:

1. The piston pump is the easiest to manufacture and if properly designed it may be operated with a direct drive. If a double acting piston pump is designed it affords the opportunity to be operated by two persons to provide extremely high flow rates. The largest problems in developing this type of pump are manufacturing the required valves and seals.
2. The diaphragm pump is more difficult to manufacture than the piston pump because of the requirement for a housing with a conical shape. The difficulty in manufacturing the valves is the same although the sealing problem is mitigated. The need for a larger diameter to achieve high flow rates necessitates the use of a lever type reciprocating drive.
3. The centrifugal pump is the most difficult to manufacture. The fabrication of both the impeller and the volute housing represents a significant challenge. Also, the design analysis leading to the specification of the dimension is obscure because it is based on a series of empirical relations. Operating the centrifugal pump at higher efficiencies requires that it be driven at high angular velocity. This requirement in turn will force the design team to develop a two-stage rotary drive, which increases the cost, complexity and the weight of the system. While much is to be learned by developing a complex system

like a centrifugal pump with a two-stage rotary drive, it is probably too difficult for the typical first design project.

REFERENCES

1. Lobanoff, V. S. and R. R. Ross, <u>Centrifugal Pumps: Design and Application,</u> Gulf Publishing Co., Houston, TX, 1985.
2. Stepannoff, A. J. <u>Centrifugal and Axial Flow Pumps</u>, 2nd Edition, John Wiley, New York, 1957.
3. Church, A. H. Centrifugal Pumps and Blowers, Krieger Publishing Co., Huntington, NY, 1972.

EXERCISES

4.1   Prepare a sketch showing all of the important components involved in a double acting piston pump.

4.2   Prepare a parts list for the pumping system described in the sketch of Exercise 4.1.

4.3   Size the components included in the pumping system described in Exercise 4.1.

4.4   Conduct a design analysis showing the flow rate from the piston pump if it is operated by two persons at a cyclic rate of 30 stokes per minute.

4.5   Extend the design analysis of Exercise 4.4 to show the forces required of the operators and the power supplied to the pump. Assume an efficiency of 75%.

4.6   Prepare an engineering brief describing the method of manufacturing each of the following components for the piston pump of Exercise 4.1:

- The pump body including the heads.
- The piston and the piston seal.
- The inlet and outlet valves.
- The foot valve.
- The piston rod and handles.
- The discharge piping.
- The inlet piping.

4.7   Prepare a sketch showing all of the important components involved in a single acting diaphragm pump.

4.8   Prepare a parts list for the pumping system described in the sketch of Exercise 4.7.

4.9   Size the components included in the pumping system described in Exercise 4.7.

4.10 Conduct a design analysis showing the flow rate from the diaphragm pump if it is operated at a cyclic rate of 45 stokes per minute.

4.11 Extend the design analysis of Exercise 4.4 to show the forces required from the operator and the power supplied to the pump. Assume an efficiency of 75%.

4.12 Prepare an engineering brief describing the method of manufacturing each of the following components for the piston pump of Exercise 4.7:

- The pump body.
- The diaphragm and the attachment plates.
- The inlet and outlet valves.
- The foot valve.
- The oscillating drive.
- The discharge piping.
- The inlet piping.

4.13 Critique the design of the diaphragm pump as depicted in Fig. 4.8. Cite the reasons for your criticism.

4.14 Determine the strain induced in a circular diaphragm with a diameter of 300 mm, if it is subjected to a stroke of 150 mm. The central plate to which the diaphragm is attached is 75 mm in diameter. Interpret this result with regard to the life of the diaphragm.

4.15 Prepare a sketch showing all of the important components involved in a single stage centrifugal pump.

4.16 Prepare a parts list for the pumping system described in the sketch of Exercise 4.15.

4.17 Size the components included in the pumping system described in Exercise 4.15.

4.18 Determine the specific pump speed for a centrifugal pump if it is driven at a speed of 400 RPM to produce 80 GPM against a head of 12 feet. Comment on the adequacy of the result as it relates to the performance of the centrifugal pump.

4.19 You have designed the impeller and the volute of a centrifugal pump with a specific speed of 1100. If the pump is to produce 90 GPM against a head of 9 feet, determine the rotary speed required driving the pump.

4.20 For the pump in Exercise 4.19, determine the efficiency of its operation.

4.21 Prepare a sketch of the rotary drive employed to power the pump of Exercise 4.19.

4.22 Size the components included in the drive of Exercise 4.21.

4.23 Perform a design analysis of the drive of Exercise 4.21 and 4.22 and indicate the input and output torque and speed for the drive.

4.24 Prepare an engineering brief describing the method of manufacturing each of the following components for the centrifugal pump of Exercise 4.15:

- The impeller.
- The housing (involute).
- The seal.
- The foot valve.
- The bearing and shaft.
- The attachment of the shaft to the impeller plate.
- The discharge piping.
- The inlet piping.

4.25 Write an engineering brief discussing the selection criteria employed in choosing the type of pump to design as a team project. Provide arguments for each of the criterion.

CHAPTER 5

FLOW MEASUREMENT

5.1 FLOW RATES

Hopefully, your pumping system will provide a stream of water to a collection point approximately eight feet above the level of the stream. One of the requirements placed on the design team is to measure the flow rate from the system and the total quantity of water delivered. Before discussing the many methods for measuring flow rates, let's define some basic terminology used in the study of fluid mechanics. Flow rates can be expressed in terms of volume, weight and mass, q_v, q_w, and q_m respectively. In all cases they refer to the quantity of fluid pumped per unit time. The only difference is the unit of measure of the quantity (i.e. volume, weight or mass).

Consider fluid flowing through a channel as shown in Fig. 5.1

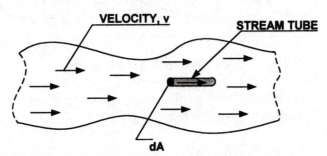

Fig. 5.1 Illustration of fluid flowing through a channel. The cross section of the stream tube is dA.

In the channel depicted in Fig. 5.1, we have placed a small imaginary stream tube with an elemental cross sectional area dA. We have also shown the flow velocity, v in the channel with a series of arrows. This velocity usually varies with position in the channel. The flow rates in the channel are given by integrating the velocity over the area of the channel:

$$q_v = \int_A v\, dA \qquad (5.1)$$

$$q_w = \int_A \gamma v\, dA = \gamma \int_A v\, dA = \gamma q_v \qquad (5.2)$$

$$q_m = \int_A \rho v\, dA = \rho \int_A v\, dA = \rho q_v \qquad (5.3)$$

q_v, q_w and q_m are the volume, weight and mass flow rates respectively.
A is the total area of the channel at the location of the imaginary stream tube.
γ is the density of the fluid (lb/ft^3) or (N/m^3).
$\rho = \gamma/g$ is the mass density of the fluid.
g is the gravitational constant (32.2 ft/s^2) or (9.81 m/s^2).
v is the velocity of the flow (ft/s) or (m/s).

In writing the extensions to Eqs. (5.2) and (5.3), we have assumed that the fluid is incompressible and that the densities are constants.

5.2 CLOSED CHANNEL FLOW

One of the difficulties in measuring flow rate is the fact that the velocity is not a constant across the channel. We must accommodate its variation with respect to position. Let's first consider the most common case of flow in a circular pipe. Due to friction on the inside wall of the pipe the flow is retarded at the edges, and the flow is a maximum at the center as shown in Fig. 5.2. The exact profile depends upon whether the flow is laminar or turbulent.

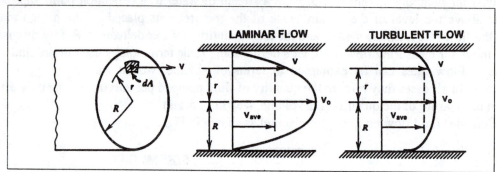

Fig. 5.2 Flow profiles in a circular pipe.

For laminar flow, a parabolic velocity profile occurs in a circular pipe as illustrated in Fig. 5.2. The flow velocity depends on the radial position r, and is given by:

$$v = v_o \left[1 - (r/R)^2\right] \tag{5.4}$$

v_o is the velocity of flow at the center of the pipe.
R is the radius of the inside wall of the pipe.

Substituting Eq. (5.4) into Eq. (5.1) and integrating yields:

$$q_v = \pi R^2 v_{ave} = A\, v_{ave} \tag{5.5}$$

$$q_w = \gamma\, A\, v_{ave} \tag{5.6}$$

$$q_m = \rho\, A\, v_{ave} \tag{5.7}$$

Where $v_{ave} = v_o/2$ and the area $A = \pi R^2$.

From these relations, it is clear that the flow rate may be determined if the size of the pipe is known and the velocity of flow at the center of the pipe is measured. The density of water is well known and has been cited previously in Chapter 3.

If the flow is turbulent, the procedure is similar although the equation for the average velocity depends on the Reynolds number. Turbulent flow is beyond the scope of this textbook, and the reader is referred to reference [1].

5.3 OPEN CHANNEL FLOW

Open channel flow occurs with one surface of the fluid free as shown in Fig. 5.3. Moreover the channel may be irregular in shape. This poses a more difficult measurement problem if one attempts to determine flow velocity v by point methods. Because of the irregular shape of the channel, the flow may vary significantly from one position to another. To accommodate these variations in flow velocity, measurements are made at many points (P_1, P_2, P_3, P_4, ... P_n) across the channel width.

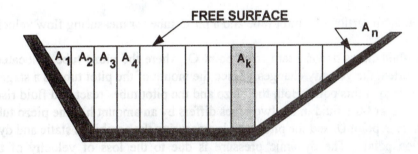

Fig. 5.3 The flow channel is subdivided into several small areas.

We measure the velocity v_k associated with the area A_k, and then determine the volume flow rate for area A_k from Eq. (5.1) as:

$$[q_v]_k = v_k A_k \qquad\qquad (a)$$

The total volumetric flow rate is determined by summing the flow rates for the incremental areas as indicated below.

$$q_v = \sum_1^n A_k v_k \qquad\qquad (5.8)$$

5.4 MEASURING FLOW RATES

There are many different methods for measuring flow rates—some are based on insertion instruments, others on the pressure produced by well-defined obstructions placed in closed pipelines, and still others on blockages placed in open channels. We will cover a few of the techniques introducing the basic instruments or concepts used with each method.

5.4.1 Insertion Instruments

Two different insertion instruments that are employed to measure flow velocity at a point—the pitot tube and hot wire anemometer are described in the following subsections.

5.4.1.1 The Pitot Tube

The use of a pitot tube inserted in a stream to measure flow velocity v is illustrated in Fig. 5.4.

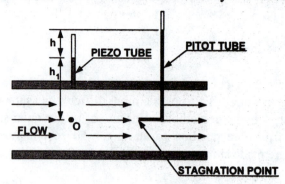

Fig. 5.4 Insertion of a pitot tube and a piezo tube for measuring flow velocity.

As the fluid flows downstream from point O, where the piezo tube is located, to where the pitot tube is inserted, the velocity changes. Since the mouth of the pitot tube is a stagnation point, the velocity goes to zero at this point. Both the piezo and the pitot tubes react and fluid rises in each tube. However, the height of the fluid in the two tubes differs by an amount h. The piezo tube is measuring the static pressure at point O, and the pitot tube is measuring the combined static and dynamic pressure at the stagnation point. The dynamic pressure is due to the loss of velocity of the flow at the stagnation point. A modified form of Bernoulli's equation enables us to relate pressure and velocity as:

$$\frac{p_0}{\gamma} + \frac{v_0^2}{2g} = \frac{p_s}{\gamma} + \frac{v_s^2}{2g} \tag{5.9}$$

where v_0 and v_s are the velocities at points O and S, respectively.
  p_0 and p_s are the static and stagnation pressures, respectively.
  γ is the density (weight per unit volume) of the fluid.

Since the stagnation velocity $v_s = 0$, the dynamic pressure p_d is given by:

$$p_d = p_s - p_0 = \frac{\gamma v_0^2}{2g} = \gamma h \tag{5.10}$$

The quantity h is the dynamic head as illustrated in Fig. 5.4. The flow velocity v_0 at the location of the piezo tube is then determined from Eq. (5.10) as:

$$v_0 = \sqrt{2g\left(\frac{p_s - p_0}{\gamma}\right)} = \sqrt{2gh} \tag{5.11}$$

It is apparent from Eq. (5.11) that the flow velocity may be determined if we measure the dynamic head h. Then Eq. 5.5 may be employed to determine the volume flow rate.

EXAMPLE 5.1

Water is flowing in a two-inch diameter delivery line with a static pressure of 1.2 psi. A pitot tube inserted into the center of the line indicates a pressure of 1.9 psi; determine the volume flow rate assuming the flow is laminar.

Solution:

Let's begin by using Eq. (5.11) to determine the flow velocity at the center of the tube.

$$v_0 = \sqrt{2(386.4)\left(\frac{1.9 - 1.2}{0.0361}\right)} = 122.4 \, \frac{in}{s} \qquad \text{(a)}$$

Recall that the average flow velocity for laminar flow is related to the flow at the center of the pipe as:

$$v_{ave} = v_0/2 = 122.4/2 = 61.2 \text{ in/s} \qquad \text{(b)}$$

Next, the average flow rate is determined from Eq. (5.5) as:

$$q_{ave} = \pi R^2 v_{ave} = \pi \, (1)^2 \times 61.2 = 192 \text{ in}^3/\text{s} \qquad \text{(c)}$$

Let's convert this result into more meaningful units:

$$192 \text{ in}^3/\text{s} \times \text{gallon}/231 \text{ in}^3 \times 60 \text{ s/min} = 49.9 \text{ GPM} \qquad \text{(d)}$$

This example shows that the pumping system is delivering about 50 GPM to a two-inch delivery pipe that is containing the flow rate at relatively low pressure.

The pitot and piezo tubes convert the measurement of flow velocity into a measurement of either pressure or head. If the pressures are low, the heads are small and can be measured with water filled tubes or manometers. However, if the pressures are high the tubes become too long to be practical, and pressure gages must be utilized for the measurement.

5.4.1.2 The Hot Wire Anemometer

The hot wire or hot film anemometer is simply an electrical resistor that is placed in a flow field. The flow velocity is measured by relating the power dissipated in the resistor to the rate of heat transfer to the surrounding fluid. The sensor for the anemometer is illustrated in Fig. 5.5.

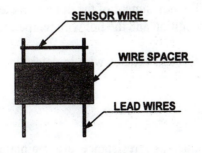

Fig. 5.5 The sensor for a hot wire anemometer.

The heat transfer from a hot wire immersed in a flow field is given by:

$$Q = I^2R = \left(A + B\sqrt{v}\right)\left(T_w - T_f\right) \tag{5.12}$$

Where Q is the rate of heat transfer
      I is the current passing through the hot wire.
      R is the resistance of the wire.
      T_w and T_f are the temperatures of the wire and fluid, respectively.
      A and B are calibration constants.

Solving Eq. (5.12), yields:

$$v = \frac{1}{B^2}\left[\frac{I^2R}{T_w - T_f} - A\right]^2 \tag{5.13}$$

In practice, either the current or the temperature is maintained at some constant value during the measurement. We suggest the use of a constant temperature for the hot wire in making the measurement of flow velocity. A Wheatstone bridge arrangement, presented in Fig 5.6, is employed in making the measurement of the current I flowing through the hot wire with a simple voltmeter.

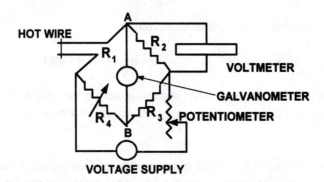

Fig. 5.6 A Wheatstone bridge circuit used to measure current flow through the hot wire (R_1).

Initially, the Wheatstone bridge is balanced with zero flow in the delivery pipe by adjusting resistor R_4 until the galvanometer shows that no current is flowing from point A to B. When fluid flow occurs, the hot wire is cooled, and its temperature T_w and resistance R_1 both decrease. This change in the resistance of the hot wire causes the bridge to become unbalanced. We restore the balance by adjusting the potentiometer to increase the supply voltage to the bridge, which increases the current flow through the hot wire. Additional current flow through the hot wire increases its temperature and resistance. The adjustment of the potentiometer is made until the bridge is restored to its initial (no flow) balance condition and the hot wire temperature and resistance equal their zero flow values.

The current passing through the hot wire is determined from the voltage $\boldsymbol{V}$ measured across arm R_2 as:

$$I = \boldsymbol{V}/R_2 \tag{5.14}$$

When the hot wire temperature and resistance are maintained at constant values by the bridge balancing procedure, Eq. (5.13) reduces to:

$$v = K\left[\left(\frac{I}{I_0}\right)^2 - 1\right]^2 \qquad (5.15)$$

Where K is a calibration constant and I_0 is the current at zero flow velocity that yields a hot wire temperature of about 120 °F.

Calibration of the system is necessary to determine the calibration constant K, and to establish the best temperature at which to operate the hot wire. Of course, it is necessary to electrically insulate the hot wire sensor and the lead wires that are exposed to the water flowing in the delivery pipe to avoid an electrical short.

5.4.2 Flow Meters

Several different flow meters are based on the idea of placing a partial obstruction in a closed pipeline and measuring the pressure drop occurring across that partial obstruction. The pressure drop across each of these obstructions is due to the change in cross sectional area, which causes a corresponding change in the velocity and pressure of the fluid flowing through the partial obstruction. We will describe two of these devices—the flow nozzle and the orifice meter in subsequent subsections.

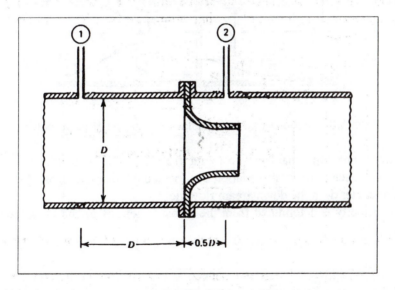

Fig. 5.7 Insertion of a flow nozzle at a flange in a pipeline for measuring flow velocity.

5.4.2.1 The Flow Nozzle

The flow nozzle as the name implies is a nozzle, which is inserted into a pipeline as illustrated in Fig. 5.7. Pressure ports are provided in the pipeline at stations 1 and 2 upstream and downstream of the nozzle. The nozzle contour is in the form of an ellipse and is specified by the ASME[1] as a long-radius flow nozzle. Measurements of the pressure p_1 and p_2 are made at ports in the wall of the pipeline located one diameter upstream and one-half diameter downstream from the nozzle. The volume flow rate is determined from:

---

[1] The ASME (American Society for Mechanical Engineers) specifies the shape of elliptical flow nozzles that are calibrated for measuring flow velocity.

$$q_v = \frac{C_v A_2}{\sqrt{1 - (A_2 / A_1)^2}} \sqrt{2g\left(\frac{p_1 - p_2}{\gamma}\right)}$$

(5.16)

Where A_1 and A_2 are the cross sectional areas of the pipe and the nozzle, respectively.
C_v is a correction factor equal to about 0.95 for low velocity flow in small pipes.

5.4.2.2 The Orifice Meter

The orifice meter is similar to the flow nozzle in that a partial obstruction is inserted in the delivery pipeline. However, the nozzle is replaced with an orifice plate, which is a flat plate with a centrally located circular hole of diameter d_0. The arrangement of an orifice plate inserted into a pipeline for flow velocity measurement is illustrated in Fig. 5.8.

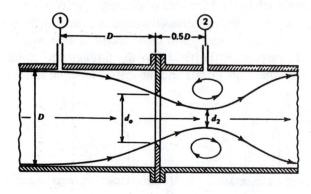

Fig. 5.8 Insertion of an orifice plate to measure flow velocities.

Station 1 is located one pipe diameter upstream from the orifice plate where the flow field is unaffected by the presence of the plate. Station 2 is located 0.5 diameter from the plate at the location of the minimum diameter d_2 in the downstream flow field.
   The flow velocity is determined from the measurement of p_1 and p_2 at the two stations from the following relation:

$$q_v = C A_0 \sqrt{2g\left(\frac{p_1 - p_2}{\gamma}\right)}$$

(5.17)

Where C is the orifice coefficient that is equal to about 0.60 for ratios of d_0/D from 0.2 to 0.4. More detailed information on the orifice coefficient is provided in reference [1]

EXAMPLE 5.2

A delivery pipeline is fabricated from plastic tubing with a two-inch inside diameter. An orifice plate with a ½ inch diameter hole is fitted into the pipeline following the procedure illustrated in Fig. 5.8. Holes are drilled into the pipeline and pressure gages are attached at Stations 1 and 2. Determine the volume flow rate if $p_1 = 3$ psi and $p_2 = 2$ psi.

Solution:

The volume flow rate is determined from Eq. (5.17) as:

Let's convert the result into units of GPM that better characterizes volume flow rate.

$$q_v = CA_0 \sqrt{2g\left(\frac{p_1 - p_2}{\gamma}\right)} = 0.60\left(\frac{\pi}{4^2}\right)\sqrt{2 \times 386 \frac{(3-2)}{0.0361}} = 17.2 \text{ in}^3/\text{s} \qquad \text{(a)}$$

$$q = 17.2 \text{ in}^3/\text{s} \times 60 \text{ s/min} \times \text{gallon}/231 \text{ in}^3 = 4.47 \text{ GPM} \qquad \text{(b)}$$

5.5 MEASURING FLOW RATES IN OPEN CHANNELS

In many applications, irrigation water flows in open ditches or channels. Pipelines are not necessary because the flow is due to small changes in elevation. Since the irrigation water flows from the source down to the field under cultivation gravity overcomes the friction constraints. There is no requirement to close the channel (as in a pipeline) and pressurize the water to deliver a specified flow rate.

One of the difficulties with the flow meters inserted in a pipeline, as described in Section 5.4, is that the flow measurement is replaced with a requirement to measure the pressure at two stations. In open channel flow, the pressures are lower and it is possible to determine flow rate with a relatively simple measurement of the head developed at the partial obstruction.

We describe three similar techniques for measuring open channel flow by placing partial obstructions in the channel and measuring the head at the location of these obstructions. These techniques include—the orifice gate, the sluice gate and the V-weir.

5.5.1 The Orifice Gate

Consider a stream of water flowing from left to right in an open channel. If we insert a gate into the channel, it acts as a dam and the water behind the gate raises until it reaches the top of the gate. It then flows over the gate into the downstream channel. Suppose we cut a circular hole into the gate as shown in Fig. 5.9. Water will flow from the hole and its exit velocity will depend on the height h of the water level behind the gate. After a short time, equilibrium will occur, the head h stabilizes, and the volume flow rate out of the orifice equals the volume flow rate down the open channel.

The volume flow rate is determined from:

$$q_v = CA_0\sqrt{2gh} \qquad (5.18)$$

Where $C = 0.61$ is the orifice discharge coefficient, and $A_0 = \pi d_0/4$ is the area of the orifice with a diameter d_0. It should be noted that the edge of the hole in the orifice gate should be sharp, otherwise the stream exiting the orifice will not contract and the discharge coefficient will be larger than 0.61.

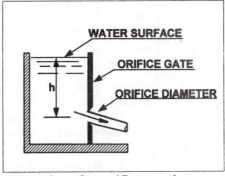

Fig. 5.9 Insertion of an orifice gate into an open channel flow stream.

EXAMPLE 5.3

Your team is preparing an orifice plate to insert in the open channel provided at the delivery site. The pumping system is to produce a flow rate of 75 GPM. Specify the diameter of the hole to be drilled into the plate and justify the specification.

Solution:

If we examine Eq. (5.18), it is clear that there are two unknowns to consider in the application of this relation—A_0 and h. Let's rewrite Eq. (5.18) and then rearrange it as shown below:

$$q_v = CA_0\sqrt{2gh} \tag{a}$$

$$h = \left(\frac{q_v}{C}\right)^2 \left(\frac{8}{g\pi^2}\right)\left(\frac{1}{d^4}\right) \tag{b}$$

For this example, the first two terms on the right hand side of Eq. (b) are constants. Let's evaluate them:

$$\left(\frac{q_v}{C}\right)^2 = \left(\frac{75 \times 231}{0.61 \times 60}\right)^2 = 224{,}070\ \frac{in^6}{s^2}$$

$$\left(\frac{8}{g\pi^2}\right) = \left(\frac{8}{386 \times \pi^2}\right) = 2.10 \times 10^{-3}\ \frac{s^2}{in} \tag{c}$$

$$\left(\frac{q_v}{C}\right)^2\left(\frac{8}{g\pi^2}\right) = 470.5\,in^5$$

Combining Eqs. (b) and (c) gives:

$$h = 470.5/d^4 \tag{d}$$

Let's determine h for selected values of d as shown in the table below:

Table 5.1
Head h for selected values of orifice diameter d
units in inches

d	d^4	h
1.0	1.00	470.5
1.5	5.06	92.9
2.0	16.00	29.4
2.5	39.06	12.0
3.0	81.00	5.8

Inspection of the table shows solution space for the determination of both the head and orifice diameter. If we were to use an orifice diameter of only one-inch the head produced is nearly 40 feet high—impossible. On the other hand, if the diameter is specified as three-inch, the head developed 5.8-inch is only slightly larger than the diameter of the orifice. It appears that either two or 2-½ inch orifice diameters are the most suitable for measuring the flow rate easily and accurately with a scale.

5.5.2 The Sluice Gate

The sluice gate is similar to the orifice gate as illustrated in Fig. 5.10. In using a sluice gate, we block the channel flow with a plate. The gate is pulled up by an amount d to open a small area A_s beneath the sluice gate.

$$A_s = d \times w \qquad (5.19)$$

Where w is the width of the opening at the sluice gate.

The volume flow rate is given by:

$$q_v = C_D A_s \sqrt{2gh_1} \qquad (5.20)$$

Where $C_D = 0.55$ to 0.60 if free flow is maintained down stream for the sluice gate.

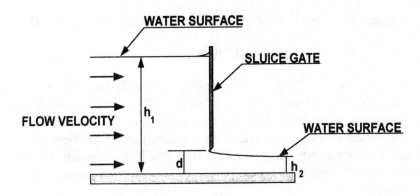

Fig. 5.10 Flow through a sluice gate.

With a constant upstream head h_1, it is evident from Eq. (5.20) that the flow rate is controlled by the area A_s. Since the width of the gate w is constant, the flow rate past the gate is adjusted by varying the height d of the opening beneath the sluice gate. The measurement of d may be accomplished with a scale or a displacement transducer located at the top of the gate.

EXAMPLE 5.4

The open channel section at the delivery site for the irrigation water, which is available to the team, is 8 inch wide by 12 inch high. Your team is expecting the pumping system to produce a minimum of 50 GPM and a maximum of 80 GPM. If you plan to hold the head h_1 on a sluice gate at a constant value of 10 inch, determine the variation in the opening of the sluice gate.

Solution:

Let's substitute Eq. (5.19) into Eq. (5.20) and solve for the depth d of the opening under the sluice gate.

$$d = \frac{q_v}{C_D \times w\sqrt{2gh_1}} \qquad\qquad (a)$$

Before substituting numerical values into Eq. (a), let's convert the volume flow rates from GPM to in^3/s.

$$q_v = \frac{50 \text{ gallon}}{min} \times \frac{231in^3}{gallon} \times \frac{min}{60s} = 192.5 \frac{in^3}{s} \qquad\qquad (b)$$

A similar conversion gives q_v = 80 GPM = 308 in^3/s.

Substituting Eq. (b) into Eq. (a) yields:

$$d_{min} = \frac{q_v}{C_D w\sqrt{2gh_1}} = \frac{192.5}{0.6 \times 8\sqrt{2 \times 386 \times 10}} = 0.456in \qquad\qquad (c)$$

For the maximum flow rate of 80 GPM, a similar computation indicates that:

$$d_{max} = 0.730 \text{ in.} \qquad\qquad (d)$$

Interpret these results—the depth of the opening beneath the sluice gate varies from about ½ to ¾ inch as the flow rate increases from 50 to 80 GPM. How accurately will your team be able to measure such small quantities? Suppose your measurement of the distance d is in error by 0.10 inch. What error is generated in the measurement of the flow rate? Would you be able to make a more accurate estimate of the flow rate if the quantity you were attempting to measure was larger? What parameters would you change to improve the accuracy of the measurement of d?

5.5.3 Weirs

A weir is very much like a sluice gate because an obstruction like a gate is placed in an open channel. The difference is that the water flows over the top of a weir and under a sluice gate. Some weirs are simple rectangular gates where the water flows at some depth d over the entire width of the weir. Another popular weir permits flow through a V-notch cut into the gate as indicated in Fig. 5.11. All of the flow is through the V-notch and the fluid level is maintained below the top of the gate. The V-notch weir is used when flow rates are small and the depth of water across a full width weir is too small to measure with accuracy.

The V-notch weir provides a more accurate measurement of q_v over a wider range of flow rates than a rectangular weir. It also has the advantage that the width of the flow section increases with the head h. The volume flow rate for the V-notch weir is given by:

$$q_v = \left(\frac{8}{15}\right)\sqrt{2g}\,tan\left(\frac{\theta}{2}\right)C_D h^{2.5} \qquad\qquad (5.21)$$

Where the discharge coefficient $C_D = 0.58$ to 0.60.

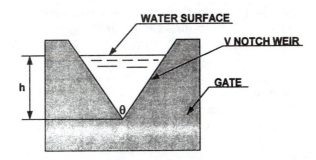

Fig 5.11 A V-notch weir with an included angle of θ.

EXAMPLE 5.5

In interpreting the solution of Example 5.4, your team decided the use of a sluice gate to measure the volume flow rate would not be sufficiently accurate. Instead, you design a V-notch weir with a gate having a width of 8 inch and a height of 10 inch. A V-notch with an included angle of 90° is cut into the gate. The altitude of the V-notch is 8 inch. The tip of the notch is 2 inch above the bottom of the open channel. Perform a design analysis that will show the effectiveness of this weir for measuring volume flow rates ranging from 50 to 80 GPM.

Solution:

Let's begin the design analysis by determining the range in the height h as the flow rate is increased from its minimum value to its maximum. We will solve Eq. (5.21) for the height of the water flowing through the V-notch at the minimum flow rate of 50 GPM.

$$h = \left[\frac{15q_v}{8\sqrt{2g}\,\tan(\theta/2)C_D}\right]^{0.4} = \left[\frac{15 \times 192.5}{8\sqrt{2 \times 386}\tan(45) \times 0.60}\right]^{0.4} = 3.42 \text{ in.} \qquad \text{(a)}$$

Similarly, at the maximum flow rate of 80 GPM, we find that:

$$h = 4.13 \text{ in.} \qquad \text{(b)}$$

In the interpretation of these results, it is clear that the V-notch weir provides an approach for determining the volume flow rate. However, the results indicate that we are only using about one-half of the 8-inch altitude of the V-notch. What should you do to improve the accuracy of the measurement of flow rate using a V-notch weir? Is the selection of θ = 90° the most appropriate choice for the included angle for the V-notch? What would be the result of decreasing this angle from 90° to 60°?

Weirs—the orifice gate, sluice gate, and the V-notch—provide an excellent approach to measuring the volume flow rate. In all cases, the flow rate depends on some head, which may be measured manually with a scale or automatically with a float activated displacement sensor that will be described later in Section 5.7. The data from the displacement sensor may be recorded with a voltmeter.

5.6 PRESSURE MEASUREMENT

In the discussion of various flow measurement devices, we noted that several of them depended on the measurement of pressure to establish the velocity of the flow. We will briefly review some of the simpler approaches to measuring pressure by covering the following devices:

- The piezo tube.
- The U-tube manometer.
- The bourdon pressure gage.
- The dead weight piston gage.

5.6.1 The Piezo Tube

The piezo tube is simply an open-ended vertical tube that is inserted into a port drilled into the wall of a closed channel (pipeline) as shown in Fig. 5.12. The pressure in the pipeline forces the fluid out into the tube. The fluid raises to a height h until its weight produces a pressure equal to the pressure in the pipeline. If the tube is fabricated from a transparent material such as plastic or glass, the height of the fluid can be measured with a scale.

The relationship between the pressure and the height of the fluid in the piezo tube is given by:

$$p = \gamma \times h \qquad (5.22)$$

Where γ is the density of the fluid. For water $\gamma = 62.4$ lb/ft^3 or 0.0361 lb/in^3.

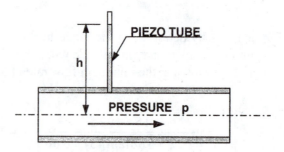

Fig. 5.12 A vertical piezo tube inserted in a pipeline carrying a fluid with a pressure p.

EXAMPLE 5.6

The height of water in a piezo tube is measured at 19.6 inches from the wall of the pipe. If the piezo tube is inserted in a pipeline with a 2-inch diameter, determine the pressure of the water flowing at the centerline of the pipe.

Solution:

Let's substitute the controlling parameters in Eq. (5.22) to obtain:

$$p = \gamma \times h = 0.0361 \times (19.6 + 1) = 0.7437 \text{ psi}$$

This example shows that a piezo tube is very useful in determining low-pressure flow in pipelines. However, when the pressure becomes larger the height of the fluid becomes so large that the tube becomes awkward. For example, a pressure of 2 psi will produce a column of water in a piezo tube that is 55.4 inch high. It is possible to increase the capacity of the piezo tube by employing mercury with a density of 0.490 lb/in^3; however, the use of mercury in measuring instruments is not usually recommended. Mercury is toxic, and we make every attempt to avoid using it in engineering applications.

5.6.2 The U-tube Manometer

The U-tube manometer is similar to a piezo tube because it is also constructed from a transparent tube of glass or plastic. Its shape, as the name implies, is in the form of the letter U. A schematic drawing of a U-tube manometer is presented in Fig. 5.13.

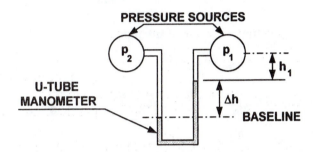

Fig. 5.13 A U-tube manometer used for measuring the difference in the pressure ($p_1 - p_2$).

One end of the tube is connected to a pressure source p_1 and the other end to a second pressure source p_2. In this case, the fluid associated with both pressure sources is water. However, the darker colored fluid in Fig. 5.13 is oil, which is less dense than water and is immiscible with it. It is possible to derive the relation for the pressure difference ($p_1 - p_2$) and the head Δh by considering the pressure developed at the baseline defined in Fig. 5.13.

On the left leg, the pressure is:

$$p_1 + \gamma_w (h_1 + \Delta h)$$

And on the right leg, the pressure at the baseline is:

$$p_2 + \gamma_w h_1 + \gamma_o \Delta h$$

Where γ_w and γ_o are the densities of water and oil respectively.

Since the pressures on both sides of the tube are equal at the baseline, we equate the relations shown above and rearrange terms to obtain:

$$p_1 - p_2 = (\gamma_w - \gamma_o)\Delta h \tag{5.23}$$

This relation shows that the U-tube manometer may be used to measure differential pressures. The requirement is that two fluids be employed in the U-tube and that these two fluids be immiscible and distinguishable. Dark colored oil is a suitable second fluid that may be used with water in the manometer.

EXAMPLE 5.7

One member of the design team suggests the use of a U tube manometer to measure the pressure difference between a piezo tube and a pitot tube. He suggests that the pitot tube be connected to one end of the manometer and the piezo tube connected to the other end. A suitable quantity of oil with a specific gravity of 0.90 is to be poured into the U-tube before the connections are made to the piezo and pitot tubes.

You are uncomfortable with this proposal since you believe that the approach using the U-tube manometer together with piezo and pitot tubes is too complicated. Perform a design analysis of the measurement system to show the feasibility of the proposal. Assume that the flow from the pump is 90 GPM, and the delivery line is two inches in diameter.

Solution:

From Eq. 5.5, we determine the average flow velocity as:

$$v_{ave} = q_v/A = (90 \times 231)/(60 \times \pi) = 110.3 \text{ in/s} \qquad (a)$$

For laminar flow, the velocity at the centerline of the pipe v_0 is:

$$v_0 = 2 \times v_{ave} = 220.6 \text{ in/s} \qquad (b)$$

The dynamic pressure acting on the pitot tube is given by Eq. (5.10) as:

$$p_d = (\gamma_w v_0^2)/(2g) \qquad (c)$$

The difference in the pressures measured by the piezo and pitot tubes is the dynamic pressure p_d. We use this fact together with Eq. (5.23) to write:

$$p_1 - p_2 = p_d = \frac{\gamma_w v_0^2}{2g} = (\gamma_w - \gamma_0)\Delta h \qquad (d)$$

Let's solve Eq. (d) for Δh to obtain:

$$\Delta h = \frac{\dfrac{\gamma_w v_0^2}{2g}}{\gamma_w - \gamma_0} = \frac{\dfrac{\gamma_w v_0^2}{2g}}{\gamma_w(1-0.9)} = \frac{\dfrac{v_0^2}{2g}}{(0.1)} = \frac{\dfrac{220.6^2}{2\times 386}}{0.1} = 630.6 \text{ in.} \qquad (e)$$

The result of the design analysis that is given in Eq. (e) shows that the U-tube manometer is not feasible. The size of the U-tube is too large to be practical. The difficulty is the small difference in the density of oil relative to that of water. The U-tube manometer would be effective in this application if the oil were replaced by mercury with a specific gravity of 13.54. If this substitution were made, Δh would be reduced from 630.6 in. to 5.07 in. Unfortunately, mercury is toxic and it is rarely used in measurement instruments today.

5.6.3 The Bourdon Pressure Gage

The bourdon pressure gage is a mechanical device fabricated from a curved tube. The free end of the curved tube undergoes a displacement when pressurized. The mechanical details of the bourdon tube pressure gage are shown in Fig. 5.14. The tube is made with an oval cross section to increase the displacement of its free end when pressure is applied.

When pressure is applied to the curved tube it tends to straighten. The displacement of the free end of the tube is transmitted through a link to a sector of a spur gear. As the gear sector rotates, its teeth mesh with a centrally located pinion causing it to rotate. A pointer, attached to the pinion gear, rotates with the pinion and indicates the pressure by its position. A dial face is marked with a calibrated scale permitting the pressure to be read directly at the location of the pointer.

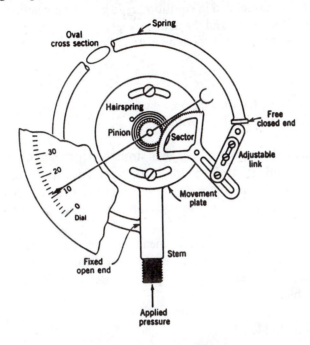

Fig 5.14 Detail of a bourdon tube pressure gage. (After ASME PTC Supplement 19.2, 1964.)

· The bourdon tube provides an accurate method of manually measuring pressure. The pressure gages are rugged and inexpensive. ANSI grade B pressure gages, with an accuracy of ± 2%, may be purchased for less than $7.00.

5.6.4 The Dead Weight Piston Gage

In our discussion of manometers and piezo tubes, we noted that they had the advantage of simplicity but the disadvantage of a very limited range. They are easy to manufacture and accurate pressure determinations can be made using a simple scale to measure the head h. However, the fact that they are limited to pressures of one or two psi if oil is used as the sensing fluid or about 15-psi if mercury is used curtails their application. The dead weight piston gage is similar in some respects to the piezo tube—it employs a tube and is simple to use. Fortunately, the dead weight piston gage is not limited to the measurement of low pressures. A schematic diagram showing the design of a dead weight pressure gage is shown in Fig 5.15.

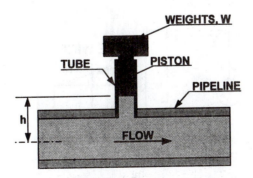

Fig. 5.15 Schematic drawing of a dead weight piston gage.

The tube inserted into the pipeline shown in Fig. 5.15 is like a piezo tube except that it also serves as a cylinder for the piston. The piston is inserted into the tube and weights are applied to the platen on the piston until it is suspended in equilibrium. The pressure acting at the base of the cylinder equals the initial weight of the piston and the added weights.

The pressure at the centerline of the pipe is given by:

$$p = \gamma h + W/A_p \qquad (5.24)$$

where W is the weight of the piston plus the added weights required to achieve equilibrium.
A_p is the cross sectional area of the piston.

The success of a dead weight piston gage is strongly dependent on the fit of the piston with the tube, which serves as the cylinder. If the piston is too tight, friction forces will develop and the gage will not respond accurately. If the piston is too loose in the cylinder, leakage past the piston will occur. If the fit of the piston in the cylinder is suitable, the piston can be rotated slowly in the tube without leakage.

The dead weight piston gage is used for measuring pressures that are essentially constant with time. It is possible to determine pressure even though small fluctuations occur if h is measured with respect to time. However, if the pressure in the pipeline varies significantly with time, the dead weight pressure gage is not suitable.

5.7 THE POTENTIOMETER

A potentiometer is a variable resistor. The simplest way of making a potentiometer is to take a length L of a very high resistance wire and attach it across a voltage source such as a battery. We provide a wiper (an electrical contact) that slides along the length of the wire as shown in Fig. 5.16. The wiper divides the slide wire and its total resistance R_w into two parts. The resistance R_x, defined in Fig. 5.16, is a function of the position x of the wiper as shown below:

$$R_x = (x/L) R_w \qquad (5.25)$$

The current flowing through the potentiometer wire is determined from Eq. (5.14) as:

$$I = V_s / R_w \qquad (a)$$

The voltage drop V_d across the resistance R_x is obtained from Eqs. (5.25), and (5.14) as:

$$V_d = I \times R_x = (V_s/R_w) \times (x/L)R_w = C_p x \qquad (5.26)$$

Where $C_p = V_s/L$ is a constant depending on the slide wire length and the supply voltage V_s.

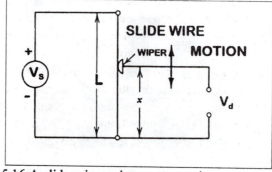

Fig. 5.16 A slide-wire resistance potentiometer.

Examination of Eq. (5.26) shows that the potentiometer converts a distance x into a voltage V. Suppose the potentiometer provides a voltage V_d that is proportional to the head h developed in one of the weirs discussed previously. How do you convert this voltage to provide a measurement of the volume flow rate? A voltmeter will measure the voltage, but not the head h or the volume flow rate q_v. It is necessary to determine a conversion factor to convert the voltage to the volume flow rate. With this conversion factor it is easy to modify an analog voltmeter to provide a direct readout in GPM by:

- Remove the cover from the analog voltmeter.
- Use the conversion factor to modify the scale to display the volume flow rate instead of voltage.

The simple slide wire resistor, shown in Fig. 5.16, is easy to understand. Unfortunately, it is rarely used because the specific resistance of common metals is too low. For instance, if we used a constantan alloy wire (45% Ni and 55% Cu) with a diameter of 0.010 in. for the potentiometer, the resistance per unit length of the wire would only be 0.25 Ω/in. A potentiometer made with a single strand of this wire would have a resistance much too low to be useful. When the resistance applied across the terminals of the battery is too low, the current drawn from the battery becomes excessively high and its life is prohibitively short. To alleviate this difficulty, high-resistance, wire-wound potentiometers are manufactured by winding the resistance wire around an insulating core as shown in Fig. 5.17. The potentiometer illustrated in Fig. 5.17a is used for linear displacement measurements.

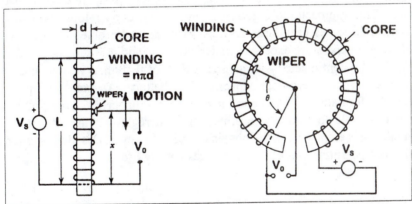

Fig. 5.17 Wire wrapped resistance potentiometers with (a) linear core and (b) cylindrical core.

Cylindrically shaped potentiometers, similar to the one illustrated in Fig. 5.17b, are used for angular measurements. Of course, cylindrical potentiometers are used to measure linear displacements by converting their rotary motion to a linear motion. An example of a conversion method is illustrated in Fig. 5.18 where an extension arm is fastened to the shaft of the potentiometer. The other end of the extension arm is fixed to a float. As the water level rises or falls, the extension arm causes the shaft of the potentiometer to rotate producing a change in the output voltage V_o from the potentiometer.

The resistance of commercially available potentiometers ranges from 10 to about 10^6 Ω depending on the selection of the wire material, its diameter and the length of the coil. If we are to employ a potentiometer in the design of our flow measuring system, the resistance of the potentiometer (its input impedance) should be relatively high to reduce the current drain on the battery. For example, suppose you have a potentiometer with a total resistance of $R_p = 100,000$ Ω and a battery with a voltage of $10V$. The current flowing through the potentiometer is determined from Ohms law [Eq. (5.14)] as:

$$I = V_s /R_p = 10/100,000 = 0.10 \text{ mA}$$

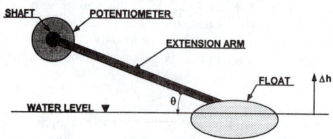

Fig. 5.18 A mechanism for converting a cylindrical potentiometer into a linear motion sensor.

This result must be interpreted before it has meaning. Is 0.10 mA satisfactory or is it a disaster? We know that quantities such as work, energy and power should be conserved. It is an easy step from the concept of conserving power to recognize that the current flow through the potentiometer should be minimized. If we examine the relation for the power dissipated by a potentiometer, as described by:

$$P = I^2 R_p = V_s^2 /R_p \tag{5.28}$$

It is evident that the power and the current are both minimized by increasing the resistance R_p of the potentiometer while fixing the supply voltage V_s. If we were to increase the value of R_p from 100 kΩ to one MΩ, the current would decrease to 0.010 mA and the power dissipated would also decrease by an order of magnitude. This approach looks good; however, if we carry it too far the input impedance (R_p) of the potentiometer becomes very large and electrical noise becomes a problem. We usually seek a compromise between generating excessive electrical noise and minimizing power dissipation.

For our 100,000 Ω potentiometer with a current flow of 0.10 mA, we can anticipate a battery life of 10,000 hours for every 1 Ampere-hour capacity of the battery. An exercise is provided that will develop your appreciation for battery life of common commercial batteries. After you have completed this exercise, you will be able to form an informed opinion regarding the importance of limiting current flow when a battery is employed as the power source. However, it is clear from inspection that selecting a high resistance potentiometer clearly enhances battery life.

EXAMPLE 5.8

Your team has developed a flow measuring system based on an orifice gate similar to that shown in Fig. 5.9. The sensor used to measure the head h is a potentiometer that is arranged with a float and extension rod as indicated in Fig. 5.18. The angle of the extension arm θ ranges from 10° when the water behind the gate is at its maximum height to 90° when the water level is at the centerline of the orifice.

The potentiometer coil is over a circular arc with an included angle of 300°. The potentiometer is adjusted to give an output voltage of zero when the float is at the 90° position. If the water level behind the gate raises the float to produce an angle θ = 20°, determine the height h, the voltage V_0, and the volume flow rate q_v. The design parameters for the system are:

- The effective length of the extension arm is 14 in.
- The supply voltage for the potentiometer is 9 V.
- The resistance of the potentiometer is 100 kΩ.
- The orifice diameter is 2.5 inch.

Solution:

Let's first determine the height h of the water above the centerline of the orifice when the extension rod makes an angle of 20°. Two drawings, which model the two states of the float, aid in visualizing the solution.

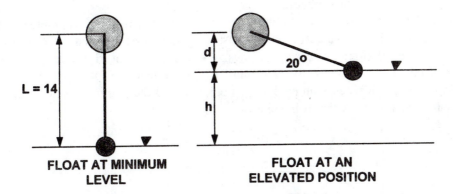

From our simple drawings, it is clear that the head behind the gate with the orifice is:

$$h = L - d = 14 - 14 \tan (20°) = 8.91 \text{ in.}$$

The output voltage V_0 is given by:

$$V_0 = (\Delta\theta/\theta_p)V_s$$

Where $\Delta\theta = 90° - 20° = 70°$ is the angle of rotation of the potentiometer as the height of the water increased, and $\theta_p = 300$ is the angle of the coil of wire in the potentiometer.

$$V_0 = (70°/300°) \times 9 = 2.1°$$

Finally, we determine the volume flow ate from Eq. (5.18) as:

$$q_v = CA_0\sqrt{2gh} = 0.61 \times \pi(1.25)^2 \sqrt{2 \times 386 \times 8.91} = 248.3 \frac{\text{in}^3}{\text{s}}$$

Converting this result into units of GPM gives the flow rate measured with the float-potentiometer system as 64.5 GPM.

5.8 VOLTMETERS

We have described several different approaches to measuring the volume flow rate from a pumping system. Some are simple and the measurement of flow rate may be made manually with a scale. Other approaches are more complex and require pressure measurements to be made. One method, the hot wire anemometer, requires the use of a Wheatstone bridge circuit, and the measurement of both resistance and voltage. Still other flow rate measurement techniques, not covered in this Chapter, require electrical measurements. For this reason, we have included a description of both digital and analog voltmeters.

5.8.1 Analog Voltmeters

There are two markedly different types of voltmeters. The first is the analog meter with its calibrated scale and a pointer. The second is the digital meter with a digital readout of the voltage. Both of these meters are available commercially as multi-meters providing a means for measuring current and resistance in addition to ac and dc voltages.

Let's consider an analog meter, which incorporates a D'Arsonval galvanometer as the sensor element as shown in Fig. 5.19. The galvanometer has a coil supported in a magnetic field with either jeweled bearings (low friction) or torsion springs. When a current flows through the coil, a magnetically induced torque causes it to rotate until it is restrained by springs that are part of the suspension system. The coil reaches an equilibrium position and the pointer indicates the reading of the current or some other quantity that is being measured.

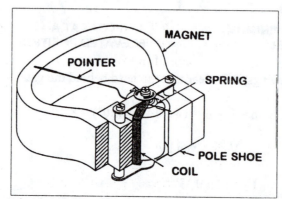

Fig. 5.19 A D'Arsonval galvanometer is used as the sensing movement in analog voltmeters.

The galvanometer is converted into a voltmeter by placing a resistor in series with the coil to control the current flow. The range and sensitivity of the voltmeter is varied by changing the value of this series resistance. Typical analog voltmeters have several scales with full-scale readings varying from 100 mV to 1000 V. These are low cost instruments ($10 to $50) with accuracies of ± 2 to 3% of full-scale. When operated on the lower voltage ranges, the input impedance of the voltmeter is relatively low and loading errors can occur because of the current that passes through the instrument. The input impedance for a typical analog voltmeter (without amplifier) in terms of ohms per volt full-scale varies from about 2000 to 20,000 Ω/V.

What is a loading error? Do we have to deal with it? You do when using an analog meter in an attempt to measure small (tens of millivolts) voltages. When analyzing a circuit like a Wheatstone bridge and developing equations for the output voltage, we assume that the voltmeter used to measure V_0 does not draw current from the bridge. Unfortunately, analog voltmeters with D'Arsonval galvanometers require a significant current to activate the movement; consequently, an error results in the measurement.

The error E due to the current drawn through the meter is given by:

$$E = \frac{R_s / R_{Imp}}{1 + R_s / R_{Imp}} \tag{5.29}$$

where R_{Imp} is the input impedance for the meter.

R_s is the impedance of the voltage source.

Let's consider an example to illustrate loading error.

EXAMPLE 5.9

Suppose that a Wheatstone bridge circuit is comprised of four resistors each 350 Ω. Also, we attempt to measure its output with a voltmeter with an input impedance of 3500 Ω. Determine the loading error.

Solution:

Since the bridge exhibits a source impedance $R_s = 350$ Ω, the ratio of the source resistance to the input impedance resistance of the voltmeter is $R_s/R_{Imp} = 350/3500 = 0.1$. From Eq. (5.29), we determine the loading error due to the voltmeter as:

$$E = 0.1/(1 + 0.1) = 0.091 = 9.1\ \%$$

This example shows that the voltmeter in this case introduces a large error—too large for most engineering applications. How do we reduce the loading error due to the voltmeter? In our design of the circuits and the selection of the voltmeter, we must make certain that the ratio $R_s/R_{Imp} < 1/100$. In other words, if the input impedance of the voltmeter is more than 100 times the output impedance of the voltage source, then the loading error will be less that one percent.

5.8.2 Digital Voltmeters

Digital voltmeters differ in many respects from analog voltmeters. First, a low resistance galvanometer is not involved; consequently, the input impedance of the digital voltmeters is very high. The digital voltmeter shown in Fig. 5.20, purchased at Radio Shack for $24.95 (plus tax), has an input impedance of 10 MΩ. This impedance is so large that loading errors are not an issue with modern digital instrumentation.

Digital voltmeters (DVM) are relatively new; they have largely replaced the analog voltmeters because they are more accurate, exhibit extremely small loading errors, have better resolution and are easier to read. The voltage is displayed with lighted numerals as shown in Fig. 5.20, rather than a pointer on a continuous scale as is the case with analog voltmeters.

The number of full digits on the display determines the range of a DVM. For example, a three-digit DVM records a count of 999. If the full-scale of the DVM were set at 1 V, the count of 999 would be recorded as a voltage of 0.999V. Many DVMs are equipped with a partial digit to extend the range. The (1/2) partial digit displays only two numbers—0 and 1. The (3/4)-digit display four numbers—0, 1, 2, 3 and 4. The value of the partial digit in extending the range of a DVM is apparent in the following example. Suppose you have a 3 digit DVM and attempt to measure a voltage of 16.66 V. Your 3-digit voltmeter would record this measurement as 16.7 V. If you made the same measurement using a 3-½ digit DVM, the reading would be 16.66 V. Since the 3-½ digit DVM provides a count to 1999, you would not lose the last digit in making the measurement.

The resolution of a DVM is determined by the maximum count displayed. For example, a 3-½ digit DVM with a maximum count of 1,999 has a resolution of 1 part in 2,000[2]. The sensitivity of a DVM is the smallest increment of voltage that can be detected. It is determined by multiplying the lowest full-scale range by the resolution. For example, the 3-½ DVM with 100 mV as the lowest (full-scale) range has a sensitivity of 100 mV x 1/2000 = 0.05 mV.

The accuracy of a DVM is usually expressed as ± x% of full-scale ± N digits. The specifications for the inexpensive Radio Shack DVM, with 3-¾ digits, indicates an accuracy ± 0.2%

---

[2] A 3 ½ DVM has a total count of 2000 because, in digital representation, we use one count for zero and 1999 for the remaining numbers shown on the display.

of full-scale plus ± 1 in the last digit. If we are operating this DVM on the 100 mV scale, the accuracy is ± (0.002 × 100 mV + 0.1 mV) = ± 0.3 mV.

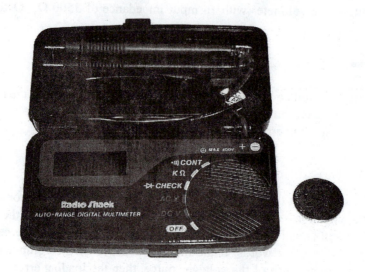

Fig. 5.20 Photograph of an inexpensive pocket size digital multi-meter.

The input to a multi-meter may be voltage dc or ac, current or resistance. In all instances, the input is immediately converted to a dc voltage as the signal enters the instrument. The voltage is amplified and then transmitted to an analog to digital (A/D) converter. The A/D converter changes the dc voltage input to a clock count by using what is called the dual slope integration method illustrated in Fig. 5.21.

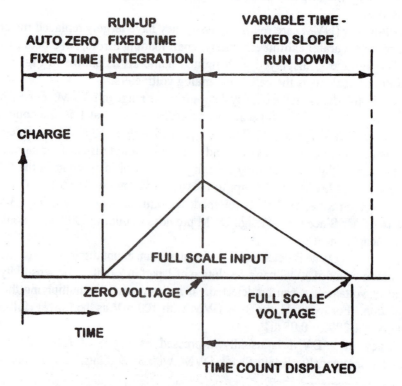

Fig. 5.21 Voltage time trace for a dual slope integrating DVM.

There are three different time controlled operations involved with the dual-slope integration technique.

1. Auto zero—where the output voltage from the integrator is zeroed for a fixed time (usually about 100 ms).

2. Run-up—the dc voltage input is integrated for a fixed time (about 100 ms). The output of the integrator circuit is a voltage, which increases with time to produce a ramp function as illustrated in Fig. 5.21.

3. Run-down—the voltage on the integrator is discharged at a constant rate with respect to time to produce a linear decreasing voltage (the run-down ramp is also shown in Fig. 5.21).

Because both the ramp-up and ramp-down voltages exhibit slopes on the graph of voltage versus time, this A/D conversion technique is known as dual-slope integration.

A counter (clock), which is initiated at the beginning of the ramp-down, runs until the voltage from the integrator goes to zero. It stops at this instant. Since this time interval is proportional to the dc voltage first applied to the integrator, the dual slope technique effectively converts voltage to a time count. The time count is presented on the digital display as the dc voltage. Counting is relatively easy to accomplish in digital electronics because extremely accurate quartz crystal oscillators are available at low cost. For example, a 10 MHz counter, capable of providing a train of well-defined voltage pulses at 0.1 μs intervals, is available at prices ranging from $1 to $3 depending on packaging and quantity ordered.

5.9 SUMMARY

We introduce fluid flow with an expression describing the volume flow rate:

$$q_v = \int_A v dA$$

and then showed that the weight and mass flow rates were related to this quantity.

Descriptions were provided for both closed and open channel flows that indicated that the velocity profile across the stream was an important parameter markedly affecting flow rate. Methods for measuring flow rates were described. The discussion began with insertion instruments and the pitot tube and the hot wire anemometer were described. Relations were developed for determining the volume flow rate from the output of these instruments, and an example was given to demonstrate computational techniques.

Flow meters were introduced where partial obstructions such a flow nozzles or orifice plates are placed in a pipeline. Since the velocity of the flow must increase in passing through these devices, a pressure drop occurs, Relations are provided to determine the volume flow rate from these pressure changes. An example was given to demonstrate the application of the relations to determining the volume flow rate.

Three methods were described for measuring flow rates in open channels—the orifice gate, the sluice gate and the V-notched weir. These methods are applicable to the determination of flow rates from the pumping systems developed in this class project. Relations are provided for the volume flow rates in terms of the head behind the various gates. Examples are provided for each method to illustrate the computation procedure used to convert from head to flow rates.

Methods of measuring pressure are described. Of the many approaches to measuring pressure, we described four—the piezo tube, the U-tube manometer, the bourdon pressure gage, and the dead weight piston gage. Equations relating the gage output to pressure are given. Also, examples are presented that will aid in the design of these instruments.

The potentiometer is introduced as a sensor for the measurement of displacement. A device incorporating a float is suggested as an approach for electrically measuring the head behind a gate. An example is presented that illustrates that the design of a system with a potentiometer is feasible for measuring flow rates in open channels.

Finally both analog and digital voltmeters are introduced as instruments to measure electrical quantities such as voltage, current and resistance. Mechanical details of the analog voltmeter are described together with loading errors that are encountered when measuring small voltages. The digital voltmeter is described with a discussion of the number of digits, the count, resolution, and accuracy. The method of converting voltage to the count of clock pulses with a dual slope integrator is described.

REFERENCES

1. Dally, J. W., W. F. Riley, and K. G. McConnell, Instrumentation for Engineering Measurements, 2nd Edition, John Wiley, New York, NY, 1993.
2. Benedict, R. P. Fundamentals of Temperature, Pressure and Flow Measurement, 2nd Edition, John Wiley, New York, NY, 1977.
3. Doeblin, E. O., Measurement Systems: Application and Design, McGraw Hill, New York, NY, 1975.
4. Northrup, R. B. Analog Electronic Circuits: Analysis and Application, Addison-Wesley, Reading, MA, 1990.
5. Tandeske, D, Pressure Sensors: Selection and Application, Dekker, New York, NY, 1991.
6. Beckwith, T. G. and R. D. Marangoni, Mechanical Measurements, 4th Edition, Addison-Wesley, Reading, MA, 1990.

EXERCISES

5.1 Determine the flow rate in terms of volume, weight and mass for water flowing in a two-inch inside diameter pipeline with velocity of 8 ft/min. Assume the flow is laminar.

5.2 Determine the flow rate in terms of volume, weight and mass for water flowing in a three-inch inside diameter pipeline with velocity of 12 ft/min. Assume the flow is laminar.

5.3 Determine the flow rate in terms of volume, weight and mass for water flowing in a one-inch inside diameter pipeline with velocity of 24 ft/min. Assume the flow is laminar.

5.4 A pitot tube is inserted into the centerline of a pipeline that is carrying a flow of 60 GPM. If the pipeline is fabricated from plastic pipe with a 2 inch inside diameter, determine the height of the water in the pitot tube. Note a piezo tube inserted near the pitot tube exhibits a height of water of 14 inches above the outside wall of the pipe.

5.5 A pitot tube is inserted into the centerline of a pipeline that is carrying a flow of 80 GPM. If the pipeline is fabricated from plastic pipe with a 3 inch inside diameter, determine the height of the water in the pitot tube. Note a piezo tube inserted near the pitot tube exhibits a height of water of 6 inches above the outside wall of the pipe.

5.6 A pitot tube is inserted into the centerline of a pipeline that is carrying a flow of 40 GPM. If the pipeline is fabricated from plastic pipe with a 1-½ inch inside diameter, determine the height of the water in the pitot tube. Note a piezo tube inserted near the pitot tube exhibits a height of water of 10 inches above the outside wall of the pipe.

5.7   Determine the velocity of flow for water in a pipeline if the height h_1 in a piezo tube is 9 inches and the height h in a pitot tube is 15 inch. Figure 5.4 gives definitions of h_1 and h.

5.8   Determine the velocity of flow for water in a pipeline if the height h_1 in a piezo tube is 6 inches and the height h in a pitot tube is 18 inch. Figure 5.4 gives definitions of h_1 and h.

5.9   Determine the volume flow rate for the water carried in the pipeline of Exercise 5.7 if the pipe has an inside diameter of 2 inch. Assume laminar flow conditions exist.

5.10   Determine the volume flow rate for the water carried in the pipeline of Exercise 5.8 if the pipe has an inside diameter of 2 inch. Assume laminar flow conditions exist.

5.11   Prepare a detailed sketch of a hot-wire anemometer with a parts list specifying the materials and size for the hot wire, the lead wires, the spacer block and the electrical insulation.

5.12   For the hot-wire anemometer of Exercise 5.11, determine the resistance of the hot wire.

5.13   If the hot-wire anemometer of Exercise 5.11 is connected into a Wheatstone bridge circuit, as shown in Fig. 5.6, determine the current flow through the hot wire if the supply voltage is set at 5 volts.

5.14   Prepare an engineering brief interpreting the results of Exercises 5.12 and 5.13 as they affect the feasibility of the design of the hot-wire anemometer proposed in Exercise 5.11.

5.15   A differential pressure gage is attached to the tubes from the pipeline with a flow nozzle, which measures the quantity $p_1 - p_2 = 3$ psi. If water is flowing in the 3-inch diameter pipeline, determine the volume flow rate. The diameter of the throat of the nozzle is 1.0-in.

5.16   A differential pressure gage is attached to the tubes from the pipeline with a flow nozzle, which measures the quantity $p_1 - p_2 = 1.8$ psi. If water is flowing in the 2-inch diameter pipeline, determine the volume flow rate. The diameter of the throat of the nozzle is ½-inch.

5.17   A differential pressure gage is attached to the tubes from the pipeline with an orifice plate, which measures the quantity $p_1 - p_2 = 2.4$ psi. If water is flowing in the 2.5-inch diameter pipeline, determine the volume flow rate. The diameter of the orifice is 1.25-inch.

5.18   A differential pressure gage is attached to the tubes from the pipeline with an orifice plate, which measures the quantity $p_1 - p_2 = 1.2$ psi. If water is flowing in the 1.5-inch diameter pipeline, determine the volume flow rate. The diameter of the throat of the nozzle is 3/8-inch.

5.19   Your team decides to exhaust the water from a hose leading from the pump into an open channel with a cross section defined by h = 12 inch, w =8 inch and L = 8 feet. Flow measurement is to take place in this open channel. Prepare an engineering brief describing the reasoning for selecting one of the following measurement methods:

- The orifice gate.
- The sluice
- The V-notched weir.

You anticipate producing 44 GPM from the pumping system.

5.20   Prepare a sketch of the design of an orifice gate to fit into the flow channel that is compatible with the pump capacity described in Exercise 5. 19. Conduct a design analysis that supports the selection of the diameter specified in your design.

5.21   Prepare a sketch of the design of a sluice gate to fit into the flow channel that is compatible with the pump capacity described in Exercise 5. 19. Conduct a design analysis that supports the selection of the opening d beneath the gate as specified in your design.

5.22   Prepare a sketch of the design of a V-notch weir to fit into the flow channel that is compatible with the pump capacity described in Exercise 5. 19. Conduct a design analysis that supports the selection of the included angle and the altitude of the notch in the weir gate specified in your design.

5.23 In conducting preliminary experiments with the orifice gate, you measure the following heads h with different diameter orifices: h = 10-inch with d = 1.5-inch; h =6-inch with d = 2.0-inch; h = 5-inch with d = 2.5-inch. Determine the volume flow rate for each of the three measurements. Prepare a list of reasons for the variations from one experiment to the other.

5.24 In conducting preliminary experiments with the sluice gate, you measure the following heads h_1 with different gate openings d: h_1 = 11-inch with d = 1.0-inch; h_1 = 8-inch with d = 1.25-inch; h_1 = 6-inch with d = 1.5-inch. Determine the volume flow rate for each of the three measurements. Prepare a list of reasons for the variations from one experiment to the other.

5.25 In conducting preliminary experiments with the V-notched weir, you measure the following heads h with a different included angle θ for the notch: h = 8-inch with θ = 60°; h = 6-inch with θ = 75°; h = 5-inch with θ = 90°. Determine the volume flow rate for each of the three measurements. Prepare a list of reasons for the variations from one experiment to the other.

5.26 The height of water in a piezo tube is measured at 11.4-inch from the wall of the pipe. If the piezo tube is inserted in a pipeline with a 2-inch diameter, determine the pressure of the water flowing at the centerline of the pipe.

5.27 The height of water in a piezo tube is measured at 9.3 inches from the wall of the pipe. If the piezo tube is inserted in a pipeline with a 1.0-inch diameter, determine the pressure of the water flowing at the centerline of the pipe.

5.28 Prepare an engineering brief describing a bourdon tube pressure gage. Include in the brief a listing of applications of the bourdon tube to measure pressure.

5.29 Prepare a sketch of a design of a dead weight piston gage for measuring pressure that varies from a minimum of 3.7 psi to a maximum of 4.3 psi. Include a parts list that describes each component in the design.

5.30 Perform a design analysis of the dead weight pressure gage of Exercise 5.29. Modify the design as required by the results indicated in the analysis.

5.31 Make a trip to an electronic supply house and examine the various types of potentiometers available and the prices for each type. If you are not able to make the real trip, which is advised, make a virtual trip. Try the supplier Digi-Key with its on-line catalog at www.digikey.com.

5.32 A supply voltage of 6 V is placed across a potentiometer. If the sensing element of the potentiometer is a circular arc with an included angle θ = 320°, determine the output voltage for a rotation of the wiper of 216°.

5.33 A supply voltage of 9 V is placed across a potentiometer. If the sensing element of the potentiometer is a circular arc with an included angle θ = 310°, determine the output voltage for a rotation of the wiper of 118°.

5.34 Prepare a drawing showing the design of a water level measuring system similar to that shown in Fig. 5.18. A variation of the head is expected to range from zero to ten inches. Include in the design a parts list identifying all of the components. Also, size each component in the measuring system.

5.35 Perform a design analysis of the float-potentiometer head measuring system developed in Exercise 5.34.

5.36 Prepare an engineering brief describing an analog multimeter that employs a D'Arsonval galvanometer.

5.37 Prepare an engineering brief describing a digital multimeter that employs a dual slope integrating circuit.

5.38 Visit an electronics supply store and examine the line of digital multimeters available. One possibility is Radio Shack, which is a commercial enterprise with thousands of retail outlets. If you are not able to make the real visit, which is advised, make a virtual visit to Radio Shack's web page at www.radioshack.com.

PART II

ENGINEERING GRAPHICS

CHAPTER 6

THREE-VIEW DRAWINGS

6.1 INTRODUCTION

Engineering graphics is a broad term that is used to describe a means of communication. We normally think of communication in terms of writing and speaking because they are more commonly employed in the normal course of our life. However, when trying to communicate design ideas, we find writing and speaking insufficient to express our thoughts. A more visual way to communicate is needed. It is more effective to present our ideas by means of drawings, sketches, pictures, and graphs of many different types. Visuals aids, such as drawings, convey our ideas quickly and with remarkable precision. It is easier to transmit and to receive information if it is conveyed in the form of drawings, sketches, or graphs.

The objective of this part of the textbook is to introduce you to visual methods of communication. You should understand the advantages of presenting information in drawings, sketches, and graphs because they are very common techniques used to communicate very complex ideas quickly and with precision. In this part of the textbook, you will begin to learn techniques for preparing orthographic projections, isometric drawings, sketches, and several different types of graphs.

There are two general approaches used in preparing drawings and graphs. The first is a manual approach where we draw the visuals by hand using a few simple drawing instruments. The second is using a computer and suitable software programs, which greatly facilitate the preparation of a drawing or graph. In this chapter, we will cover the manual methods for preparing drawings. In a different textbook, we introduce a computer-aided design (CAD) software program used in preparing engineering drawings. In still another chapter, we describe EXCEL, which is a spreadsheet program that is used in computation and in preparing several different types of different graphs. Finally, we describe a graphics presentation program, PowerPoint, employed to prepare slides and overhead transparencies for oral presentations.

6.2 VIEW DRAWINGS

Let's begin by considering the block-like object shown in Fig. 6.1. We could describe this object as a rectangular block with a slot cut from its top. However, this written description is vague because we have not conveyed the relative proportions of the block, the precise location of the slot, or the size of its features. In preparing an engineering drawing, it is essential to communicate proportion, exact location, and size of every feature of the object, which we are describing. The objective of the drawing is to convey sufficient information for the object to be produced by anyone capable of reading a drawing. We use pictorial drawings and/or multi-view drawings to quickly and accurately convey this information to the reader. The drawing shown in Fig. 6.1 is a pictorial, which we will use to aid in describing multi-view drawings.

The pictorial (isometric) drawing in Fig. 6.1 is shown with three arrows. These arrows represent the directions of observation of the object which show the front, top and side views. When considered individually, each view is a two-dimensional rendering of the three-dimensional object. Of course, two-dimensional drawings are incomplete, because they show only the information observed in one of several possible views. Nevertheless, the two-dimensional views are important because we can show objects to a true scale on a sheet of ordinary paper or the screen of a computer monitor. We avoid the problem of incomplete information inherent in two-dimensional drawings by presenting a sufficient number of views to completely locate and size each feature of the object.

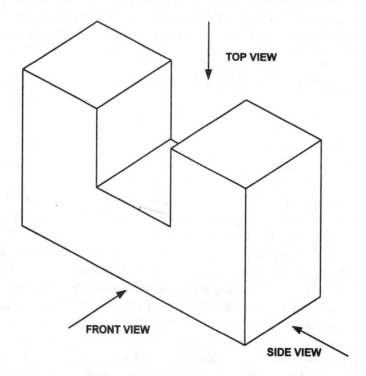

Fig. 6.1 Pictorial drawing showing a rectangular block with a slot. The three viewing directions present the front, top, and side views.

A three-view drawing of the rectangular block presented in Fig. 6.1 is illustrated in Fig. 6.2. The single pictorial drawing of the block is now represented with three different two-dimensional drawings representing the front, top, and side views. These three drawings completely define the proportions of the rectangular block, its size, and the location and size of the slot.

The arrangement of the views is important. The front view is placed in the lower left corner of the paper, the top view is directly above it, and the side view is to the right of the front view. This arrangement is important because it permits us to prepare the three-view drawing using orthographic projection. With orthographic projection, we project dimensions from one view to another. For example, we measure width and height of the block, draw the front view, and project construction lines upward to the top view. The width of the object is identical in both the top and front views. We also project construction lines to the right from the front view to define the height of the side view. The front and side views both show the height dimensions of the block. We can also project from the top view to the side view if we draw a 45° construction line from the right hand corner of the front view as shown in Fig. 6.3. We then project construction lines to the right from the top view until they intersect the 45° construction line. We turn them downward to show the depth of the block on the side view. All of the construction lines used to prepare Fig. 6.2 are illustrated in Fig. 6.3. Orthographic projection is very important because it saves you time in preparing your drawings. It reduces the number of time consuming measurements that you must make in preparing the different views necessary to completely describe any given part. The construction lines also help you in forming visual images of the locations of each feature on the adjacent views.

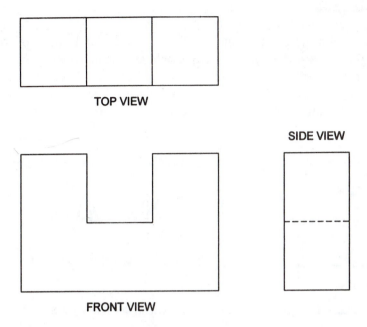

Fig. 6.2 A three-view drawing of the object shows the front, top, and side views.

One more point to consider before we leave Fig. 6.2. Did you notice the dashed line on the side view? Why were all of the lines solid except that one? We follow a convention in preparing engineering drawing for the use of line styles and weight. Solid lines represent edges of features that are observed in a particular view. As we observe the top of the block shown in Fig. 6.1, we can see the four sides of the rectangle and the two edges of the slot. We use solid lines then to represent all six edges visible in the top view. However, when we view the object from the side only the four edges outlining the rectangle are visible. Viewing the right side, we cannot observe the slot. If we look at the pictorial or the top and front views, we know that the slot exists. To show the slot on the side view, we use a dashed line to represent a line hidden from view. Drawings differ from photographs because hidden features can be shown on the appropriate view with dashed lines.

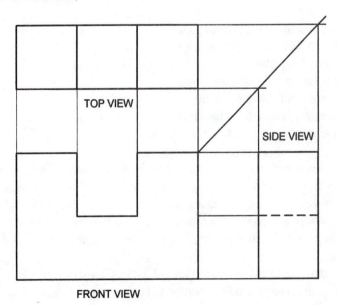

Fig. 6.3 Construction lines used to convey dimensions from one view to another.

6.3 LINE STYLES

We use different line styles and different line weights (thickness and darkness of the lines) in preparing engineering drawings as illustrated in Fig. 6.4. The weight and the thickness of the lines are controlled by the selection of the pencil lead and the sharpness of the point. The degree of hardness is governed by variable amounts of clay that is mixed with graphite to form pencil lead. Hard leads have a small amount of clay while soft leads have higher clay content. The scale from hard and light to soft and dark is given below:

Hard and Light $\Leftarrow$ 6H–5H–4H–3H–2H–H–F–HB–B–2B–3B–4B–5B–6B $\Rightarrow$ Soft and Dark

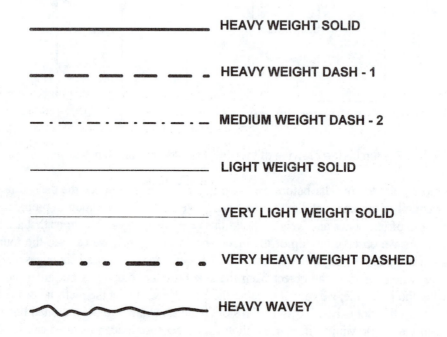

HEAVY WEIGHT SOLID

HEAVY WEIGHT DASH - 1

MEDIUM WEIGHT DASH - 2

LIGHT WEIGHT SOLID

VERY LIGHT WEIGHT SOLID

VERY HEAVY WEIGHT DASHED

HEAVY WAVEY

Fig. 6.4 Different line styles used in view drawing.

   a. Heavyweight, solid-lines representing edges
   b. Heavyweight, dashed-lines representing hidden lines.
   c. Medium-weight, long-dash then short-dash lines for centerlines.
   d. Lightweight solid-lines for dimensioning.
   e. Very thin, light-lines for construction.
   f. Very heavyweight, dashed-lines for section cuts.
   g. Heavy wavy-lines to indicate a break in the view.

We use 4H to 6H leads with a very sharp point to draw construction lines. These lines are so light and thin that they will not show when the drawing is reproduced. Light-weight dimension lines are drawn so they will remain visible when the drawing is copied. We use an H or F lead to give the correct relative darkness and a sharp point with a very slightly rounded tip. Medium-weight centerlines are drawn with a HB or B lead with a rounded point. Heavy-weight lines, both solid and dashed, are drawn with B or 2B lead with a rounded point.

The very soft leads—3B to 6B are usually employed by artists to prepare sketchings, but not by engineers in their sketches. Engineers handle a completed drawing much more than an artist. In handling, the soft lead smears degrading the appearance of the drawing and detracting from the quality of the copies usually made from the originals.

We recognize that many of you will be short on pencils, but perhaps team members can share resources. Please do not use ball-point pens for drawing. There are two problems with ball-point pens. First, you have no control over the width of the line because the line width is determined by the diameter of the ball in the pen. Second, you may make errors in preparing the drawing. The lines drawn with a ball-point pen cannot be erased; you eventually complete a drawing that is sloppy and difficult to read. Work in pencil and use a clean, white, vinyl-eraser to eliminate all traces of your errors. Strive hard to produce an accurate, clean, and error-free drawing.

While we are on the topic of pencils and erasers, let's list some other simple tools that you will want to acquire.

1. Transparent ruler with scales in both inches and millimeters.
2. Two triangles—30/60 and 45/45 degrees.
3. Protractor
4. Compass
5. Templates for circles, ellipses, etc.

This is a short list. We have not included a complete set of drawing instruments or a drafting board with a T-square. Our purpose is not to teach drafting, but to introduce you to the essentials of engineering graphics. A complete set of drafting tools is not necessary to learn the early lessons in graphics. We encourage you to invest the small sum needed to acquire the items listed above. You will find them useful in many other courses during your program in engineering.

Finally, a suggestion is made for the paper to use in preparing your drawings. We recommend National Brand engineering paper 8-1/2 by 11 in size. It is light green in color to relieve the strain on your eyes. It also has a grid (five squares to the inch) on the backside of each sheet. The grid shows through to the front side providing guidelines that are helpful in projecting the construction lines needed to prepare three-view drawings. The paper erases well, produces good copies, and is pre-punched for a three-ring binder.

Equipped with the proper pencils, eraser, tools and paper, you are ready to learn techniques for drawing several different features normally encountered in preparing multi-view drawings of engineering components.

6.4 REPRESENTING FEATURES

We will demonstrate the techniques for drawing various features in three-view drawings. The examples selected include a block with a slot and a step, a block with a tapered slot, and a block with a step and a hole.

6.4.1 Block with Slot and Step

The rectangular block in Fig. 6.1 has one feature—a slot. We encounter many different geometric features in depicting parts that we design, including holes, curved boundaries, steps, and tapers. A pictorial of an object with both a slot and a step, presented in Fig. 6.5, serves as our next example. Let's prepare a three-view drawing of this object using the methods of orthographic projection. Examine the pictorial starting with the front view.

As you examine the front view in Fig. 6.5, visualize the outline of the block. Even with the step and the slot, the outline of the block (the outside edges) is a rectangle. Draw the rectangle, representing the outline of the front view, in the lower left hand corner of your quadrille paper. We will not worry about exact dimensions of the rectangle at this stage; however, try to maintain the proportions shown in Fig. 6.5. When drawing the rectangle in the front view, extend the vertical lines upward into the region of the top view and the horizontal lines to the right into the region of the side view. These are construction lines; you should keep them thin and light. Now examine the step in the upper right of the front view of the block.

Note the intersection of the planes defining the step with the front plane of the block. These intersections produce two new edges that are visible in the front view. Draw the location of these edges with the horizontal and the vertical lines shown in the upper right hand corner of the rectangle (see Fig. 6.6). Extend the vertical line used to locate the step in the front view upward and the horizontal line to the right. Finally, locate the slot on the left side of the pictorial drawing in Fig. 6.5. Again note the planes defining the slot and observe that they intersect the front plane to form two vertical edges. We draw two vertical lines in the correct location to represent these edges. Project these lines upward into the region of the top view with construction lines. Okay! We have completed the front view and are ready to draw the top view.

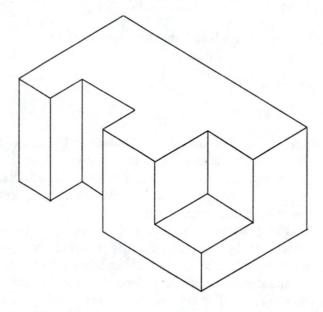

Fig. 6.5 A pictorial drawing of a block with a slot and a step.

Look downward at the pictorial drawing. Again, the outline observed is rectangular in shape; however, the rectangle is not perfect since the slot interrupts it. Looking downward, we observe that the edge defining the outline of the rectangle across the width of the slot is missing. It is possible to observe the bottom plane at the location of the slot. Draw the outline of the rectangle in the location of the top view of Fig. 6.6. Use the construction lines that exist at the location of the top view to establish the width dimensions. An open section in the rectangle is evident at the location of the slot. Next, show the depth of the notch in the top view by drawing the three lines that represent the edges formed by the three vertical planes of the slot with the top plane of the block.

Next, examine the step. The step has two vertical planes that intersect the top plane to form the defining edges. We draw two lines representing those edges in our drawing to complete the top view. If you made full use of the construction lines projected from the front view into the region of the top view, it was easy to draw the top view. It was only necessary to locate the position of three horizontal lines that defined the depth of the slot, step, and rectangular block. These depth dimensions are carried to the right with construction lines.

The final view is of the right side. It is possible to draw a left side view, but the convention is to draw the right view. After having completed the front and top views, with the orthographic projection of construction lines, the side view is easy to prepare. Draw a construction line at 45° so that it intersects the horizontal construction lines from the top view. Then draw a second set of construction lines downward from these intersection points into the region of the side view. The side view now contains seven construction lines. The three horizontal lines projected from the front view give the height of the step and the rectangular block. The four vertical lines projected from the top view give the depth of the slot, step, and block. We have the location and dimensions of all of the lines in the side view from our orthographic projections. No measurements are necessary. To complete the side view, we darken select portions of the construction lines that represent edges in the side view. Again, examine the side of the pictorial drawing in Fig. 6.5 and observe that the rectangular outline of the block is completed. Also note the edges produced by the intersections of the planes forming the step with the plane of the right side of the block. Draw these two edges in the upper left corner of the side view (see Fig. 6.6). From a visual perspective, our drawing of the right side is complete; however, a multi-view drawing often carries more information than what we can observe directly. We know from the front and top views that a slot exists. Our projection lines from the top view to the side view show the depth of the slot. Even though we cannot "see" the slot in the side view, we represent its presence and its location with a dashed (hidden) line. The three-view drawing of a rectangular block containing a slot and a step is complete as illustrated in Fig. 6.6.

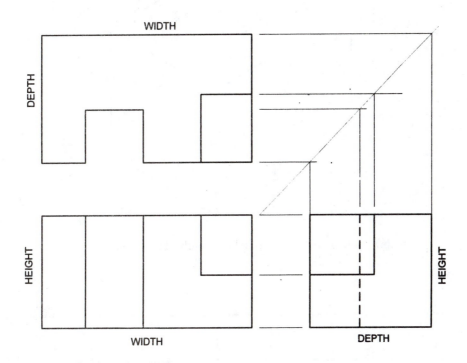

Fig. 6.6 A three-view drawing of the block with a slot and a step.

6.4.2 Block with Tapered Slot

Let's consider drawing still another feature—a block incorporating a tapered slot as illustrated pictorially in Fig. 6.7. Begin a three-view drawing with the front view in the lower left corner of your drawing paper as shown in Fig. 6.8. Examine the front view in Fig. 6.7 and observe the edges produced by all the planes intersecting the front plane. Drawing these edges gives us four vertical lines and three horizontal lines. The tapered surface intersects the front plane giving an edge shown as a horizontal line in the front view. The top surface is represented by still another edge visible in the front view. We project all of these lines into the remaining two views with construction lines.

Examine the top view and observe that the rectangular outline of the block remains intact. Observe three edges inside the rectangular outline. The edge due to the intersection of the taper plane with the top surface produces a horizontal line, and the two vertical planes, defining the width of the slot, intersect the top surface to produce two vertical lines in the top view. Project the horizontal lines in the top view to the right so that they intersect the 45° construction line shown in Fig. 6.8. Then project these three lines downward from the intersection points on the 45° line into the side view.

The six construction lines projected into the side view provide the dimensions needed to complete this view. Clearly, it is evident from Fig. 6.7 that the rectangle outlining the side view is intact. In fact as we view the right side, only a simple rectangle formed by the four edges is visible. The information regarding the tapered surface, evident in the pictorial drawing and the front view, is not visible in the side view. We add information, showing the location of the tapered surface on the side view, by using the dashed (hidden) lines.

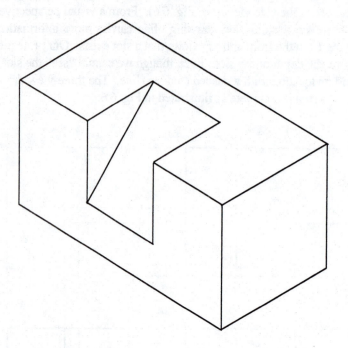

Fig. 6.7 A pictorial drawing of a block with a tapered slot.

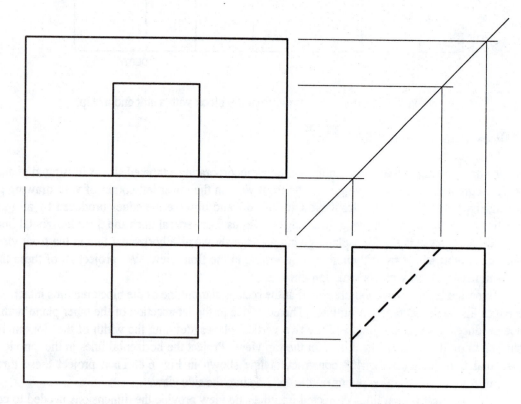

Fig. 6.8 A three-view drawing of the block with a tapered slot.

6.4.3 Block with Step and Hole

A pictorial drawing of a block with a step and a hole is illustrated in Fig. 6.9, and a three-view drawing of this object is presented in Fig. 6.10. The purpose in showing the fourth example of a three-view drawing is to illustrate the method for representing holes and curved boundaries in engineering drawings. Assume in this discussion that you understand the techniques used to draw the outlines of the three-views. The only question remaining concerns drawing the curved boundary that outlines the step and the hole that is drilled through the step.

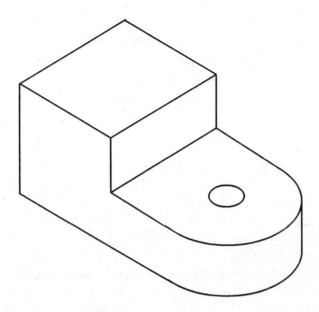

Fig. 6.9 A pictorial drawing of a block with a step and a hole.

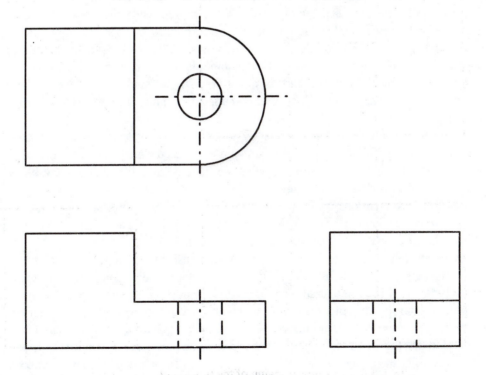

Fig. 6.10 A three-view drawing of the block with a step and a hole.

The circular boundary outlining the step is evident only in the top view (see Fig. 6.10). The front and side views give no evidence of the presence of a curved boundary because the vertical curved surface does not produce intersections (edges) on the front or side planes. We produce the curved surface in the top view by using a compass with its point located in the center of the hole. The radius is set on the compass so that the circular arc is tangent to the horizontal lines forming the outline of the block in the top view. It is advisable to draw the arc with the compass prior to drawing the horizontal lines that are tangent to the ends of the arc.

The compass is also employed to draw the hole in the top view. Note the two orthogonal lines defining the center of the hole. These are centerlines that locate the position of the hole relative to other features on the block. The presence of the hole is also depicted in the front and side views even though it is not visible in either of these views. We use dashed (hidden) lines to locate the edges of the holes. We also show the centerline of the hole in the front and side views. Note the long-dash, short-dash line used to denote the centerlines in all three-views. The centerlines are very important when we dimension the drawings because we locate holes with the centerlines and not the their edges.

6.5 DIMENSIONING

In the previous section, we prepared three-view drawings maintaining proportionality, but without concern for the dimensions. This approach was a simplification taken to clarify the discussion of three-view drawings. In actual practice, the dimensions are critical because they control the size of the component and the location of all of its features. Let's examine dimensions from two different points of view. First, as the individual preparing the drawing, you must know (or decide) upon the dimensions required to completely define the component. If you are the designer of a new component, it is necessary to start with a blank sheet of paper and assign the dimensions that you believe will optimize the design of the component. If you are redesigning an existing component, it is possible to start with the drawing of that component and modify existing dimensions to refine the design. The point here is that you (as the person preparing the drawing) must know the dimensions. It is your responsibility to include on your drawing all dimensions necessary for some stranger, perhaps in some other country, to fabricate the component.

Next consider the second point of view—as the person using the drawing. This person may manufacture the part, may assemble the product that incorporates the part, or may repair the product. Engineering drawings serve many purposes and are used by several people not involved in the design. The three-view drawing defines this component without ambiguity. The dimensions give its size precisely. The component as defined by the drawing and its dimension can be made by anyone in the world. The component must be interchangeable with another component manufactured previously by another plant in another country.

Let's begin to learn about dimensioning by adding dimensions to the drawing of the block with a slot shown in Fig. 6.2. We have copied this drawing and added dimensions as indicated in Fig. 6.11. Examine the front view of this drawing noting that we have defined the width of the block as 3. This dimension is inserted in a break in the dimension line. No units are given except for a notation in the drawing block that all dimensions are in inches, mm, ft, etc. Arrowheads that point to the two extension lines terminate the dimension line. The arrowhead is long (about 1/8 in.) and thin. The extension lines are separated from the view drawings by a small gap (about 1/16 in.).

In the example presented in Fig. 6.11, we have dimensioned the block using inches as the unit of measure. The fact that we have defined the width as 3 indicates to the person manufacturing the block that the width can be $3 \pm 1/64$ in. The tolerance on the width is implied or is given in the drawing block. If you, as the designer, wish to specify a block manufactured with more accuracy, you would specify the dimension using decimals. For example, if you provided the dimension as 3.00, the person manufacturing the part would understand that the width of the block would have to be between 3.00 ± 0.01 in. If you require still more accuracy, specify the dimension as 3.000; the block must be produced with a width between 3.000 ± 0.005 in. Remember—there is a great difference between 3, 3.00 and 3.000 in the accuracy of the block and tooling required to manufacture it. There is also a great difference in the cost. As we tighten the tolerances, the cost of manufacturing a component increases significantly.

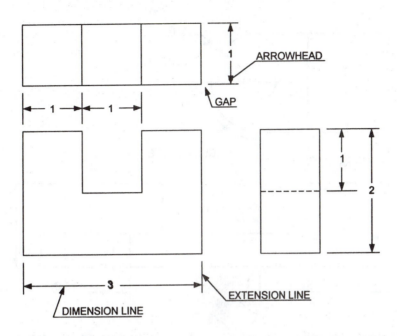

Fig. 6.11 Dimensioning example showing the dimension line, extension line, arrowhead and gap.

Let's return to Fig. 6.11 and examine the dimensioning of the top and side views. We have already shown the width on the front view, but we still must define dimensions for the depth and height of the block and the location and size of the slot. Note, the top view is used to give dimensions for the width and depth, and the side view is used to give dimensions for the height and depth. In the top view of Fig. 6.11, we have specified the depth of the rectangular block as 1. We have also specified the dimensions locating the slot and defining its width on this view. The dimensions of the height of the block and the height of the slot are provided on the side view.

We have completed the dimensioning of the block and the slot. Every dimension necessary to manufacture the component has been specified in the drawing. Also the drawing has not been over-dimensioned. (We have not specified the same dimension twice). The choice of where we place the dimensions is somewhat arbitrary. We can dimension the width on either the front or top views, the depth on the top or side views and the height on the front or side views. Two choices for the location of each dimension are possible. We usually select between one of the two views in defining a particular dimension to keep the drawing clear and uncluttered.

6.5.1 Dimensioning Holes and Cylinders

To demonstrate techniques for dimensioning either holes or cylindrical boundaries, let's add dimensions to the drawing shown in Fig. 6.10. Again start with the front view and locate the centerline of the hole as 3.20 from the left edge of the block. A hole is always located by dimensioning between suitable edges to the hole's centerline. Use centerlines for dimensioning because the drill employed to make the hole is inserted into the component at the location of the centerline.

We choose not to dimension the width in the front view because we prefer to specify the radius of the curved (cylindrical) boundary in the top view. Examine the top view and note the radius of the curved boundary is specified as 1.20 R. The R is added to the dimension to indicate that it is given in terms of the radius and not the diameter. In specifying the radius R, we have in effect defined the width of the block as the sum of 1.20 + 3.20 = 4.40. If we had specified the total width of the block in the front view, the drawing would have been over-dimensioning.

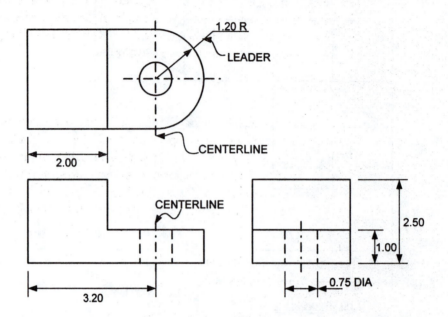

Fig. 6.12 Dimensioning example showing centerlines, leader, and radius and diameter indicators.

We show the location of the step from the left edge of the block as 2.00 in the top view. The height dimensions of the features are given in the side view. We have indicated the total height of the block as 2.50 and the height of the step as 1.00 from the bottom edge of the block. We have dimensioned the diameter of the hole in the side view as 0.75 DIA. We dimension holes in terms of their diameter because the drill sizes used in manufacturing the holes are given in terms of diameter and not radius.

Have we completed the dimensions as shown in Fig. 6.12? You might question if we have specified the depth of the block since it has not been indicated in the side view. However, we specified the depth with the radius of the curved boundary in the top view. The depth is equal to two times the radius R or 2.40.

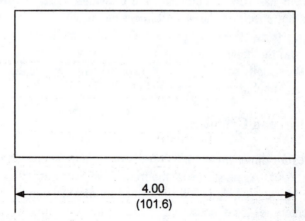

Fig. 6.13 Illustration of a convention used for dual dimensioning.

Again, we have not shown units in the dimensioning. The units are defined in the drawing block along with other information pertaining to the component. The use of two standards for units, the U. S.

Customary and the SI systems cause some difficulty in preparing and reading drawings. Some drawings are prepared using U. S. Customary units of in. or ft, and others are prepared using the SI system with units given in mm, cm or m. Sometimes, when the drawing is to be used by many people from different countries, both systems are employed in the dimensioning. An example of dual dimensioning is given in Fig. 6.13. In this case we do not break the dimension line because the U. S. Customary unit is placed above the line and the SI unit is placed below the line and enclosed with parentheses.

6.6 DRAWING BLOCKS

The drawing block serves a very important function on an engineering drawing. In the previous section, we referred to the fact that the units for the dimensions were specified in the drawing block. The unit of measurement used in preparing the drawing is only one of the many facts presented in the drawing block. As illustrated in Fig. 6.14, the drawing block is located in the lower right hand corner adjacent to the border of the drawing. A typical drawing block conveys important information pertaining to all drawings produced by a certain company and to the individual drawing. Information commonly shown in the drawing block includes:

1. The name of the company issuing the drawing.
2. The name of the part that the drawing defines.
3. The scale used in preparing the drawing.
4. Tolerances to be employed in manufacturing the part.
5. Date of the completion or the release of the drawing.
6. Material to be used in manufacturing the part.
7. Heat treatment of the part after manufacturing.
8. Units of measurement to be used in manufacturing the part.
9. Initials of the individual preparing the drawing.
10. Initials of the individual checking the drawing.
11. A unique drawing number to identify the drawing.

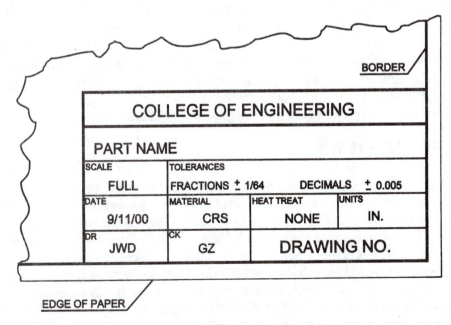

Fig. 6.14 An example of the information presented in a drawing block.

Let's examine each of these items. The name of the company is self-evident. It is illustrated in Fig. 6.14 with the College of Engineering. In designing a part, identify it with some name such as bracket, support, shaft, etc. and include this name on your drawing. The part name should be brief (no more than two or three words).

Next you will indicate the scale used in preparing the drawing. The example shown in Fig. 6.14 indicates that we used full scale in drawing the three-views of the part. The scale that we select depends on the size of the part and the size of the paper that is used for the drawing. In a typical college class, we are constrained to standard size copy paper (8-1/2 by 11 inches) because of size limitations imposed by available laser printers. Suppose the part that you are designing is 12 inches wide, 6 inches high and 1 inch deep. Will the three-view drawing of this part fit on this standard size paper? The answer to this question is a clear NO. We need to scale down the part dimensions on the drawing so that it will fit on a single sheet of paper. In this case you would probably use ¼ scale. With ¼ scale the front view of the part would be 3 inches wide by 1 ½ inches high. You would dimension the drawing using the actual sizes of the part (i.e. you would show the width as 12 inches although it is only 3 inches on the scaled down drawing). You indicate to those reading the drawing that you have reduced the size of each view by a scaling factor of ¼.

Tolerances that are to be employed in manufacturing the part are specified in the drawing block. Many companies have standardized their tolerances and these limits are preprinted in drawing blocks that are incorporated on the company's drawing paper. The date of completion of the drawing is also shown. In some instances, the drawing release date is used instead of the completion date. The release date identifies when the design engineers turn over the ownership of the design of a product to the operations (production) function, which is responsible for producing the product.

The material from which the part is to be fabricated is often identified in the drawing block. In the example shown, the abbreviation CRS indicates that cold rolled steel will be used in manufacturing the component. Sometimes the heat treatment of the component, if required, is identified in the drawing block. Heat treatment is a multi-step process employed to enhance the strength and hardness of a component. Usually heat treatment follows the machining of a part; we indicate whether or not heat treatment is required to assist those in charge of operations in controlling the flow of parts in the production process.

The units used in the drawing are given in the drawing block. We can use U. S. Customary units, SI units, or dual units. We cannot use mixed units. (Note the difference between mixed and dual units). U. S. Customary units are in terms of inch (in.) or foot (ft). SI units are in terms of millimeter (mm) or meter (m).

Those responsible for the drawing are identified. The individual who has prepared the drawing signs the block with his or her initials. The drawing is checked for completeness and accuracy, and the individual checking the details also initials the drawing block. Finally, the drawing is given a number. In a large company, which may release thousands of drawings each year, the engineering records department controls the drawing numbers. This department maintains a numbering system that insures that each part or component has a unique drawing number. The engineering records department also organizes the numbering system to group together all of the drawings needed to build a specific product.

6.7 ADDITIONAL VIEWS

Some of the parts that we design are complex and additional views may be needed to ensure that the drawings are properly interpreted. We are not restricted to three-view drawings. For very simple parts, we can adequately describe the object with one or two views. There is no need to waste time drawing additional views. For complex parts that are more difficult to visualize, we can provide additional views to clarify the drawing. These extra views can show either external surfaces or internal sections. If it helps to visualize the object, we can add back, bottom and left side views to the more commonly employed front, top and right side views. To draw these additional external views, we simply rearrange the layout of the views on the drawing as shown in Fig. 6.15. This drawing, which shows five views of a multiply tapered block, illustrates the proper arrangement to show additional external views. It also maintains the advantages of orthographic projection. Clearly, the extra views help us visualize the complex shape of this block.

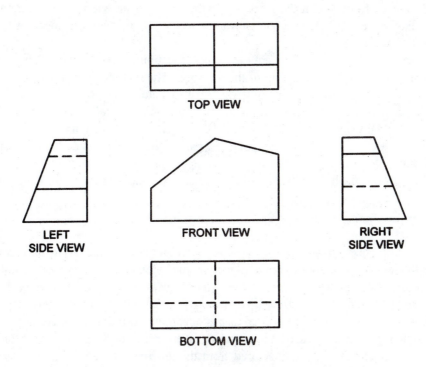

Fig. 6.15 A five-view drawing of a complex block.

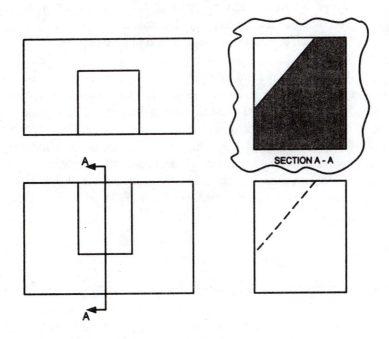

Fig. 6.16 A section cut and the corresponding section view.

An addition internal (sectional) view is often more useful in clarifying a drawing than an additional external view. The concept of a sectional view requires us to mentally slice the object into two pieces with a cutting plane. We then open the part and view the internal section revealed by the cut. An example of the sectioning technique is illustrated in Fig. 6.16 where adding a sectional view has modified Fig. 6.8. On the front view drawing in Fig. 6.16, a very heavy dashed line is shown to indicate the location of the cut. The cut line is turned through 90° at both ends and arrowheads are drawn to indicate the direction of the view.

The arrowheads are each labeled with the letter A. In this case, we are viewing the section cut from the right side. Accordingly, we expect the section view to closely resemble the right side view. We then draw the section that is revealed by the cut.

The section view may be placed at any convenient location on the drawing. We have placed it in the open space in the upper right hand corner of the drawing. The section view is usually encircled with a wavy line to distinguish it from the usual three-views. We also match the section view with the cut by using the letters A-A for identification. In some drawings, we might make several cuts each revealing a different section with each sectional view identified by A-A, B-B, C-C, etc. In the sectional view, we show the view revealed by the cut with cross-hatching. The remaining part of the view (outside the cut section) is drawn without cross-hatching. Prepare a section view of the component shown in Fig. 6.12. Make the section cut along the centerline through the hole in the top view, and then examine the section from the right side.

6.8 SUMMARY

We have introduced engineering graphics by describing techniques for preparing multi-view drawings. Multi-view (usually three-view) drawings are used to communicate the relative proportions of a component and the exact size of all of its features. The drawing techniques are covered in detail with a few simple examples. We hope that you will invest the time necessary to learn these techniques because learning to draw and read three-view drawings is an essential means of communicating in engineering.

Dimensioning is a significant aspect in preparing an error free drawing. All of the dimensions must be included so that the part can be manufactured from the drawing. Often the part is manufactured in another city, or state or even another country. The person manufacturing the part does not know you and cannot ask questions to clarify the ambiguities in the drawing. The drawing and all of its dimensions must completely define the component. An important implication of the dimensioning is the tolerances required in fabricating the part. As you write the numbers for the dimensions, you implicitly assign the tolerances. There is a great difference between 3, and 3.000 in the precision required in manufacturing and in the cost of the component.

We introduced drawing blocks and indicated the type of information that they normally convey. Remember drawing blocks are not unique and you will see differences in the information presented by different companies. We also introduced the concept of scale in this section. The scale used in preparing a drawing is your decision. Select the scale factor so that the three-views fit the paper without crowding. Sometimes you will scale down the view drawings so they will fit on the sheet. Other times you will scale up the views of a very small part so that very small features are clearly represented.

The coverage of three-view drawings presented in this chapter is very brief. If you need additional information, we recommend the complete textbook by James Earle [1] for an in depth discussion.

REFERENCES

1.   Earle, J. H., Engineering Design Graphics, 4th edition, Addison Wesley, Reading, MA 1983.

EXERCISES

6.1   Prepare a three-view drawing similar to the one shown in Fig. 6.2 except change the width of the notch from 1 inch to 1 ½ inch.

6.2   Prepare a three-view drawing similar to the one shown in Fig. 6.2 except change the width of the notch from 1 inch to 2 inch.

6.3   Prepare a three-view drawing similar to the one shown in Fig. 6.2 except change the depth of the object from 1 inch to 1 ½ inch.

6.4   Take a piece of clay or foam plastic and use a razor knife to manufacture a block with the shape shown in Fig. 6.6.

6.5   Prepare a three-view drawing of a block with a taper notch like that shown in Fig. 6.8. Select the dimensions yourself, but be consistent from one view to another with these dimensions.

6.6   Prepare a three-view drawing similar to that shown in Fig. 6.10 except increase the diameter of the hole to 1 ¼ inch.

6.7   Prepare a three-view drawing similar to that shown in Fig. 6.10 except decrease the diameter of the hole to ½ inch.

6.8   Prepare a three-view drawing similar to that shown in Fig. 6.10 except increase the depth of the object from 2.4 to 3.0 inch.

6.9   Dimension the drawing that you prepared for Exercise 6.1.

6.10 Dimension the drawing that you prepared for Exercise 6.2.

6.11 Dimension the drawing that you prepared for Exercise 6.3.

6.12 Dimension the drawing that you prepared for Exercise 6.5.

6.13 Dimension the drawing that you prepared for Exercise 6.6.

6.14 Dimension the drawing that you prepared for Exercise 6.7.

6.15 Dimension the drawing that you prepared for Exercise 6.8.

6.16 Design a drawing block that your team will employ to identify their drawings.

6.17 Prepare a five view drawing of the object shown in Fig. 6.5.

6.18 Prepare a five view drawing of the object shown in Fig. 6.7.

6.19 Prepare a five view drawing of the object shown in Fig. 6.9.

6.20 Prepare a drawing with a section view for the block defined in Fig. 6.1.

6.21 Prepare a drawing with a section view for the block defined in Fig. 6.5.

CHAPTER 7

PICTORIAL DRAWINGS

7.1 INTRODUCTION

Pictorial drawings are three-dimensional illustrations of a machine component, structure, or an object. For a person trying to visualize some object, the pictorial drawing is the most effective means to convey its size and shape. Some people have difficulty placing the three standard (front, top and side) views together to "see" the object. Pictorials drawings assemble the three-views on a single sheet providing a three-dimensional rendering facilitating visualization. Because pictorials are so easy to visualize, they are often used for catalogs, maintenance manuals, and assembly instructions.

Three different types of pictorials are in common usage:

- Isometric
- Oblique
- Perspective

A simple cube with three different types of pictorials is illustrated in Fig. 7.1. The isometric pictorial is drawn with its three axes spaced 120° apart. The term isometric means "equal measurement" indicating that the three sides are all scaled by the same factor relative to their true length. Parallel lines defining edges on the object are also parallel on the isometric drawing. Drawing paper with isometric axes is available in well-stocked office and drafting supply stores. We encourage you to use it because it greatly facilitates the preparation of an isometric pictorial.

Oblique pictorials are drawn with the front view in the x-y plane. We project oblique lines, which represent the z-axis, at some angle—often 45°. However, the angle used for the oblique lines can vary from 0 to 90°. Parallel lines defining edges on the object are also parallel on the oblique drawing. If the true length of the lines is employed in scaling all three sides, it is known as a cavalier oblique pictorial. The cavalier oblique style is frequently used, but the resulting pictorial is distorted. The distortion is due to the depth dimension, which appears to be too long.

The perspective is a pictorial drawing that represents what we actually see. Artists usually draw or paint using the perspective style. Engineers sometimes represent their designs in this style; however, this is the most difficult of the three types of pictorials to master. In perspective drawing, we do not have a well-defined coordinate system. Parallel lines tend to converge to a vanishing point as they recede from the observer. Scales used on the different axes are different in order to foreshorten the lines located some distance from the picture plane. The use of converging lines instead of parallel lines and the foreshortening of select dimensions give the drawing perspective. The prospective drawing looks like a photograph of an object.

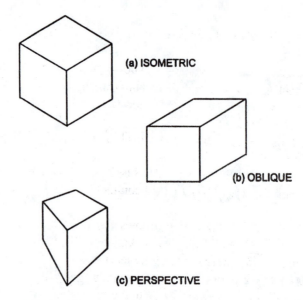

Fig. 7.1 A cube represented with isometric, cavalier oblique, and perspective pictorials.

7.2 ISOMETRIC DRAWINGS

In this discussion of the three forms of pictorial drawings, we will introduce the axes used to frame the drawing, the direction of viewing the three-dimensional object, and the dimensions used for the width, height and depth. Let's begin with the isometric pictorial, shown in Fig. 7.2, that employs axes, which make 120° with each other. The axes divide the paper into the three zones utilized to present three-views. If the axes form the letter Y with the vertical line oriented downward from the two branches, we are looking downward at the object. From this perspective, we visualize the top view in the region between the branches of the Y as shown in Fig. 7.2. The front view is displayed in the region to the right of the vertical axis, and the left-side view is located in the region to the left of this axis.

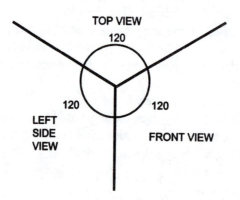

Fig. 7.2 Isometric axes at 120° angles give three regions for the front, top and left-side views.

Okay! We have defined the axes and the direction of viewing the object in Fig. 7.2. Let's discuss the dimensions used on the drawing for the width, height and depth. The word isometric means equal measurement; hence, it is clear that the same scale is used along all three axes. To illustrate the techniques used to prepare isometric drawings, consider a simple rectangular block with width W, height H, and depth D. We have prepared an isometric pictorial of this rectangular block in Fig. 7.3. To draw an isometric pictorial of a rectangular block:

1. Use a 30° triangle to draw the isometric axes identified with the numbers 1, 2, and 3 as shown in Fig. 7.3.
2. Using a scale, measure a length of H down the vertical and establish point A.
3. From point A, draw two additional lines (numbers 4 and 5) parallel to lines number 1 and 2.
4. Along line 5, measure the width W locating point B. Similarly measure the depth D along line 4 to locate point C.
5. From points B and C, draw the vertical lines 6 and 7 that intersect lines 1 and 2, and position points E and F.
6. From point F, draw line 8 parallel to line 2, and from point E draw line 9 parallel to line 1.
7. Lines 8 and 9 intersect at point G to complete the isometric drawing.

The isometric pictorial that we have drawn shows the left-side view, the top view, and the front view because we are viewing the object from above looking from the left to the right. The origin of the isometric coordinates is positioned at the upper left hand corner of the rectangular block.

The procedure for preparing isometric drawings is easy to implement; the lines are all parallel to the isometric axes, and the measurements are all to the same scale.

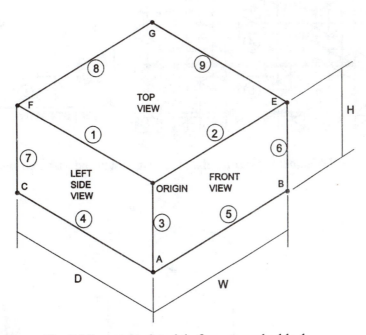

Fig. 7.3 Isometric pictorial of a rectangular block.

The example of a rectangular block was too easy. Let's try a more complex geometry as shown in Fig. 7.4. Examine the three-view drawing of an odd shaped block illustrated in Fig. 7.4. It is an unusual three-view drawing since it presents the left-side view instead of the more conventional right-side view. We present the three-view drawing in this manner because the isometric drawing shows the same three-views. For the simple rectangular block, we placed the origin of the isometric axes at the upper left hand corner of the block (front view). For the more complex geometry, we again place the origin at the same location as indicated in Fig. 7.4.

The isometric pictorial, presented in Fig. 7.4, is drawn on isometric paper with only a straight edge to guide the lines. If you have a steady hand, you can sketch the lines without the straight edge. Isometric paper gives many evenly spaced lines parallel to the isometric axes eliminating the need for the 30° triangle and the scale. We begin at point B and draw plane 1 in the region of the left view. We do not draw the entire left-side view, because there is a taper to the block that complicates this view in the isometric representation. We leave the left-side view incomplete, and move to the top view and draw plane 2. Both planes 1 and 2 are clearly defined in the three-view drawing and are easy to construct on the isometric pictorial. Again we do not complete the top view because the step and the taper complicate the geometry. We move next to the front view and add plane 3, which defines the depth of the step. We now return to the top view and draw plane 4 as shown in Fig. 7.4. Now that the top view is complete in the isometric drawing, and the height of the step is established, it is easy to draw plane 5. Note, plane 5 is on the surface formed by the taper; it does not lie in a plane formed by the isometric axes. We locate this non-isometric plane by first drawing the four well defined isometric planes on the pictorial. The location of plane 5 is then clearly established by the position of planes 1 and 4.

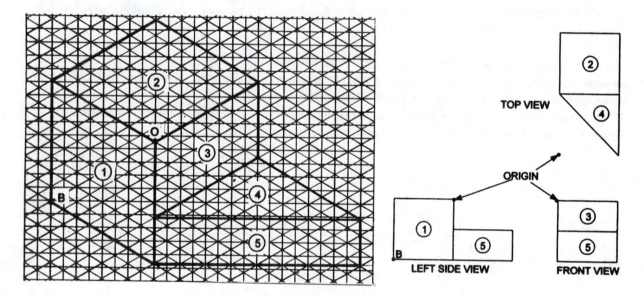

Fig. 7.4 Isometric pictorial of a block with a step and a tapered side.

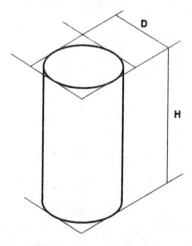

Fig. 7.5 Isometric pictorial of a right circular cylinder of diameter D and height H.

You now understand the procedure for drawing isometric pictorials of simple rectangular blocks and more complex blocks with both a step and a taper. Let's next consider a cylinder of diameter D and height H and illustrate it with an isometric pictorial. To draw the cylinder, lay out two isometric axes located a distance H apart using light construction lines as indicated in Fig. 7.5. On the upper set of axes, draw a square in the top plane with the length of the sides of the square equal to the diameter of the cylinder. Then select an isometric ellipse (35° - 16') from an ellipse template, and draw an isometric ellipse so that it is tangent to each side of the square drawn in the top plane. (If you do not have one of these handy templates, sketch the ellipse in by hand). Note that the ellipse is tangent to the isometric axes at four points as indicated in Fig. 7.5. Move down to the second set of isometric axes, which represent the bottom plane. Take the handy template and draw an isometric ellipse again. This time draw only the front half of the ellipse, since that portion is all that is visible when we view the cylinder from above. The two ellipses are joined with vertical lines at their outer most points to complete the cylinder. The construction lines in Fig. 7.5 remain to aid in understanding the procedure used in preparing this drawing. To finish, erase the construction lines and shade the cylinder to enhance the visual effect of a three-dimensional object. Shading and shadows will be discussed later in this chapter when we refer to this isometric drawing of a cylinder again.

7.3 OBLIQUE DRAWINGS

Oblique pictorials and isometric pictorials are similar, because both use parallel lines in constructing the three views. The difference between isometric and oblique pictorials is in the definition of the axes. In oblique drawings, we use a x, y, and z coordinate system as shown in Fig. 7.6. The three coordinate axes divide the sheet into three regions for drawing the front, top and right-side views. With the axes defined as shown in Fig. 7.6, we are viewing the object from above and observing from right to left. The z-axis, which is the receding axis in Fig. 7.6, is drawn with a 45° angle relative to the x-axis; however, other angles such as 30° or 60° are often employed to represent the receding axis.

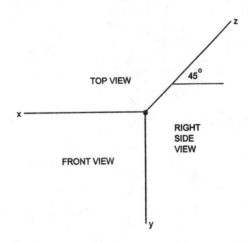

Fig. 7.6 Oblique axes use the x, y, and z coordinate system.

Three different types of oblique drawings are frequently used for pictorials as illustrated in Fig. 7.7:

1. Cavalier oblique is drawn with the receding axis at any angle from 0 to 90°, but the measurements along all three axes are the same scale.
2. Cabinet oblique is drawn with the receding axis at any angle from 0 to 90°, but the measurements along this axis are half-scale.
3. General oblique is drawn with the receding axis at any angle from 0 to 90°, but the measurements along all this axis varies from half to full-scale.

If we examine the cube represented by the three types of oblique drawings in Fig. 7.7, it is evident that full-scale (true length) measurements are used in the front view in all three types of oblique pictorials. The difference among them is the scale used along the receding axis. In cavalier oblique, the full-scale measurements are made along the receding axis; however, this scale produces a drawing that is out of proportion. The cube does not look like a cube.

In cabinet oblique, half-scale measurements are made along the receding axis. The resulting drawing is in better proportion than the cavalier oblique, but sometimes it appears that the depth dimension along the receding axis is too short. We prefer the general oblique where the measurement on the receding axis can be varied from half-scale to full-scale. The scale is adjusted between these limits to give what appears to the eye to be the correct proportions.

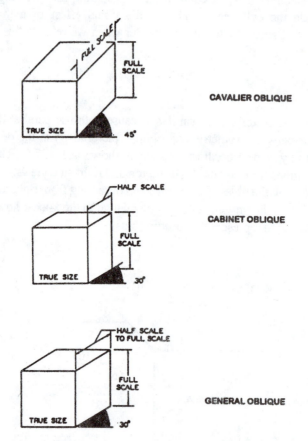

Fig. 7.7 A pictorial of a cube drawn with cavalier, cabinet and general oblique techniques.

To illustrate the procedure followed in drawing an oblique pictorial, examine the three-view drawing of a pair of the connected rectangular blocks as shown in Fig. 7.8. To begin an oblique pictorial, draw the x, y, z-axes using the lower left hand corner as the origin (point O). First, draw the part of the front view (plane 1) corresponding to the front of the large block. Next, draw part of the top view (plane 2) of the large block. In drawing the front view, use true lengths (on plane 1) to layout the width and height of the large block. On the top view, establish the depth of the large block, maintaining proportion by using a scaling factor of ¾.

Note, the smaller block obstructs the right-side view of the large block. We handle this obstruction by drawing the small block. Using the information in the three-view drawing, we can place plane 3 on the oblique pictorial. Plane 3 is located by working from the back edge in the top view. The width and height measurements required to locate and size plane 3 in the oblique pictorial are true lengths, but the depth is again scaled by ¾.

After we have drawn plane 3 locating the small block in the pictorial, it is easy to draw plane 4 by referring to the right-side view in the three-view drawing. Complete the drawing of the small block by dropping vertical lines down from the three corners in plane 3 and by closing the sides that form planes 5 and 6.

The resulting oblique pictorial clearly captures the relative proportions and positioning of the two blocks. A comparison of the pictorial with the three-view drawing demonstrates the advantages of the pictorial in visualizing the object—the pictorial is much more effective. The three-view drawing is employed for the precise definition of size and location of all of the features of a component or structure. We dimension the three-view drawing and use it in the shop for manufacturing. Usually the pictorial is not dimensioned and is not used as a substitute for a detailed three-view drawing.

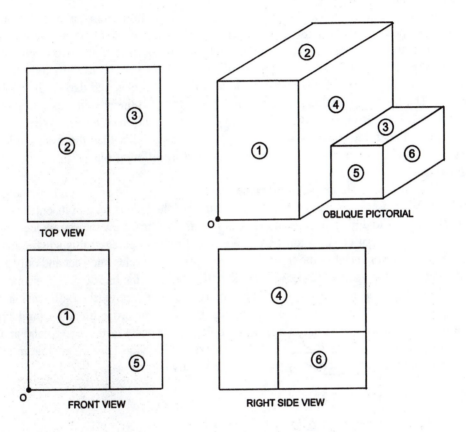

Fig. 7.8 A three-view drawing of a connected pair of rectangular blocks together with an oblique pictorial of the same object.

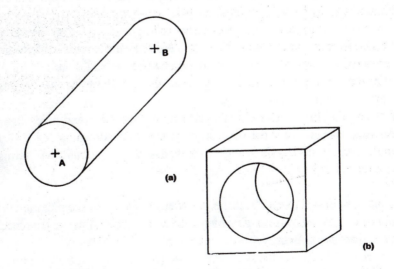

Fig. 7.9 Examples of oblique pictorials.

For two final examples of preparing oblique pictorials, consider the drawing of a cylinder or a block with a circular hole as shown in Fig. 7.9. It is easy to draw a cylinder or a circle on an oblique pictorial providing the required circles are placed on either the front plane or any plane parallel to the front plane. Since both the width and height dimensions are true length and the x, y-axes are orthogonal on the front view, the circle is not distorted into an ellipse. We can draw the circle with a compass or a circle template. This is a significant advantage.

To represent the cylinder shown in Fig. 7.9a, we draw two circles centered at points A and B. Both points A and B lie on the z (receding) axis, which has been drawn at a 45° angle to the x-axis. The circles are drawn in with construction lines with full-scale diameters. The spacing of the two circles along the z-axis is ¾ scale relative to the length of the cylinder. Draw two lines parallel to the z-axis tangent to the circles for the sides of the cylinder. Draw the front circle with a heavy line, and darken the visible portion of the back circle. The resulting pictorial shows a somewhat distorted cylinder. The distortion is due to the fact that we have drawn the back circle with the same diameter as the front circle. When observing an actual right circular cylinder, the back circle would "appear" to be smaller than the front circle. We will address the apparent difference in the size of features located in different planes in the next section on perspective drawing.

The final example of oblique pictorial, presented in Fig. 7.9b, shows a simple rectangular block with a central hole. We have drawn the block so that the circle of the hole is seen in its entirety in the front view. The question is how to draw the circle in the back plane? We first draw the square locating the back plane in the pictorial with light construction lines. We then find the center of this square and draw the second circle with the same diameter as the front circle. Note, the two circles intersect and only a portion of the circle on the back plane is visible. Darken the construction lines on the back circle over that part of the arc that is visible through the hole. The resulting pictorial is an effective three-dimensional drawing showing the appearance of a hole through a rectangular block. Our artist friends would critique the drawing, because it too distorts the visual image of the object. The scale in the back plane and the front plane are the same producing distortion of the image. This distortion is usually acceptable in engineering drawing, but we can improve our pictorials by preparing perspective drawings.

7.4 PERSPECTIVE DRAWINGS

Perspective drawings are pictorials that represent observations made with our eyes or a camera. This method of illustration is critical to the success of an artist making either sketches or paintings. Engineers also use perspective drawings particularly when preparing visuals for those who are not trained to read our more conventional three-view drawings. The tools used are the same as those described previously except for adding a thumbtack and a piece of string to the list.

There is one very significant difference between perspective drawings and isometric or oblique drawings. In both isometric and oblique drawings, the lines defining the edges are parallel to the axes; however, in perspective drawing, the lines defining some of the edges are not parallel. Drawing parallel lines distorts the drawing, because parallel lines appear to converge as they recede in space. The best illustration of this fact is a pair of railroad tracks, shown in Fig. 7.10. Looking down the tracks, you see that the tracks converge to a point. Also the railroad ties and the trees and poles lining the track appear to become shorter. Everyone knows that the tracks are parallel. What's going on?

To understand what is happening, it is necessary to define four terms used in describing prospective drawing.

1. A picture plane is the surface (i. e. the sheet of paper) of the pictorial. The edges of the paper represent the window through which you "see" a three-dimensional object that is the subject of the perspective drawing.
2. A horizon line divides the sky and the land (or the sea) if you are outdoors. The horizon line is at the level of your eyes and will change with your elevation. In a room where the true horizon

cannot be located, because the walls of the room block our view, we assume a horizon line at the elevation of our eyes.

3. A viewing point and direction of view depends on the location of eyes relative to the object. You can look directly at the object, from left to right, right to left, downward, upward, etc. What you see changes markedly depending on these parameters. Look at an object from a window, and change where you stand and the direction of your view. Does the view of the object change?

4. A vanishing point is where parallel lines converge to a point as they recede into the distance. You can clearly identify the vanishing point in Fig. 7.10 where the tracks appear to meet.

Fig. 7.10 A photograph of railroad tracks showing the visual effect of parallel lines that appear to converge and the foreshortening of objects in a distance.

7.4.1 One Point Perspective Drawings

Depending on the view, an object can be represented with a one, two, or three-point perspective. Let's start with the simple one-point perspective and illustrate the approach by drawing a simple rectangular block. In one-point perspective, you place the front view in the picture plane and show its true width and height as illustrated in Fig. 7.11. Then a construction line is drawn to represent the horizon. The location of this line depends on the viewing point and the viewing direction. In Fig. 7.11, you are viewing the block straight on (not from the right or the left); however, you are above the block. Your eyes look downward and observe the top surface of the block. The elevation of your eyes relative to the top of the block is taken into account by raising the horizon line. Next, locate the vanishing point on the horizon line at the center point behind the front view, because we are looking straight on at the block. Draw construction lines from the vanishing point to the top corners of the block in the front view as shown in Fig. 7.11. The back edge on the top view is drawn parallel to the front top edge to establish the depth of the block. Note that the back edge is much shorter in length than the front edge. The shortening of the lines on the recessed planes give an illusion of the third dimension. The edges at the side are darkened to complete the pictorial. Again these edges are converging giving an illusion of depth on the two-dimensional sheet of drawing paper.

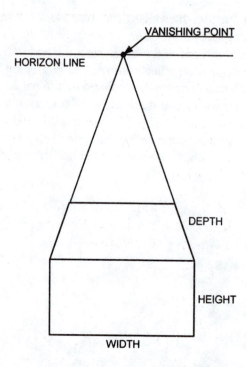

Fig. 7.11 An example of a pictorial drawing with one-point perspective.

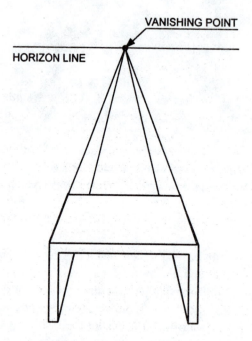

Fig. 7.12 A one-point perspective drawing of a coffee table.

Another example of one-point perspective is the drawing of a coffee table presented in Fig. 7.12. Again the front of the table is drawn to scale in the picture plane. The width and height dimensions are true lengths. Coffee tables are low, so when we stand it is necessary to look downward to view the table. In this example, we are centered relative to the table and looking directly at its top surface. To develop a perspective view with this direction of viewing, draw a horizon line at eye level aligning the vanishing point with the center of the table. Next, draw light construction lines from the top outside corners of the table to the vanishing point. We also draw construction lines from the lower inside corners of the table legs as shown in Fig. 7.12. These construction lines define two triangles. We use the larger of these two triangles to provide the length of the back top edge of the table. Next, we draw the edges of the tabletop to complete our perspective rendering of the top view. The smaller of the two triangles is used to guide us in drawing the bottom edges of the legs, which are visible under the table.

In a normal perspective drawing, we would erase the construction lines, the vanishing point and the horizon line. We have not erased them in Fig. 7.12, because they were used to illustrate the procedure followed in making a one-point perspective drawing.

7.4.2 Two-point Perspective Drawings

The one-point perspective drawing is useful when we view an object straight on with its front view shown in the picture plane. However, if the object is rotated so that neither the front or side view is in the picture plane, as illustrated in Fig. 7.13, a two-point perspective is required.

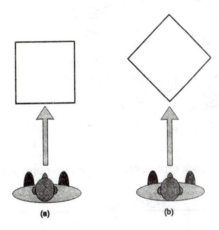

Fig. 7.13 Directions of viewing control the number of vanishing points used in perspective drawing.
a. Straight-on—one-point perspective. b. Angle view—two-point perspective.

Again consider a rectangular block to illustrate the approach followed in drawing a two-point perspective pictorial. From Fig. 7.13, it is evident that only one vertical edge of the block lies in the picture plane. We will start with this fact and proceed step by step to illustrate the block with a two-point perspective drawing.

- Draw a vertical line, as indicated in Fig. 7.14, to establish the edge between the left-side view and the front view.
- Draw the horizon line to reflect the fact that we are standing above the block looking downward onto the top surface.
- Place two vanishing points on the horizon line. We have spaced the vanishing point located to the right of the vertical line (VP - R) farther from the vertical line than the vanishing point on the left-side (VP - L), because we are viewing the block at a slight angle from the right toward the left.
- Measure the true length of the vertical line (line 1) and label its ends with the letters A and B.
- Draw four construction lines connecting points A and B with VP - R and VP - L.
- Draw vertical lines (2 and 3) to locate the back edges of the left side and the front of the block. The ends of these lines, which are located by the construction lines, are labeled (C, D) and (E, F).
- Draw top edge lines (4, 5) by connecting points F, A and D.
- Draw bottom edge lines (6, 7) by connecting points E, B and C.

- Draw construction lines from point F to VP - R and from point D to VP – L. Then locate point G at the intersection of these two lines.
- Draw sides (8, 9) by connecting points F, G, and D.

The two-point perspective drawing of the rectangular block is complete as shown in Fig. 7.14. The procedure may seem long and tedious, when it is described in a sequence of steps, but it is not difficult. When you understand the basic concepts of establishing the true length of the vertical line in the picture plane and the vanishing points on the horizon line, the remaining steps are routine.

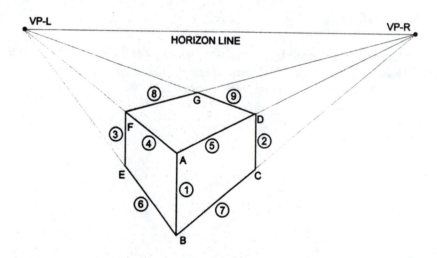

Fig. 7.14 A rectangular block represented with a two-point perspective drawing.

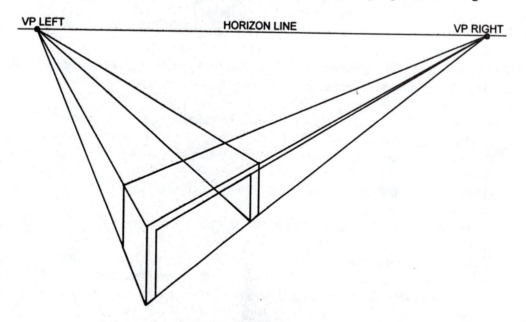

Fig. 7.15 A two-point perspective drawing of a coffee table.

To complete the discussion of the two-point perspective, let's again consider the coffee table. This time, we will use a two-point perspective drawing to illustrate the table. Begin by placing the vertical edge of the table leg in the picture plane as indicated in Fig. 7.15. The edge of the leg is shown in true length on the picture plane. Establish the horizon line and the vanishing points to provide the direction of the view that you are showing. Draw construction lines from the ends of the vertical line to the vanishing points.

Vertical lines are then drawn to show the width and the depth of the table. The position of these vertical lines is not established by measurement, because both the width and depth dimensions are not true length in this two-point perspective. Place the vertical lines in a position that maintains, "to the eye," the correct proportions of the table.

When the vertical lines that bound the left-side view and the front view are drawn, the remainder of the drawing is easy to complete. Construction lines are drawn from the ends of these vertical lines to the vanishing points. The construction lines outline the table. It is only necessary to darken those segments of the construction lines that define the visible edges of the table.

7.4.3 Three-point Perspective Drawings

Three-point perspective drawings are used when the object being illustrated is very tall. Architects drawing a city view with skyscrapers would use three-point perspectives and taper the building as it extends into the sky. Engineers usually deal with smaller objects that usually can be represented in pictorials with either one or two-point perspective drawings. For this reason, we will not describe the methods used in preparing three-point perspectives. However, if you are interested in learning much more about perspective drawing, we recommend the excellent book by Powell [2].

7.5 SHADING AND SHADOWS

Adding shading and shadows to our pictorial drawings makes them appear more realistic. Let's first distinguish between shading and shadowing. If you light the object, some surfaces are exposed to this light and other surfaces are in the shade. There is a difference in the intensity of light reflected from these surfaces. Those in the shade are darkened slightly. When an opaque object blocks the light, a shadow is produced. The shadow occurring on the floor plane is shown as a dark area in the drawing. We show an example of shading and shadowing of a right circular cylinder in Fig. 7.16. In this example, parallel light rays are illuminating the cylinder from the upper left. (They are included on the drawing only to show the logic for determining the shade and shadow regions). The right half of the cylinder is in the shade and is darkened slightly. A shadow is cast on the plane of the floor upon which the cylinder rests. The depth of the shadow is the same as the depth of the cylinder on the isometric pictorial. The length of the shadow is dependent on the direction of the parallel rays of light. Extending the lines representing the light rays defines the edge of the shadow.

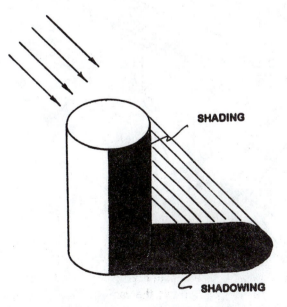

Fig. 7.16 Shading and shadowing of an isometric pictorial of a right circular cylinder.

Let's consider another example of shading and shadowing in a two-point perspective drawing of a cube. The drawing, presented in Fig. 7.17, is similar to that shown in Fig. 7.14, so we will assume that you can draw the cube in the two-point perspective pictorial. In Fig. 7.17, the construction technique is demonstrated for determining the exact size of the shadow when the light is coming from a point source.

Suppose you have completed the two-point perspective drawing of the cube and are ready to shade and shadow the drawing. The process is first to select the location of the light source. It is your choice, but you place the light source well above the horizon either to the left or the right of the cube. Second, you must select the vanishing point for the shadow. Again it is your choice as long as it is directly below the light source and in the ground (floor) plane. The reasons for these two constraints are evident. The shadow must vanish when the light source is directly overhead, and the shadow must always lie on the ground (floor) plane. In Fig. 7.17, we selected the vanishing point of the shadow on the horizon line; however, it could have been placed anywhere on the ground plane directly under the light source.

Let's consider a step-by-step procedure for locating the shadow outline shown in Fig. 7.17:

1. Draw the light rays (1, 2, and 3) from the source through three of the top-forward corners of the cube.
2. Draw two construction lines (4 and 5) from the vanishing point of the shadow through the two bottom-side corners.
3. Locate points A and B at the intersections of the lines (1 and 6) and (3 and 7).
4. Draw lines 6 and 7 from the left and right vanishing points through points A and B.
5. The intersection of lines (6 and 7) locates point C.
6. Darken segments of the construction lines (8, 9, 10 and 11) to give the shadow outline.
7. Erase all of the construction lines and darken the shadow region within the outline.

We have completed the shadow, but the front and left-side views of the cube are in the shade. To complete the pictorial, these two sides should be darkened slightly indicating they are in a shadow.

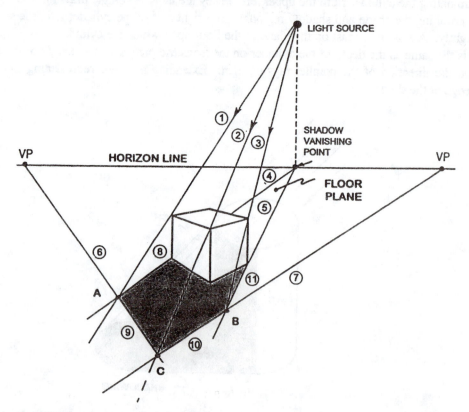

Fig. 7.17 Technique for determining shadow outline on a two-point perspective drawing.

7.6 SUMMARY

Three different pictorial drawings—isometric, oblique, and perspective—have been introduced. Both isometric and oblique pictorials are widely used in engineering to present three-dimension illustrations on a two-dimensional format (a sheet of drawing paper or a computer monitor). The isometric and oblique pictorials are easy to draw because lines are parallel, and most, if not all, of the measurements are to scale. However, exact scaling of dimensions and parallel lines tend to distort these three-dimensional illustrations. The eye or camera "sees" dimensions to different scales, and parallel lines as parallel only if the view depicted is on the picture plane. If the view is on any other plane behind the picture plane, the dimensions are foreshortened. If the parallel lines are on views with receding axes, they tend to converge. We ignore these distortions in engineering drawings when we prepare isometric or oblique pictorials.

Perspective drawings provide a more realistic representation of three-dimensional objects. Artists almost always use perspective concepts in their sketches or paintings. In engineering, we usually employ one-point or two-point perspectives to produce realistic three-dimensional drawings that are used in catalogs and maintenance manuals. The techniques of perspective drawing require that we understand four quantities: the picture plane, the horizon line, vanishing point or points, and viewing direction. The techniques for drawing perspectives with converging lines and foreshortened dimensions follow directly from the definitions of these four quantities.

Shading and shadowing is an added technique for making pictorial drawings even more realistic. We have indicated methods for determining the outline of shadows either from point light sources or collimated light sources that produce parallel light rays.

REFERENCES

1. Franks, G. Pencil Drawing, Walter Foster Publishing, Laguna Hills, CA 1988.
2. Powell, W. F. Perspective, Walter Foster Publishing, Laguna Hills, CA 1989.
3. Earle, J. H., Engineering Design Graphics, Addison Wesley, Reading, MA 1983.

EXERCISES

7.1 What are the three types of pictorial drawings? Why do we use pictorial drawings in engineering applications?

7.2 Prepare an isometric pictorial drawing of a rectangular block 10 mm wide by 15 mm deep by 50 mm high. What scale should you use in preparing this drawing? Why?

7.3 Prepare an isometric pictorial drawing of a right circular cylinder 15 mm in diameter by 35 mm long. What scale should you use in preparing this drawing? Why?

7.4 Prepare an oblique pictorial drawing of a rectangular block 20 mm wide by 25 mm deep by 70 mm high. What scale should you use in preparing this drawing? Why?

7.5 Prepare an oblique pictorial drawing of a right circular cylinder 25 mm in diameter by 45 mm long. What scale should you use in preparing this drawing? Why?

7.6 Prepare an isometric pictorial drawing of a rectangular block with a centrally located circular through hole. State the dimensions of the block and the diameter and orientation of the hole.

7.7 Prepare an oblique pictorial drawing of a rectangular block with a centrally located circular through hole. State the dimensions of the block and the diameter and orientation of the hole.

7.8 Prepare an isometric pictorial drawing of the wind machine that your team is developing.

7.9 Prepare an oblique pictorial drawing of the wind machine that your team is developing.

7.10 Prepare a prospective pictorial drawing of the J. M. Patterson building. View the building from the intersection of the two frontage streets.

7.11 Draw a pictorial illustration of a sphere with shading and shadowing.

CHAPTER 8

TABLES AND GRAPHS

8.1 INTRODUCTION

On many occasions we find ourselves with a collection of numerical data that we must present at a design briefing or in an engineering report. At other times, we may have extensive results from a mathematical relationship that must be expressed in a meaningful format. We have two choices in presenting numerical data. We can show the numbers in tabular form or on a graph. Both methods of presentation, the table or the graph, have advantages and disadvantages. The method that you select will depend on your audience, the message, and the purpose of the presentation.

In this chapter, we will describe the technique for arranging data in a table using Microsoft's Word 2000 as the word processing program. In Chapter 9, we describe data entry into a spreadsheet, which is also a convenient method for preparing tables. We then introduce five common methods for representing data in the form of graphs, which include:

1. Pie charts
2. Bar charts
3. X - Y graphs.
4. Semi-log graphs
5. Log - log graphs

Each type of graph is used for a different purpose and selection of the correct type is essential in communicating effectively. For instance, the pie chart is used to show distributions—who gets the most and the least. The bar chart is used to compare one set of data with another. X – Y graphs, the most frequently used type of graphing in engineering, show how Y varies with X. When Y varies by very large amounts with small changes in X, we often use semi-log graphs, which are capable of covering a very large range of Y values. When both X and Y vary over very large ranges, we display the results graphically by using a log-log representation.

8.2 TABLES

In Chapter 1 we described the responsibilities of Mechanical Engineers in their first position and presented statistical data in a table. Let's reproduce this data in tabular format to show you the ease with which a table may be prepared in Word 2000. We first prepare the table heading by selecting the center alignment, identifying the table number, and typing the title. We may choose a bold font for the table heading to attract the reader's attention.

Table 8.1
Responsibilities of Mechanical Engineers in Their First Position

Okay. We are now ready for the body of the table. With the mouse, point to the menu row and click on "Table". Wait a moment and an extended drop-down menu appears. We have a choice of whether or not we want to show grid lines in the table. Let's click on the "Gridlines" icon and use grid lines as guides in constructing the table. The gridlines aid us in entering the data and they are also helpful to the reader. If we do not like them they can be removed after the data is entered. Next, click on the "Insert" then "Table" shown on the drop down menus. A dialog box that appears on the screen, prompts us to enter information defining the table. It is necessary to understand the arrangement of the table (how many columns and rows). To represent the data for the responsibilities of mechanical engineers, it is necessary to establish three columns and 15 rows in the arrangement of the table. The initial arrangement is not critical, because we can always add or delete either columns or rows later. Finally, we select auto-format in the dialog box and permit Word to control the initial width of the columns. Click the OK button, and an outline of the table appears on the monitor. If the grid lines are light gray when the table first appears on the monitor, they will not print as they are intended only to guide your typing. If you prefer the grid lines to delineate the rows and columns when the table is printed, click on Format in the menu and select Borders and Shading. A dialog box appears on the screen and you click on the "Grid" button; then choose the style and width of the grid lines outlining each cell. The result of our first trial after we have typed the data into the correct cells is shown below.

ASSIGNMENT	% TIME	% TIME
Design Engineering		40
Product Design	24	
Systems Design	9	
Equipment Design	7	
Plant Engineering/ Operations/ Maintenance		13
Quality Control/ Reliability/ Standards		12
Production Engineering		12
Sales Engineering		5
Management		4
Engineering	3	
Corporate	1	
Computer Applications/		4
Basic Research and		3
Other Activities		7

The result of our first attempt is fair, but it could be improved. The auto-format selection divided the width of the page into thirds for the column width. This partitioning is not the best choice considering the wasted space associated with the numerical entries and the double rows needed to accommodate one of the assignments. The first column, titled **ASSIGNMENT**, needs more width and the 2nd and 3rd columns are too wide. Let's modify the table by changing the column widths. We accomplish this by clicking on Table in the menu row and selecting "AutoFit" from the pull down menu. From the adjacent pull down menu that appears, select "To Contents". The appearance of the table has improved and it requires less space on the page. Unfortunately, it is not centered. To center it, click Table from the menu row, click on "Table Properties", and select the "Table" tab from the dialog box. From the Table dialog box, under the choice for alignment, select "Center" and click the OK button. The improved presentation is presented in Table 8.1. As a result of our modifications, the table looks very professional.

Table 8.1
Responsibilities of Mechanical Engineers in Their First Position

ASSIGNMENT	% TIME	% TIME
Design Engineering		40
Product Design	24	
Systems Design	9	
Equipment Design	7	
Plant Engineering/ Operations/ Maintenance		13
Quality Control/ Reliability/ Standards		12
Production Engineering		12
Sales Engineering		5
Management		4
Engineering	3	
Corporate	1	
Computer Applications/ Systems Analysis		4
Basic Research and Development		3
Other Activities		7

Tables are most effective in presenting precise numerical data. We do not need to crudely estimate a number by reading from a curve on an X-Y graph. If we have a reason to show our results with six significant figures, it is very easy to do so. A government agency that probably produces the most widely read tables in the United States is the Internal Revenue Service. They prepare tax tables each year that precisely show our tax obligations. We will present another example later that shows the advantage of a table in conveying numerical data with many significant figures. We will also show the application of spreadsheets to produce both graphs and tables in another Chapter 9.

8.3 GRAPHS

We use graphs to visualize the data in both reports and presentations. The graphs are not intended to give precise results since we use tables for that purpose. The graphs show our audience a trend, a comparison, or a distribution—quickly and effectively. There is no need for the audience to study the data to develop an understanding. Your graph shows the bottom line and eliminates the time and effort required for the reader to analyze the data.

There are many different ways of presenting numerical data in a graphical format. The distribution of goodies is usually shown with pie charts, because we all quickly relate to getting a good size share of the pie. Comparisons among several quantities are made with bar charts with the height of each bar indicative of the magnitude of the quantity being compared. Linear graphs, where we plot a dependent parameter Y as some function of an independent parameter X, are frequently employed in engineering presentations. The data plotted can be generated with mathematical functions when we know the function Y(X). However, if the mathematical relation is not known, we conduct experiments and measure Y as we systematically change X. In both cases, the X-Y graphs represent the numerical data and show a trend—increasing, neutral or flat, decreasing, oscillating, etc. In some instances, the variations in X, Y or both of these quantities is extremely large; therefore, it is not possible to clearly show the trends over the entire range of either X or Y on a graph with linear scaling. In these cases, we use a non-linear format, namely semi-log or log-log to present the large ranges in the data. Let's consider five types of graphs and learn manual techniques for preparing them. Later in Chapter 9, we will show how a spreadsheet program is used to prepare these same graphs.

8.3.1 Pie Charts

A pie chart is used to show how some quantity is distributed. In Table 8.1, we showed the various assignments for mechanical engineers starting their careers. Let's represent this same data in a pie chart as shown in Fig. 8.1. We begin by drawing a circle to represent the pie, and then we decide what fraction of the pie corresponds to each assignment. This determination is easy if we recall that the whole pie contains an included angle of 360°. The slice of pie representing design engineering is a fraction of the whole pie, i. e. (360°)(0.40) = 144°. The size of the pie slices in terms of degrees for all of the engineering assignments are listed below:

1. Design engineering   (360°)(0.400) = 144°.
2. Plant engineering, operations and maintenance   (360°)(0.130) = 47°.
3. Quality Control, reliability and standards   (360°)(0.121) = 43°.
4. Production engineering   (360°)(0.121) = 43°.
5. Sales engineering   (360°)(0.050) = 18°.
6. Management   (360°)(0.040) = 14°.
7. Computer applications and systems analysis   (360°)(0.040) = 14°.
8. Basic research and development   (360°)(0.030) = 11°.
9. Other activities   (360°)(0.070) = 25°.

We construct the pie chart by using a protractor to layout these angles on the circle. The pie slices are then labeled as shown in Fig. 8.1. When examining the pie chart, we can visualize the importance of design, because the design slice is huge. We can also see the importance of producing products since these activities—plant operations, quality control, and production—combine to produce still another huge slice. The remainder of the pie, which is less that a quarter, is divided between, sales, management, computers and systems, research and development, and other activities. Of course, these activities are important, but they represent opportunities for a much smaller fraction of the engineers beginning their careers. Compare the effectiveness of the data as presented in Table 8.1 and the pie chart of Fig. 8.1. The pie chart, with a visual representation, quickly leads the reader to the conclusion that design and production are where most opportunities exist for entry-level positions in mechanical engineering.

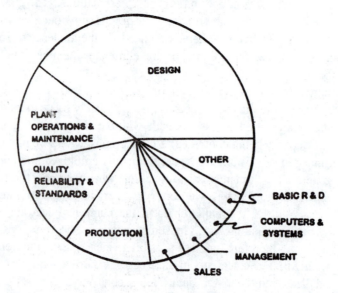

Fig. 8.1 A pie chart showing initial assignments in industry for beginning engineers.

8.3.2 Bar Charts

Bar charts can also be used to show distribution, but they are better suited for illustrating comparisons. As an example of a bar chart, consider the results of a study by the Council of Chief State School Officers on the percentage of high school students graduating with three or more years of secondary mathematics and science in 1982 and 1994. We have prepared a bar chart showing the comparison in Fig. 8.2. The visual effect enhances the comparison. At a glance, we can see that many more students were taking mathematics and science in 1994 than in 1982. The second glance indicates that only about a third of the students were in math and science in 1982 while more than half of them were taking math and science in 1994. Finally, we quickly observe that math was slightly more popular than science in both 1982 and 1994. Three glances and the reader understands the data, because you have presented it in a visual format that effectively carries the message.

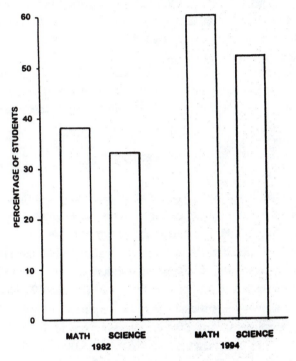

Fig. 8.2 A bar chart showing the percentage of high school students taking math and science in 1982 and 1994.

The preparation of the bar graph is easy. In constructing Fig. 8.2, we used an engineering paper with a barely visible grid that facilitated establishing a suitable scale along the ordinate. Only a straight edge was employed in drawing the bars; the grid lines visible on the engineering paper were used to scale each bar. The time required to prepare the bar chart was about 10 minutes.

8.3.3 Linear X-Y Graphs

X-Y graphs are probably the most frequently employed graph used in engineering because it illustrates trends. The curves are produced by plotting Y, the dependent variable, along the ordinate as a function of X, the independent variable, which is displayed along the abscissa. Connecting the points with a curve or line segments indicates the trend in Y with changes in X. Graphing X-Y can be accomplished using linear scales for both X-Y, a linear scale for X and a log scale for Y, or log scales for both X and Y. We will cover all three of these methods of producing X-Y graphs.

Since numerical results will be necessary for our examples of X-Y graphs, we will generate our numerical data using a very important equation that relates to money, and how you may choose to accumulate or spend it. Let's begin by assuming that you are going to accumulate some money. Sounds good! Suppose your rich uncle purchased a mutual fund for you at your birth, and plans to give you the proceeds of the fund on your 21st birthday. How much will you have accumulated when the big day arrives? The following relation gives the sum S accumulated:

$$S = P(1 + i)^n \qquad\qquad (8.1)$$

where S is the sum accumulated.
      P is the amount of the initial investment.
      i is the interest rate for the compounding period.
      n is the number of periods over which the interest accumulates.

Suppose your rich uncle invested P = $5000 in a mutual fund that guaranteed an annual interest rate I = 10 %, and the interest is compounded semi-annually. We will also assume that uncle avoided the tax collectors[1] (Federal, State and Local). Let's calculate exactly how much money you can expect to collect on your 21st birthday. We need to compute S in Eq. (8.1) for P = $5000. The number of compounding periods that will occur from your birth until you are 21 years of age is 42, because the fund compounds the interest earned twice each year. The interest rate I = 10% on an annual basis, but the interest rate i is only 5% per the semi-annual compounding period. Substitute these numbers into Eq. (1) to obtain:

$$S = \$5000(1 + 0.05)^{42} = \$5000(7.761587) = \$38,807.94$$

Wow! This is wonderful. You can buy a very nice car on the good uncle, or defer this option and continue to compound interest accumulating an even larger sum. We have evaluated Eq. 8.1 with a spreadsheet program (EXCEL) for a total of 80 periods. The results are shown in Table 8.2.

In examining Table 8.2, we find that the uncle's principal of $5,000 has grown over the years. The table gives accurate values at the end of each compounding period. For example, at the end of the 20th period, your tenth birthday, the principal in the mutual fund was $13,266.49. Note that the table conveys the exact numerical value with as many significant figures as required. If you maintain the fund until you are 40 years old, accumulating interest for 80 periods, you would have a total of $247, 807.21. Compounding interest has produced a significant growth from the $5,000 seed-money planted by uncle.

Let's represent the data shown in Table 8.2, as an X-Y graph. We first decide that the sum S is the dependent variable represented along the ordinate (the Y axis), and that your age is the independent parameter displayed along the abscissa (the X axis). We draw the X and Y-axes on our engineering paper (which has a faint grid) and apply a scale to each axis, as shown in Fig. 8.3. Scaling is very important because it determines the size of our graph. If the scale is too small, the graph looks like a postage stamp and you have wasted an opportunity to show the graph to full advantage. However, if the scale is too large, you cannot plot all of the data on the paper. To scale the graph correctly, examine the range of the data to be covered for both X and Y. In this case, the range for X varies from 0 to 21 and Y varies from $5,000 to $38,807.94. We have selected a scale of 1 inch equal to 5 years along the X-axis, and a scale of 1 inch equal to $5000 along the Y-axis. The range displayed for X was from 0 to 25, and for Y from 0 to $40,000. We add numbers to each axis, beginning with zero at the origin and increment in steps of 5 along both axes. Captions are added to both axes to remind the reader of the definitions of X and Y. This scaling results in the graph shown in Fig. 8.3; it is approximately 5 by 8 inches in size, which fits nicely on a single page with sufficient space for the figure caption and the margins.

---

[1] Avoidance of federal taxes is possible, if uncle places the funds in a Roth account.

Table 8.2						
Accumulated Sum from an Investment of $5000.00						
Interest Rate of 10% Compounded Semi-annually.						
No Taxes Paid on Interest Earned						
Periods	Multiplier		Sum	Periods	Multiplier	Sum
0	1.0000	$	5,000.00	40	7.0400	$ 35,199.94
1	1.0500	$	5,250.00	41	7.3920	$ 36,959.94
2	1.1025	$	5,512.50	42	7.7616	$ 38,807.94
3	1.1576	$	5,788.13	43	8.1497	$ 40,748.33
4	1.2155	$	6,077.53	44	8.5572	$ 42,785.75
5	1.2763	$	6,381.41	45	8.9850	$ 44,925.04
6	1.3401	$	6,700.48	46	9.4343	$ 47,171.29
7	1.4071	$	7,035.50	47	9.9060	$ 49,529.86
8	1.4775	$	7,387.28	48	10.4013	$ 52,006.35
9	1.5513	$	7,756.64	49	10.9213	$ 54,606.67
10	1.6289	$	8,144.47	50	11.4674	$ 57,337.00
11	1.7103	$	8,551.70	51	12.0408	$ 60,203.85
12	1.7959	$	8,979.28	52	12.6428	$ 63,214.04
13	1.8856	$	9,428.25	53	13.2749	$ 66,374.74
14	1.9799	$	9,899.66	54	13.9387	$ 69,693.48
15	2.0789	$	10,394.64	55	14.6356	$ 73,178.15
16	2.1829	$	10,914.37	56	15.3674	$ 76,837.06
17	2.2920	$	11,460.09	57	16.1358	$ 80,678.92
18	2.4066	$	12,033.10	58	16.9426	$ 84,712.86
19	2.5270	$	12,634.75	59	17.7897	$ 88,948.50
20	2.6533	$	13,266.49	60	18.6792	$ 93,395.93
21	2.7860	$	13,929.81	61	19.6131	$ 98,065.73
22	2.9253	$	14,626.30	62	20.5938	$ 102,969.01
23	3.0715	$	15,357.62	63	21.6235	$ 108,117.46
24	3.2251	$	16,125.50	64	22.7047	$ 113,523.34
25	3.3864	$	16,931.77	65	23.8399	$ 119,199.50
26	3.5557	$	17,778.36	66	25.0319	$ 125,159.48
27	3.7335	$	18,667.28	67	26.2835	$ 131,417.45
28	3.9201	$	19,600.65	68	27.5977	$ 137,988.32
29	4.1161	$	20,580.68	69	28.9775	$ 144,887.74
30	4.3219	$	21,609.71	70	30.4264	$ 152,132.13
31	4.5380	$	22,690.20	71	31.9477	$ 159,738.73
32	4.7649	$	23,824.71	72	33.5451	$ 167,725.67
33	5.0032	$	25,015.94	73	35.2224	$ 176,111.95
34	5.2533	$	26,266.74	74	36.9835	$ 184,917.55
35	5.5160	$	27,580.08	75	38.8327	$ 194,163.43
36	5.7918	$	28,959.08	76	40.7743	$ 203,871.60
37	6.0814	$	30,407.03	77	42.8130	$ 214,065.18
38	6.3855	$	31,927.39	78	44.9537	$ 224,768.44
39	6.7048	$	33,523.76	79	47.2014	$ 236,006.86
40	7.0400	$	35,199.94	80	49.5614	$ 247,807.21

The data consisting of X and Y coordinates, shown in Table 8.2, locate the points that are plotted on the graph. We have used a drop compass to draw the small points, but the points can be placed by free hand if you do not have this instrument. We have only plotted seven points to establish the curve. The number of points plotted depends on the smoothness of the function being graphed. In this case, Y varies monotonically with respect to X. We can draw the curve accurately with only six to eight points because the function is relatively smooth and without peaks or valleys.

Inspecting Fig. 8.3 clearly shows the trend of the accumulation sum S with respect to time (age). The sum increases continuously with time and the rate of the increase (the slope of the curve) is also increasing. The visual effect of the X-Y graph is dramatic. At a glance it is apparent that you are becoming rich. The only question is how rich? If we examine the ordinate and abscissa, we can estimate the sum accumulated for a given age, but the estimate may be in considerable error. The faint grid lines used in plotting the data do not show in the copy presented in Fig. 8.3. We lose accuracy in reading our graph without these grid lines.

If we seek to prepare an X-Y graph that is effective in showing trends, while retaining some accuracy in reading the scales, a good quality graph paper with a fine pitch grid should be employed. An example of this graph, plotted to the same scale on a graph paper with 20 grid lines to an inch, is shown in Fig. 8.4. With these fine pitch grid lines, we can read the scale to:

$$\pm (1/20)(5) = \pm 0.4 \text{ years of age}$$

$$\pm (1/20)(\$5,000) = \pm \$400 \text{ for the sum S.}$$

The resolution of the scale, possible with good quality graph paper, is a big improvement over the graph without visible gridlines. However, the best accuracy in reporting numerical data is obtained with a tabular format.

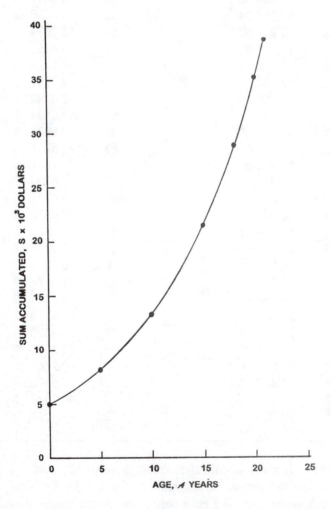

Fig. 8.3 A X-Y graph showing the sum accumulated with time from an initial investment of $5,000.

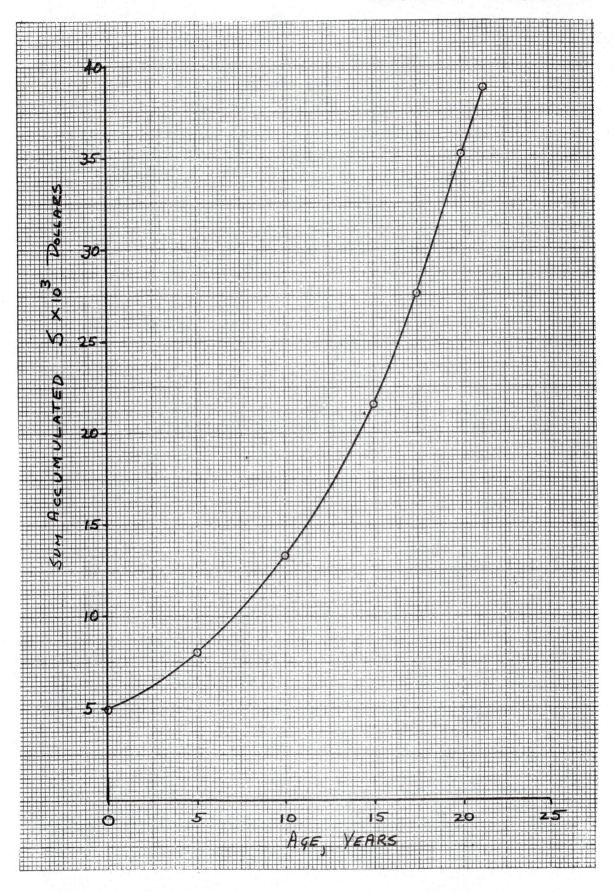

Fig. 8.4 An X-Y graph with fine pitch gridlines used for showing both trends and numerical values.

8.3.4 X-Y Graphs With Semi-log Scales

The results presented in Table 8.2 indicate that you can accumulate some very serious money from a relatively small initial investment if it is invested for a sufficiently long time. If we try to plot the larger numbers associated with the sum S in Table 8.2 on the linear graph of Fig. 8.4, we would go off-scale. We could re-scale, but we would lose resolution along the ordinate where the range in S is very large. A better approach is to use semi-log paper in preparing the graph. Semi-log paper, shown in Fig. 8.5, has a linear scale (in this example along the X axis) and a log scale (along the Y axis). Looking closer at the Y-axis indicates that the graph paper is scaled with three log cycles over the 10 inches of the paper that are ruled. We have selected three log cycles because it gives the range needed to cover the range of values for the data presented in Table 8.2. Note semi-log graph paper is commercially available with different numbers of cycles on the log axis.

Let's scale both axes on our semi-log graph paper. The X-axis is scaled with one inch equal to 10 years covering the range from 0 to 50 years. The Y-axis scale is imposed upon us by the selection of the paper. (We selected three-cycle paper so we may cover a range of three decades). The first decade is in thousands (10^3 to 10^4), the second in tens of thousands (10^4 to 10^5), and the third in hundreds of thousands (10^5 to 10^6). Zero cannot exist on a log scale because the $\log_{10}(0)$ is not defined. In our graph in Fig. 8.5, we have placed the origin at Y = $2000 and have provided space for a caption and margin at the bottom of the semi-log paper. It is evident that the maximum value of Y, which we can show on our three-cycle paper, is $1,000,000 or 10^6.

We have labeled the Y scale at locations of 1 and 5 as designated on the log scale of the preprinted graph paper. The number is multiplied by 10^3, 10^4, or 10^5 depending on the cycle involved. Next, we place points on the graph at the coordinate locations defined in Table 8.2. In this example, we may draw a straight line through the data points. The reason for the linearity of the relation for the sum S on a semi-log graph paper is evident, if we take the log of both sides of Eq. (8.1) to obtain:

$$\log_{10}(S) = \log_{10}(5000)(1 + 0.05)^n$$

$$\log_{10}(S) = \log_{10}(5000) + [\log_{10}(1 + 0.05)]\,n \qquad (8.2)$$

Examination of the right hand side of Eq. (8.2) shows that $\log_{10}(S)$ is linear in the number of periods of accumulation (n). The intercept of the straight line with the Y-axis (when n = 0) is $5000. The slope of the straight line is $\log_{10}(1 + 0.05)$ that is the coefficient of n.

The use of semi-log paper has two advantages. First, it permits us to cover a very large range in the quantity represented along the Y (log) axis. We have used a three-cycle paper that covered a range from 2000 to 1,000,000. If the need existed, we could cover even a larger range by selecting a semi-log paper with more log cycles. The second advantage of semi-log representation is that it converts certain non-linear (exponential) functions, such as Eq. (8.1), into linear relations that are much easier to interpret and extrapolate.

8.3.5 X-Y Graphs With Log-log Scales

When you encounter power functions of the form:

$$Y = AX^k \qquad (8.3)$$

it is advantageous to represent both X and Y on log scales in preparing an X-Y graph. Taking the log of both sides of this power function, we obtain:

$$\log_{10}(Y) = \log_{10} A + \log_{10} X^k = \log_{10} A + k \log_{10} X \qquad (8.4)$$

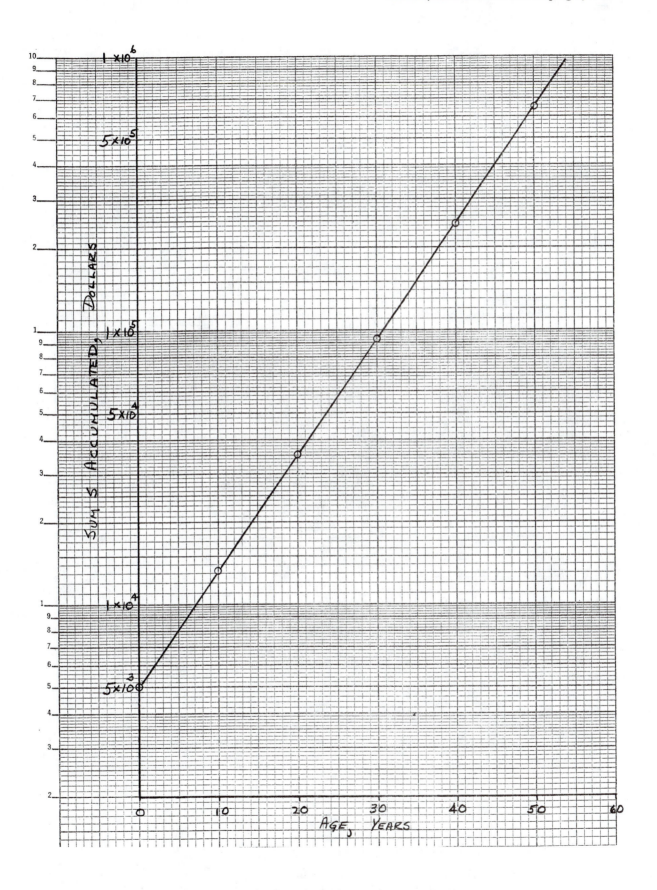

Fig. 8.5 A semi-log representation of the sum S accumulated with time.

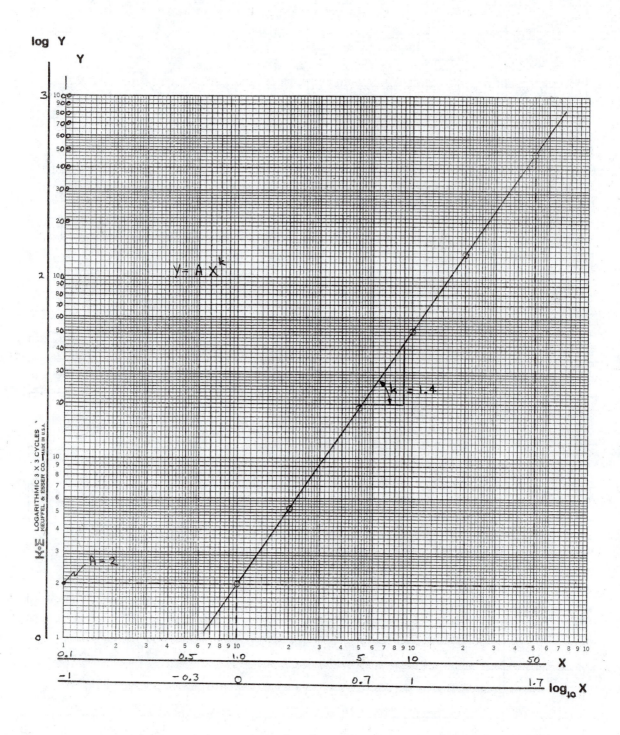

Fig. 8.6 A log-log graph of a power function $Y = AX^k$.

With the logarithmic form of the function $Y = AX^k$, the use of log-log graph paper is ideal. Representing both X and Y on a log scale gives us the opportunity to visualize the constants A and k in the power function. To illustrate this fact, we have taken as an example the function $Y = 2X^{1.4}$ and described it graphically in Fig. 8.6. In this graph, X varies from 0.1 to 100 and Y from 1 to 1000. We have also shown the scales for $\log_{10} Y$ and $\log_{10} X$ in Fig. 8.6 to emphasize the difference between X and $\log_{10} X$ and Y and $\log_{10} Y$.

There are three advantages for using log relations and log-log scales on graphs:

1. The capability of linearizing non-linear power functions that often arise in engineering.
2. The ability to plot wide ranging numerical data.
3. A power relation graphed on log-log paper gives a straight line that is easy to interpret and extrapolate.

If the data is from experimental studies, it is often necessary to establish the constants A and k in a power relation like that shown in Eq. (8.3). The graphical procedure for determining these constants is given in Fig. 8.6. We find the constant A by setting X = 1 and reading the corresponding Y value as indicated with the dashed lines in Fig. 8.6. We find the constant k by determining the slope of the line. In finding this slope, it is necessary to work with the small numbers taken from the log scales. Accordingly, the slope k is given by:

$$k = (\log_{10} Y_2 - \log_{10} Y_1) / (\log_{10} X_2 - \log_{10} X_1)$$

$$k = (2.68 - 0.30)/(1.7 - 0) = 1.40. \tag{8.5}$$

The log or log-log graphs are very important because many processes, widely used in engineering, exhibit non-linear behavior which may be modeled with either exponential or power functions. The behavior of these processes is usually best represented by graphs with either one or both scales expressed in terms of logarithms.

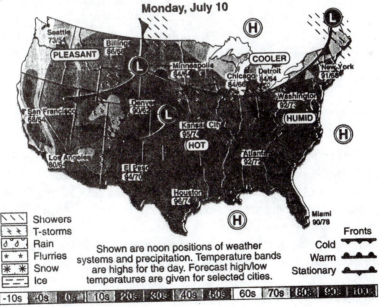

Fig. 8.7 A geographical graph showing the weather prediction for the U. S[2].

---

[2] The National Weather Service Forecasting Office provides daily weather maps of the U. S. on Web site http://www.nnic.noaa.gov/cgi-bin/page?pg=netcast.

8.3.6 Special Graphs

There are many special graphs that have been developed to help the reader quickly absorb and understand numerical data. This brief chapter on Tables and Graphs does not provide a complete description of the many special and clever graphs that have been devised. However, we will introduce one of these, the geographical graph, which is based on a map of the world, the U. S. or some state. Superimposed on the map are contour lines dividing the entire geographical region into sub-regions. Either the contour lines or sub-regions are labeled to show the geographical distribution of some quantity across the mapped region. An example of a geographical graph, presented in Fig. 8.7, shows the weather prediction by the National Weather Service for the U. S. for Monday, July 10, 2000.

8.4 SUMMARY

We have introduced tables as an approach for reporting data. They have the advantage in presenting precise numerical data. If you need six-figure accuracy, it is easy to obtain in a tabular representation of the data. We also outlined the procedure for producing very presentable tables using Word 2000.

Many different types of graphs are used to aid in visualizing data. Graphs are much less precise than tables, but they more effectively show trends, comparisons, and distributions. We first introduced the pie chart, which as the name implies, is a circular area that is divided into pie like segments to indicate distribution of some quantity. Techniques for determining the size of the slices of pie were described.

We then introduced the bar chart that is used to compare the magnitudes of two or more quantities. The bar chart was illustrated and the methods used in its construction were discussed.

Trends are best illustrated using line graphs including the linear, X-Y linear graphs, X-Y graphs with semi-log scales, and X-Y graphs with log-log scales. Linear X-Y graphs are used to show the trend of one variable with respect to another when the range of both variables is not too great. X-Y graphs are presented on semi-log graph paper when one variable has a very large range while the other variable has a more limited range. Log-log graph paper is employed to represent data fields where both variables change over very large ranges. We have provided examples for these three different types of graphs and have described techniques used in their construction.

Finally, we have discussed special graphs that make use of maps to convey the concept of geography in the presentation of data. The map presented to demonstrate this technique is the weather map. It shows the use of contour lines dividing the U. S. into temperature zones.

EXERCISES

8.1 Write a memo for a group of young engineers starting their careers with your company that explains the policy of the engineering department relative to reporting data with tables or with graphs.

8.2 Prepare a table using a word processing program that shows the SAT scores for entering freshman at your College of Engineering for the time period from 1980 to 1999. Report both the math and verbal scores for women and men. Also tabulate the number of women and men in each class. The Office of the Dean of Engineering should be able to provide you with the data that is needed.

8.3 Use the data from Exercise 8.2 to construct a bar chart comparing the scores of women and men for the years 1989 and 1999.

8.4 Use the data from Exercise 8.2 to construct two pie charts showing the percentage of men and women enrolled as freshmen in the College of Engineering in 1980 and 1999.

8.5 Prepare a table similar to the one presented in Table 8.2 showing the sum S with time for the following investment parameters. Use P = \$1000, I = 8%, with interest compounded quarterly. Determine the sum after each compounding period for a total of 20 years.

8.6 Prepare an X-Y graph showing the sum with respect to time using the data generated in Exercise 8.5.

8.7   Prepare a semi-log graph showing the sum determined in Exercise 8.5 on the log scale and the time on the linear scale.

8.8   Evaluate the power function $Y = 1.8 \, X^{1.75}$ and plot the result on a log-log graph. Let X vary from 1 to 100. Try to find log-log paper with a suitable number of cycles for both axes. Confirm that $A = 1.8$ and $k = 1.75$ from your graph.

8.9   Evaluate the power function $Y = 3.2 \, X^{1.44}$ and plot the result on a log-log graph. Let X vary from 1 to 100. Try to find log-log paper with a suitable number of cycles for both axes. Confirm that $A = 3.2$ and $k = 1.44$ from your graph.

8.10 Using the Internet print a copy of today's weather map for the U. S.

PART III

SOFTWARE APPLICATIONS

CHAPTER 9

MICROSOFT EXCEL–2000

9.1 INTRODUCTION

Microsoft® EXCEL–2000, a spreadsheet program, is an extremely important tool in engineering because it is useful in many different applications. We will describe techniques for three of these applications— preparing a parts list, performing calculations, and drawing graphs. There are several different spreadsheet programs on the market, namely EXCEL® by Microsoft, QUATTRO PRO® by Borland, and LOTUS 1 • 2 • 3® by Lotus. All of these programs have a similar format and they all have essentially the same capabilities. If you learn to use one program, it is relatively easy to change to a different program with a modest investment in time. We have selected EXCEL–2000 because it is a recent release, many universities have adopted it, and you will probably have access to it on the computer net during your tenure in college.

The purpose of this chapter is to show you how to navigate on the EXCEL spreadsheet, to make tables, perform calculations, prepare charts and graphs, print the output, and save a file. You should recognize that these objectives are limited. We are concentrating on developing your entry-level skills. Hopefully, you will find the spreadsheet tool important enough to develop a higher skill level with independent study. More complete and detailed treatments are given in the references listed at the end of this Chapter.

9.2 THE SCREEN

You load EXCEL from WINDOWS by clicking on the EXCEL icon and the program opens with the spreadsheet display shown in Fig. 9.1. Let's explore this display. At the top you will note six rows, then below in the center is a big table that is the spreadsheet proper, and finally two more rows at the bottom of the display.

9.2.1 The Six Top Rows or Bars

The top most row confirms that you have opened the Microsoft EXCEL program. The three buttons to the far right side of the title row are common to all Microsoft® programs.

The left hand button with the dash symbol places the program aside (but does not close it) so that you can work in a different program. The middle button is a toggle permitting you to either expand the size of the display on the screen or to shrink it. The right button with the X symbol closes both EXCEL and the file upon which you are working. Do not use the X button if you intend to use EXCEL again during your current session. Instead, use the dash button and set the active program aside. Click on each of these three buttons until you understand their functions.

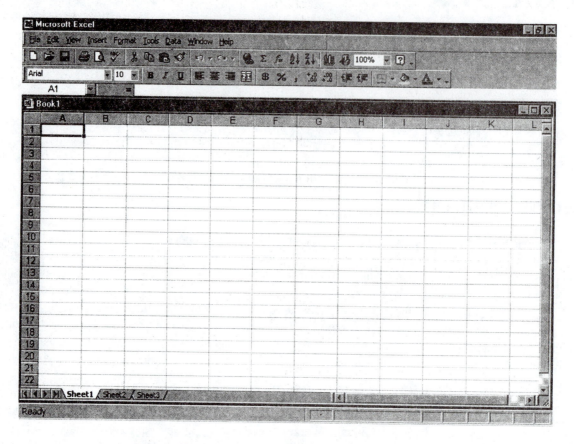

Fig. 9.1 The screen display for Microsoft EXCEL 2000.

9.2.1.1 The Menu Bar

The second row from the top is called the menu bar; it lists nine different features included in the program beginning with File and ending with Help. The menu bar is similar to that found in other Microsoft® programs. Using your mouse, select one of the menu items and click. A pull-down menu is displayed. We will not list the items on each of these pull-down menus because you will learn faster by clicking than by reading. The purpose of each of the menu entries is listed below:

1. File is to open or close, save, or print a file or to exit EXCEL.
2. Edit is to change entries that you have already made.
3. View is to change the size of the screen or to add toolbars, a header or footer, etc.
4. Insert is to add cells, rows and columns to the spreadsheet.
5. Format is to change the height of rows and the width of columns.
6. Tools are to check spelling, audit calculations and develop macros.
7. Data is to provide a means for sorting, filtering and forming data entries.
8. Window is to enable you to open a new window.
9. Help is to provide answers to questions that arise as you attempt to use the program.

9.2.1.2 The Standard Toolbar

The third row from the top exhibits the standard toolbar with 21 buttons. The standard toolbar incorporates many icons common to other Microsoft® applications. The pictures on the icons give a clue as to the purpose of each button; however, if you are in doubt, use the mouse and point to a particular button. Do not click. Wait for a moment and a small label appears giving you the function of each button. Again, point and wait until you understand the function of each icon.

9.2.1.3 The Formatting Toolbar

The fourth row down is the formatting toolbar with 20 buttons. Again many of its icons are common to other Microsoft® applications. Icon labels on each button give us a hint of the purpose of each button, but if we are in doubt, point to the button and a label will appear defining the function. You should try using the two buttons to the far left of the format toolbar to change the type of font and its size.

9.2.1.4 The Cell Locator and Formula Toolbar

The cell locator and formula toolbar is the fifth row down from the top of the display. It provides information about the active cell location and its contents. The small window to the left defines the active cell location on the spreadsheet (A1 is the active cell when you first open a new book or file). The longer window on the right, called the formula bar, is activated when you make an entry into a cell. The numbers or letters that you have entered into the cell appears in this window. If you wish to edit a cell entry without completely retyping it, activate the cell in question by clicking on it, point with the mouse to place the cursor in the formula block, click there, and then edit the entry.

Note the = sign just to the left of the long window. When you click on the = sign, EXCEL interprets the cell entry as a formula. The = sign effectively turns on the computational capabilities of the program.

9.2.1.5 The Title Bar

The sixth and last row from the top is the title bar. When opening a new file, the title Book 1 is displayed on the left side of this bar. When you save this file and rename it, the new name appears in the title bar replacing Book 1. On the far right side of the tile bar, we find the three buttons for setting aside, changing the size of the display, and closing the file. These buttons refer to the file; whereas, the three buttons in the topmost row refer to the program.

Okay! We have briefly defined the top six rows, but it is necessary for you to develop an understanding of the menu items and the icons by pointing and clicking. Do not be concerned with the beeps from the computer—that is part of the learning process.

9.2.2 The Spreadsheet

Most of the screen is occupied by the spreadsheet, which is simply a big table with many columns and rows. The columns are labeled across the top with letters A, B, C,.. L and beyond. The rows are labeled down the left side with numbers 1, 2, 3, ….. 22 and beyond. Each little block, called a cell, is identified with its coordinates (e. g. in Fig. 9.1, the cell A1 is outlined with a border indicating it is active). Note, we place the letter (column location) first and the number (row location) second. The cells are locations in a very large table (spreadsheet) where we enter numbers, text or formulas.

We can move about on the spreadsheet using either the mouse or the keyboard. Since the mouse usually permits the cells to be identified more rapidly, it is the preferred method for locating the cursor over a particular cell. When the pointer is on the spreadsheet, its location is identified with a large plus sign. We may use the mouse to point to any cell on the screen; however, if the cell of interest is not visible on the screen, the scroll buttons or bars located to right or below the spreadsheet are used to bring this cell into the field of view. When the big plus sign is pointed at the correct cell—click; the cell becomes active and is ready to receive the data we enter. We can also move the active cell by using the arrow keys, the tab key, the shift-tab keys, and the control-arrow keys. Try them and note how the active cell moves about the spreadsheet.

The spreadsheet is much larger than it looks on the screen, which shows only columns A through L and rows 1 through 22. (The number of columns and rows visible on the screen will depend on the size of the monitor. A 17-in. monitor was used in recording the image shown in Fig. 9.1). Explore down on the spreadsheet by holding the control + the arrow-down key, and note that the number of the last row is 65,536. Depressing the control and the arrow-right keys moves the active cell to the far right of the spreadsheet, and indicates a column heading of IV which is the 256th column. We have a total of $(256)(65,536) = 16,777,216$ cells on the spreadsheet. Our spreadsheet is huge—let's hope we never have to use all of the available cells!

9.2.3 The Bottom Rows

In the next to last row from the bottom of the spreadsheet, we have three tabs identifying sheet numbers 1, 2 and 3. When opening EXCEL and starting a file, we begin with a workbook for a given project. The program automatically establishes many sheets (pages) in our workbook, although only three are visible to us. If our project requires more than three worksheets, we add more by clicking on Insert, which is located on the menu bar. Then from the drop-down listing we click on Worksheet to add another worksheet. We may enter data and perform calculations on several different worksheets in this file by clicking on the tabs to move from one sheet to another. If you right click on the buttons to the left of these tabs, a drop-down menu allows you to select an active sheet number.

The row at the very bottom of the spreadsheet is called the status bar. It provides a brief statement about the function of any button that is being activated and other information about the status of the program. In Fig. 9.1, the status message is "Ready" indicating that the EXCEL program is prepared to accept your data entry in cell A1.

9.3 PREPARING A PARTS LIST

Let's begin to learn techniques for using EXCEL by preparing a parts list for a chain drive to be used in the speed changer for a human powered centrifugal pump. Load the program from **WINDOWS** by clicking the Start button, select Programs and then Microsoft EXCEL. The screen display should resemble that shown in Fig. 9.1, indicating that we are ready to start. We recommend that you title the file before beginning to develop the parts list. Click on File and select Save As. The Save As dialog box page is displayed as shown in Fig. 9.2. Type in a suitable file name for the project (PARTS LIST) and click the save button. The file name is now displayed on the top row of the spreadsheet.

We begin the parts list by typing its title in cell A1. Since the title should be obvious and easy to read, we use bold Arial font and increase its size to 16 point. Skip a row and position the cursor on cell A3 to begin organizing the section headings as shown in Table 9.1.

We have five section headings (columns) in the parts list:

A. A part number—every unique component has its own number.
B. The name of the item.
C. A description of the item or a drawing number defining the part.
D. The quantity of parts that will be required to build a single prototype.
E. The unit price for the item.

When we type these column headings in cells A3, B3, C3, D3, and E3, it becomes clear that some of the columns are too wide and others too narrow to accommodate the text. To adjust the width of column A, point with the mouse to the line between A and B on the column headings. When the pointer is positioned correctly, you will observe a short vertical line with arrows to the left and right. Press the left mouse button and drag the column boundary either to the left or the right to change the width of column A. Move the position of the cursor to a position on the column headings between column B and C; repeat the process to size the width of column B. Continue to adjust all of the column widths until they are the proper width to accommodate the data included in the parts list.

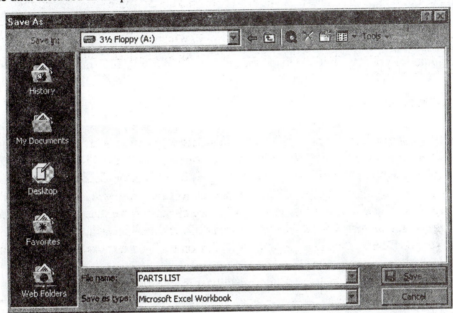

Fig. 9.2 Dialog box for naming a file and saving it.

Table 9.1
Parts List Example

PARTS LIST FOR THE CHAIN DRIVE				
PART NO	NAME OF ITEM	DESCRIPTION OR DRAWING NO.	QUANTITY	PRICE
1	LARGE SPROCKET	CATALOG NO. 6236K27 McMASTER CARR	1	$ 23.48
2	PEDAL ASSEMBLY	LOCAL BICYCLE SUPPLY HOUSE	1	$ 15.55
3	CHAIN ANSI NO. 35	CATALOG NO. 6261K531 McMASTER CARR	4 ft	$ 6.76
4	SMALL BRACKET	CATALOG NO. 6280K111 McMASTER CARR	1	$ 5.04
5	SPROCKET SHAFT	CATALOG NO. 88934K42 McMASTER CARR	1	$ 4.49
6	SLEEVE BEARINGS	CATALOG NO. 6391K214 McMASTER CARR	4	$ 3.76
7	FRAME SIDES	WOOD BARS-------DWG. NO. 100-01	2	$ 1.20
8	FRAME ENDS	WOOD BARS-------DWG. NO. 100-01	2	$ 0.60
9	SEAT	LOCAL BICYCLE SUPPLY HOUSE	1	$ 8.50
10	SUPPORT	WOOD STRUT-------DWG. NO. 100-02	1	$ 0.60
11	SET SCREWS	1/4 - 20, 1/2 LONG, ALLEN HEAD	2	$ 0.85
12	WOOD SCREWS	NO. 8, 1-1/2 IN. LONG, FLAT HEAD, PHILLIPS	8	$ 0.80

In our example parts list, we have shown 12 part numbers. You may have more items because the number of parts depends on the complexity of the design. We type the necessary information for the parts list into the table that is provided by the spreadsheet. Our only concern in this process is that we have activated the correct cell before making an entry. If one of our entries is too long, we can change the column width at any time to accommodate the description. However, it is important to use brief descriptions because the parts list should fit the page.

When you have completed the entries, work on the spreadsheet to improve its appearance. We have centered the part numbers, the quantities, and the prices within the respective columns. We center the entries by marking all of the cells in a given column (point and drag with the mouse and watch a blue background color fill the column). When the blue covers the column entries you wish to mark, lift the left hand button on the mouse and move the pointer to the formatting toolbar and click on the centering button. On the column for price, we initially made the entries with the default format; however, prices are in terms of dollars. We changed the format of the column by marking it, and then clicked on the $ button located on the formatting toolbar to show the prices in terms of a currency.

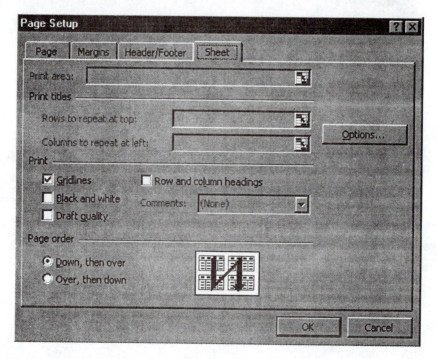

Fig. 9.3 Dialog box for page setup used to adjust margins, page layout and gridlines.

We are ready to print the parts list, but we want to check to see if it fits the page, has the correct margins and is readable. We click on File then Print Preview, to examine the appearance of the spreadsheet when it is printed. To change its appearance, click on Setup button that is located on the bar at the top of the page preview screen. The dialog box that appears is presented in Fig. 9.3. From the four tabs available choose the sheet tab. From this dialog box, point and click on the open square for gridlines so that the cell boundaries (gridlines) are visible when the spreadsheet is printed. Printing the cell gridlines makes the parts list much easier to read. Next, click on the page tab, and select either the portrait or landscape mode. The default setting is the portrait mode because it is more common, but if the spreadsheet is much wider than it is long, the landscape mode is preferred. You can also adjust the size and the quality (dots per inch) of the printing on this page. Finally, click on the margin tab and change the margins so that the spreadsheet is positioned properly on the page.

When you have made all of the necessary modifications on the print preview screen, the spreadsheet should look professional (that means very good). Next, click on the print button, and from the dialog box that appears (see Fig. 9.4) select the number of copies you require, and print the spreadsheet. Your parts list should compare favorably to the one shown in Table 9.1.

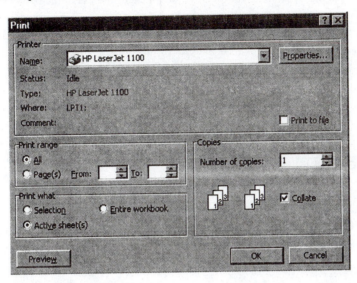

Fig. 9.4 Dialog box used in printing the spreadsheet.

9.4 PERFORMING CALCULATIONS

One of the most important advantages of any spreadsheet program is its capability to perform calculations. EXCEL performs not just one calculation, but a whole sheet full of them. Moreover, EXCEL performs the calculations almost instantaneously and without error. The results can be displayed in two different ways— as an accurate table or a suitable graph. In this section, we will show you two simple examples illustrating the computing power of EXCEL and illustrate the results in tabular format.

9.4.1 The Rich Uncle

For the first example demonstrating the computational power of EXCEL, consider your rich uncle that was introduced in Chapter 8—Tables and Graphs. Recall that we discussed the accumulation of compound interest, which was earned from his initial investment of $5000. We will now describe the results of this analysis, presented in Table 9.2, which is a record of the growth of the investment.

Let's recall the relation for the sum of the accumulation and the input data from the previous chapter where:

$$S = P(1 + i)^n = PM \qquad (9.1)$$

Uncle contributed P = $5,000.00, and the interest rate, (i) for the six month compounding period was 5%. (The annual interest rate was 10%). The sum, S accumulated for a total of 80 compounding periods (40 years) is to be determined. Recall, n is the number of compounding periods and M is a multiplier dependent on n.

Load EXCEL and Save As—Compound Interest—to establish the file name. In the results shown in Table 9.2, we titled the spreadsheet, and entered text describing the parameters that control the results in the first five rows (A1 to A5). We skip a row and define the headings for Periods, Multiplier and Sum in row 7. Initially, we define only three columns (A, B and C). We place the active cell at location A8 and begin by entering the number of the period starting with zero. We could type all 40 entries into column A, but it is easier to fill in the entries automatically.

To permit EXCEL to do the work, position the cursor on the zero entry in cell A8 and drag to mark the A column from cell A8 to cell A48 to mark the location for the required number of cells for 40 entries.

Next, click on Edit, select Fill, and then select Series. A dialog box is displayed as shown in Fig. 9.5 than enables you to select Column indicating that the input data (1 to 40) is filling entries into column A. You also select Linear because the series that you will use to fill column A is a linear series (i.e. 1, 2, 3, etc.). At the bottom of the dialog box, note two small windows—one for Step and the other for Stop. We enter 1 for the Step value and 40 for Stop value and then click on the OK button. The program fills all of the cells from A8 to A48 with numbers starting with 0 in cell A8 and increasing in steps of one until it stops at the number 40 in cell A48. The Fill command listed in the Edit menu item saves time when it is necessary to enter a long series of numbers.

Table 9.2
Accumulation over a Forty-year Period.

DETERMINING THE RUNNING SUM						
FOR AN INITIAL INVESTMENT OF $5000.00						
ANNUAL INTEREST RATE OF 10%						
COMPOUND INTEREST SEMI-ANNUALLY						
NO TAXES PAID						
Periods	Multiplier	Sum		Periods	Multiplier	Sum
0	1.0000	$ 5,000.00		40	7.0400	$ 35,199.94
1	1.0500	$ 5,250.00		41	7.3920	$ 36,959.94
2	1.1025	$ 5,512.50		42	7.7616	$ 38,807.94
3	1.1576	$ 5,788.13		43	8.1497	$ 40,748.33
4	1.2155	$ 6,077.53		44	8.5572	$ 42,785.75
5	1.2763	$ 6,381.41		45	8.9850	$ 44,925.04
6	1.3401	$ 6,700.48		46	9.4343	$ 47,171.29
7	1.4071	$ 7,035.50		47	9.9060	$ 49,529.86
8	1.4775	$ 7,387.28		48	10.4013	$ 52,006.35
9	1.5513	$ 7,756.64		49	10.9213	$ 54,606.67
10	1.6289	$ 8,144.47		50	11.4674	$ 57,337.00
11	1.7103	$ 8,551.70		51	12.0408	$ 60,203.85
12	1.7959	$ 8,979.28		52	12.6428	$ 63,214.04
13	1.8856	$ 9,428.25		53	13.2749	$ 66,374.74
14	1.9799	$ 9,899.66		54	13.9387	$ 69,693.48
15	2.0789	$ 10,394.64		55	14.6356	$ 73,178.15
16	2.1829	$ 10,914.37		56	15.3674	$ 76,837.06
17	2.2920	$ 11,460.09		57	16.1358	$ 80,678.92
18	2.4066	$ 12,033.10		58	16.9426	$ 84,712.86
19	2.5270	$ 12,634.75		59	17.7897	$ 88,948.50
20	2.6533	$ 13,266.49		60	18.6792	$ 93,395.93
21	2.7860	$ 13,929.81		61	19.6131	$ 98,065.73
22	2.9253	$ 14,626.30		62	20.5938	$ 102,969.01
23	3.0715	$ 15,357.62		63	21.6235	$ 108,117.46
24	3.2251	$ 16,125.50		64	22.7047	$ 113,523.34
25	3.3864	$ 16,931.77		65	23.8399	$ 119,199.50
26	3.5557	$ 17,778.36		66	25.0319	$ 125,159.48
27	3.7335	$ 18,667.28		67	26.2835	$ 131,417.45
28	3.9201	$ 19,600.65		68	27.5977	$ 137,988.32
29	4.1161	$ 20,580.68		69	28.9775	$ 144,887.74
30	4.3219	$ 21,609.71		70	30.4264	$ 152,132.13
31	4.5380	$ 22,690.20		71	31.9477	$ 159,738.73
32	4.7649	$ 23,824.71		72	33.5451	$ 167,725.67
33	5.0032	$ 25,015.94		73	35.2224	$ 176,111.95
34	5.2533	$ 26,266.74		74	36.9835	$ 184,917.55
35	5.5160	$ 27,580.08		75	38.8327	$ 194,163.43
36	5.7918	$ 28,959.08		76	40.7743	$ 203,871.60
37	6.0814	$ 30,407.03		77	42.8130	$ 214,065.18
38	6.3855	$ 31,927.39		78	44.9537	$ 224,768.44
39	6.7048	$ 33,523.76		79	47.2014	$ 236,006.86
40	7.0400	$ 35,199.94		80	49.5614	$ 247,807.21

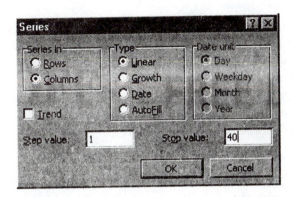

Fig. 9.5 The dialog box for filling a column or row with a series of numbers.

Next, let's locate column B and activate cell B8. We want to calculate the multiplier, M in this column. The multiplier, M is a number that increases with the number of the compounding periods. It is given by:

$$M = (1 + i)^n = (1.05)^n \qquad (9.2)$$

With cell B8 active, we strike the = key, which notifies EXCEL that a formula will be placed in the active cell. Then we type the formula, which in this case the entry is =(1.05)^A8. We use the cell address A8 because it contains the value of n for the column of calculations being considered. We use the ^ symbol to indicate that we are raising (1.05) to a power. Just to the left of the formula bar, note the buttons labeled X, ✔ and f_x. After you have entered and carefully checked the formula, click the ✔ button and examine cell B8. It should read 1, because the entry =(1.05)^A8 = $(1.05)^n$ = $(1.05)^0$ = 1.

Okay! You have successfully employed EXCEL to compute the first multiplier M. Let's determine the other 40 multipliers by copying the first one. Activate B8, and then click the Copy button on the toolbar. You have just stored the formula =(1.05)^A8 in temporary memory; it will be stored there until it is replaced the next time you use the Copy command. Next, mark (select) the B column starting with B9 and continuing to B48 by dragging the pointer down the column. Check that the region is blue where you want to Copy the multiplier term and then click on the Paste button. The cells B8 to B48 show the multiplier M superimposed on the blue background color of the column. Clicking on any cell outside this region eliminates the blue background. In performing the calculations down the B column when using the Copy—Paste commands, EXCEL modified the formula =(1.05)^A8. To check the actions of the program, activate cell B9—note the formula has changed to =(1.05)^A9. EXCEL modifies the formula to accommodate the changes in n as we move down the column. In this instance, it was essential for EXCEL to make this modification. However, in the next example, this type of modification leads to errors. To avoid errors in some extended calculations, we will show the technique necessary to control the way EXCEL modifies the formulas in the Copy—Paste command sequence in an example presented later in this Chapter.

Inspecting the results for the multiplier M in column B, we note that the number of decimals is not consistent from cell to cell. Mark all of the cells in column B, click on either one of the two decimal buttons on the formatting toolbar, and observe the number of decimals change. Click on one or the other of these buttons until you display the multiplier with four digits after the decimal point.

To determine the sum, S that is accumulated with time, it is necessary to multiply the multiplier M by the initial investment of P = 5000. We select cell C8 and type =5000*B8. The symbol * is used to indicate multiplication. When we click on the check button on the formula bar, the entry in cell C8 changes from =5000*B8 to 5000. EXCEL noted that cell B8 contained the number 1, performed the calculation (5000)(1), and displayed the result in cell C8. When we wish to incorporate a cell address into a formula, it can be typed or we can point to the cell with the correct address. After pointing at the cell, click the left mouse button to enter its contents into the formula. In the point and click approach, one of the math symbols (+, −, *, /, ^) **must precede the point and click process.** We activate cell C8 then Copy and Paste this result to the column of cells from C9 to C48. This action computes the sum accumulated at each interval during the 20-year period of the investment (40 interest periods).

Review the quantity in cell C8, and note that the result 5000 is not in the correct format. It is more appropriate to show the results as a currency indicating the sum accumulated from Uncle's investment is in terms of dollars. To change the format:

1. Mark the entire column from C8 to C48 by dragging the cursor down the column.
2. Check that the blue background covers only this region.
3. Point and click on the $ button located on the formatting bar.
4. Clear the blue background by clicking on any cell outside the blue region.

When we converted the C column into a currency format another problem occurred. The results for several cells changed into #######. This symbol indicates that the column width is not sufficient to display the result. To widen column C, we point to the column-heading row at a location between columns C and D. When a line with the double arrows replaces the pointer, press the left mouse button and drag to the right to widen column C. With a wider column the result in cell C8 reads $5,000.00.

We have a result for the sum S in cell C8. Let's arrange for EXCEL to complete the calculations. Activate cell C8 because it has the required formula, and click on the Copy button on the toolbar. Next mark the cells C9 through C48 to locate where the formula is to be applied. The blue background verifies that the region is marked correctly. Next, click on the Paste button and EXCEL applies the formula to each of the cells marked in the C column calculating the results in each cell. Scroll down the column and check to ascertain of the results are displayed correctly. If you encounter #######, it is necessary to widen the column. The result in cell C48 should read $35,199.94.

We have computed the accumulated sum S for 40 periods, but we originally stated that S was to be determined for 80 periods. Why did we stop at $n = 40$? We divided the calculation into two parts so that the output, when printed, would fit on one page. It is easy to extend the calculation to $n = 80$ by using the Copy command.

To continue our work, mark the block with corners at A7 and C8 and click the Copy button. We now have our headings and the initial formula in temporary storage (memory). Move the cursor and activate cell E7. We skip column D to provide space in the table between the two lists of results. After activating cell E7, click on the Paste button and the results from the A7 to C8 block are copied (with modifications for the shift in columns) into the E7 to G8 block. Look at cell E8 and note that the period shown is $n = 0$. Let's change that value to 40 by editing the number displayed on the formula toolbar. When cell E8 is changed to 40, the results in E9 and G9 are identical with previous results for the period $n = 40$. This fact implies that the program in EXCEL has adjusted the formulas to account for the fact that we have changed the columns in which the calculations are made. Activate cells F8 and G8 and compare the formulas in them with those in cells B8 and C8. Have you noted the changes automatically made by EXCEL as we copied formulas from one column to another?

To complete the table, activate cell E8 and fill in the entries for $n = 40$ to 80. We again use the automatic number generator in EXCEL. Click on Edit, select Fill, and then Series. On the dialog box, click on Column, Linear, and indicate a Step value of 1 with a Stop value of 80. Click on OK and then scroll down the spreadsheet to make certain that the correct entries have been generated for column E. Next, mark cells F8 and G8 and Copy them. Mark the F8 – G48 block and Paste the formulas from F8 and G8 into this block. Clear the blue background and scroll down the G column. You will probably note the symbol ####### in many of the cells. Widen column G until all of the results are shown in the currency format. As a check, the result $247,807.21 should appear in cell G48.

We are now ready to print the table showing all of the results. Click on File, then on Print Review. The screen shows the spreadsheet as it will appear when printed. We note that it could be improved with gridlines and a larger left hand margin. To show the gridlines, click the Set-up button. From the dialog box, which appears on the screen, select the sheet tab. Click on the gridlines square and note that a check appears indicating the gridlines will be printed. To adjust the margins, click on the margin tab. Then simply point, click and drag the left margin line until the table is positioned correctly in the sheet shown in the print review screen. Next, click on the OK button to return to the preview of the spreadsheet. Finally, click on the Print button to print the spreadsheet.

We will return to this spreadsheet later when preparing a graph of the results.

9.4.2 Strain on a Simply Supported Beam

In many engineering applications it is necessary to determine the strain developed when a beam is subjected to loads. Beams are long slender members that serve to carry transverse loads. You have seen many beams when driving under bridges that support highway overpasses. When a load is applied in the center of a beam, and its two ends are simply supported, the relationship for the maximum strain developed at the load point on the beam is given as:

$$\varepsilon = 3 \ W_u \ S/(2bh^2 \ E) \qquad\qquad (9.3)$$

where: W_u is the load (weight) in pounds or Newton.
S is the span of the beam in inch or meter.
b is the depth of the section of the beam in inch or meter.
h is the height of the section of the beam in inch or meter.
E is the elastic modulus in psi or Pascal.

Let's program this relation in EXCEL to determine the strain induced in the beam at a location under the load point. In examining Eq. (9.3), note that the strain ε depends on five variables—S, b, h, E and W_u. For this example, let's fix three of these variables at the values shown below:

S = 4 in., b = ¼ in. and E = 10.6 x 10^6 psi. (An aluminum alloy).

We will treat the weight, W_u and the height, h as variables, and explore solution space for the strain, ε. Let's consider the load, W_u increasing from zero to a maximum of 4 lb_f in steps of 0.25 lb_f. Also, consider the height h of the cross-section of the beam varying from 0.01 in. to 0.07 in. in steps of 0.01 in. To determine the strain, we display the variable W_u in column A, and the other variable, height, along the 8th row. This arrangement of the spreadsheet is shown in Table 9.3.

We have selected the ranges for the load and the height to cover the region of interest in the design of sensing beam for a small scale that is intended to weigh letters and small packages[1]. The maximum weight in the specifications for this scale is 4 lb_f. Also, when the prototype of the scale is built, it will be calibrated using a number of small weights to verify the scale's accuracy. The results of this spreadsheet analysis are essential in verifying the calibration.

In the spreadsheet illustrated in Table 9.3, we have entered the data for the height of the beam in row 8, columns B to H. In rows 1 to 4, we have provided the title of the spreadsheet and information related to the analysis of the strain developed in the beam. On row 6, the equation used in determining the strain is given. Both rows 5 and 7 are blank to provide space above and below the equation. In row 9, we have entered the heading for W_u and its units and repeated the data for the height of the beam together with its units. The spreadsheet is now organized with an arrangement that displays the two variables and their units of measure.

---

[1] This product is described in <u>Introduction to Engineering Design: Book 3, Postal Scales</u>, College House Ent., 1998.

Table 9.3
Strain in a Simply Supported Beam

DETERMINE STRAIN							
FOR A SIMPLY SUPPORTED BEAM							
SPAN S =4 in., DEPTH b = 1/4 in., ELASTIC MODULUS E = 10.6 x 10^6							
WEIGHT Wu AND HEIGHT h ARE VARIABLES							
STRAIN = 3WuS/(2bh^2E)							
HEIGHT	0.01	0.02	0.03	0.04	0.05	0.06	0.07
Wu (lb.)	h = 0.01 in.	h = 0.02 in.	h = 0.03 in.	h = 0.04 in.	h = 0.05 in.	h = 0.06 in.	h = 0.07 in.
0.00	0.000000	0.000000	0.000000	0.000000	0.000000	0.000000	0.000000
0.25	0.005660	0.001415	0.000629	0.000354	0.000226	0.000157	0.000116
0.50	0.011321	0.002830	0.001258	0.000708	0.000453	0.000314	0.000231
0.75	0.016981	0.004245	0.001887	0.001061	0.000679	0.000472	0.000347
1.00	0.022642	0.005660	0.002516	0.001415	0.000906	0.000629	0.000462
1.25	0.028302	0.007075	0.003145	0.001769	0.001132	0.000786	0.000578
1.50	0.033962	0.008491	0.003774	0.002123	0.001358	0.000943	0.000693
1.75	0.039623	0.009906	0.004403	0.002476	0.001585	0.001101	0.000809
2.00	0.045283	0.011321	0.005031	0.002830	0.001811	0.001258	0.000924
2.25	0.050943	0.012736	0.005660	0.003184	0.002038	0.001415	0.001040
2.50	0.056604	0.014151	0.006289	0.003538	0.002264	0.001572	0.001155
2.75	0.062264	0.015566	0.006918	0.003892	0.002491	0.001730	0.001271
3.00	0.067925	0.016981	0.007547	0.004245	0.002717	0.001887	0.001386
3.25	0.073585	0.018396	0.008176	0.004599	0.002943	0.002044	0.001502
3.50	0.079245	0.019811	0.008805	0.004953	0.003170	0.002201	0.001617
3.75	0.084906	0.021226	0.009434	0.005307	0.003396	0.002358	0.001733
4.00	0.090566	0.022642	0.010063	0.005660	0.003623	0.002516	0.001848

To begin the programming of the equation for the strain, let's fill column A with information about the weight W_u. Type zero in cell A10 and keep this cell active. Mark column A by clicking and dragging from Cell A10 to cell A26. Then click Edit on the menu bar, select Fill and Series. On the series dialog box, select Column, Linear, and enter a Step value of 0.25 and Stop value of 4. Finally, click the OK button and check to determine if the numbers automatically entered into column A are correct.

Next, activate the B10 cell and type the formula for the strain as:

$$=3*A10*4/(2*(1/4)*10.6*10^6*(B8)^2) \qquad (a)$$

This entry is correct for cell B10, but it will give us significant difficulties later when we try to use the Copy and Paste commands to extend the calculation of strain to other cells in the spreadsheet. Let's proceed by copying the formula in cell B10 into cells B11 to B26. To copy the equation into other cells in column B, position the cursor on cell B10 and click on Copy. Then mark the column from B11 to B26. Next, click on Paste. When we examine our results, they are clearly in error. To troubleshoot the problem, activate cell B11 and examine the formula displayed in the formula bar. You observe the following relation:

$$=3*A11*4/(2*(1/4)*10.6*10^6*(B9)^2) \qquad (b)$$

Unfortunately this relation is not correct. In the Copy-Paste operation, EXCEL modified our formula indexing the entries for row locations in both columns A and B by 1. This indexing was correct for the entries in column A, but the indexing produces an error in the entries in column B. Instead of indexing the entries from column B, we want to use the data in cell B8 for all the calculations in column B. To prevent unwanted indexing, we modify the formula in cell B10 to read:

$$=3*A10*4/(2*(1/4)*10.6*10^6*(B\$8)^2) \qquad (c)$$

The $ symbol before the number 8 locks the value contained in the cell B8 into the equation. When we perform the Paste operation and EXCEL indexes down the B column in modifying the formula, the value of the entry in cell B8 is used in all of the calculations. We now Copy cell B10 and Paste down the B column from B11 to B26 to generate the correct results for the strain corresponding to a beam with a height of 0.01 in.

The Copy and Paste operation worked well on column B after we corrected the B$8 entry. Let's try to copy the modified formula in cell B10 into the row of cells C10 to H10. If you Copy, Paste, and examine the formulas in these cells, you will note that EXCEL has not provided the correct relations. Again you must modify the entry in cell B10 to properly perform the Copy and Paste operation along row 10.

When we Copy the contents of cell B10 along a row to the right, the cell address that defines the weight—A10 changes to B10, C10, D10 etc. Of course, the content of cells B10, C10, D10, etc. have nothing to do with the weight applied to the beam. Since it is always necessary to multiply by the weight which is listed in column A, it is imperative that the A column designation remain fixed in the Copy-Paste operation. To modify our entry in cell B10, we type:

$$=3*\$A10*4/(2*(1/4)*10.6*10^6*(B\$8)^2) \qquad (d)$$

The $ symbol before A locks column A into the formula; it remains fixed as we change columns in the Copy-Paste operation.

Now that our entry into cell B10 has been corrected to permit us to Copy and Paste along both the rows and columns, we proceed. Copy and Paste the formula in cell B10 to cells C10 to H10. Then Copy this row (C10 to H10) and Paste to the block C11 to H26. The program executes these calculations and displays the results as shown in the previous spreadsheet. For the maximum load of 4.00 lb$_f$, the strain varies from 0.090566 for a beam 0.01 in. high to 0.001848 for a beam 0.07 in. high.

You should always use your calculator to manually check a few values of the strain to insure that errors were not made in programming the equation into the EXCEL spreadsheet. Finally, the spreadsheet does not indicate the units for the strain. In this case, the absence of units does not present a problem because strain is a dimensionless quantity.

You may also want to change the number of decimals used in reporting the results. We usually use six decimal places to represent strain. This is a very large number of decimal places to carry, but strains are usually small and often two or three of these decimal places are expended carrying zeros. You can check the spreadsheet results for the strain to verify this fact. You may change the number of decimals by marking the block containing the results (B10 to H26). Then click on one or the other of the increasing or decreasing decimal keys on the formatting toolbar until the results are presented with six decimal places.

Try these examples and understand and appreciate the power of EXCEL in performing not one but an entire sheet full of calculations. Use EXCEL to solve the homework problems assigned in Physics and Mathematics courses. You can determine the single answer usually sought in an assigned problem and then easily explore solution space in EXCEL by treating one or two of the quantities in the controlling equation as variables. Try exploring solution space. Understand the importance of determining all reasonable solutions to a well-posed problem in a single analysis.

9.4.3 Entering Formulas

In the two previous sections, we have entered formulas by hand using the well known mathematical symbols +, −, * , / and ^ to add, subtract, multiply, divide and raise to a power, respectively. For engineers with a good understanding of the mathematical operations, entering the formulas by hand and making use of these mathematical operators is usually the preferred approach. However, EXCEL has an embedded formula palette to aid in entering and editing formulas in a worksheet. The formula palette, shown in Fig. 9.6, appears when you have activated a cell and pressed the = key.

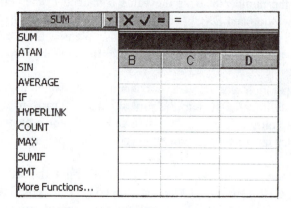

Fig. 9.6 The formula palette.

The formula palette appears on the screen with SUM as the default operation. It will aid you in adding a column or row of numbers and placing the result in the activated cell. Exercises are given at the end of the Chapter to provide the opportunity to use the SUM command. Several other commonly employed operators appear, as shown in Fig. 9.6, when you click on the ▼ button. A much more extensive list of operators is available by clicking on the bottom line of the listing—More Functions.

The Paste Function dialog box, which appears on the screen, provides a technique for determining all of the functions that have been programmed into EXCEL. In this course, the Math and Trig functions are the most important. The Paste Function dialog box is presented in Fig. 9.7. It includes a very complete listing of functions beginning with ABS (the absolute value) to TRUNC (an operator which truncates a number with a decimal or fraction to its integer value).

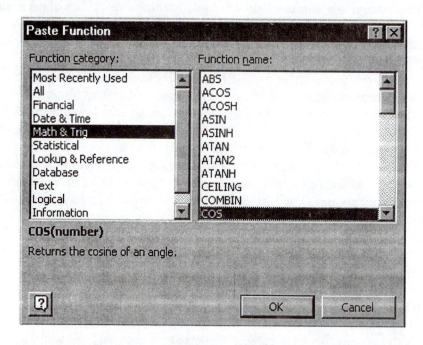

Fig. 9.7 The Paste Function dialog box with the Math and Trig operators activated.

Let's consider using the cosine function to illustrate the technique for the Paste Function dialog box. Suppose we are to determine the cosine of a 45° angle and to place this result in cell A8. Click on cell A8 and either type = or click on the equal sign on the formula bar. Click the arrow button ▼ to the right of the SUM button and then click More Functions to bring up the Paste Function dialog box. Click on the

Math & Trig line under the Function category and then select the COS function as shown in Fig. 9.7. The COS dialog box, presented in Fig. 9.8, appears, and you enter the argument of the cosine in the box labeled number. It is important to recognize that the argument of the cosine **must be** given in **radians**. To convert the 45° to radians, we divide 45 by 180 and multiply that quantity by π. Unfortunately EXCEL cannot handle π without a fuss. We must type PI() or its decimal equivalent 3.14159265358979. The parentheses () following PI are essential because EXCEL recognizes PI alone as text not the number corresponding to π. After the entry for the argument is complete, the result for the cos (45°) is shown at the bottom of the dialog box. With mathematical functions, it is imperative that the argument be enclosed with parentheses as shown in Fig. 9.8

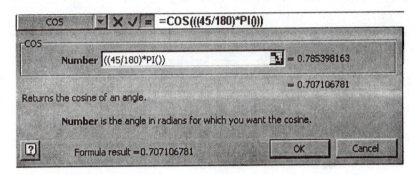

Fig. 9.8 The COS dialog box is typical of the aids available for math and trig functions.

9.5 GRAPHS WITH EXCEL

EXCEL can be employed to produce several types of graphs easily and quickly after you have learned the procedure. To show you the techniques used for producing graphs, several examples are used to demonstrate the capabilities of EXCEL in preparing pie, bar, and X-Y graphs. Let's begin by discussing the pie chart.

9.5.1 Pie Charts

In the previous chapter on Tables and Graphs, we demonstrated a manual technique for preparing a pie chart. If you remember our previous discussion, this chart is used primarily to show the distribution of some quantity. The larger the piece of the pie—the larger the share of the distribution. To illustrate how to make a pie chart in EXCEL, we open a new book and title it—PIE CHART—using the Save As selection under File on the menu bar. We use the top four rows for the title and row six for the column headings. The tabular data, that describes the distribution of time in various assignments for Mechanical Engineers in their initial positions in industry, is entered in cells B7 to B15. These entries are shown in the spreadsheet presented in Table 9.4.

The data in Table 9.4 is interesting, but the information presented does not carry a visual impact. There is no need for great precision in reporting these numbers; we are not even using a single decimal place. Clearly, the tabular format is not appropriate. The pie chart presented in Fig 9.9 is a much better method for representing this information. It would be much more effective if it was printed in color.

This pie chart looks very good. How did we generate it? After the numerical data on the distribution has been entered in cells B7 to B15, we mark them. Then we click on the chart wizard button on the standard toolbar. The chart wizard presents a series of four dialog boxes that guide us through the steps necessary to convert the data displayed on the spreadsheet into a chart or graph of one type or another. The first dialog box, presented in Fig. 9.10 guides us in the selection of the type of chart or graph that we choose to represent the data entered on the spreadsheet. We find that EXCEL offers many options from which to select. We select the pie chart noting that they can be constructed with six different options. Let's select the three dimension pie chart, shown in Fig. 9.10 by clicking on its symbol and then click on the Next > button which brings the second dialog box to the screen.

Table 9.4
Listing of Data for Pie Chart

PIE CHART	
RESPONSIBILITIES OF MECHANICAL ENGINEERS	
FIRST ASSIGNMENT IN PERCENT TIME	
ASSIGNMENT	PERCENT
1. DESIGN ENGINEERING	40
2. PLANT ENGINEERING, OPERATIONS, MAINTENANCE	13
3. QUALITY CONTROL, RELIABILITY, STANDARDS	12
4. PRODUCTION ENGINEERING	12
5. SALES ENGINEERING	5
6. MANAGEMENT	4
7. COMPUTER APPLICATIONS, SYSTEMS ANALYSIS	4
8. BASIC RESEARCH AND DEVELOPMENT	3
9. OTHER ACTIVITIES	7
TOTAL	100

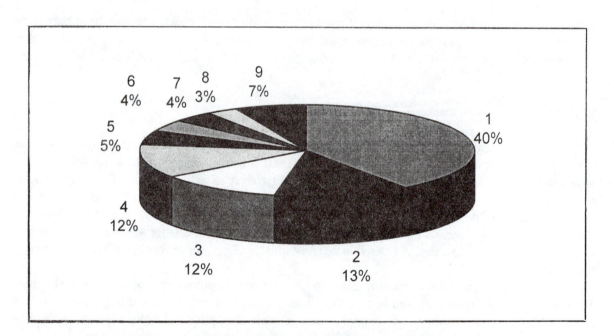

Fig. 9.9 The Pie chart shows the responsibilities of Mechanical Engineers in their first assignment.

The second dialog box, titled Chart Source Data, is presented in Fig. 9.11. Its purpose is to identify the range of data to be employed in constructing the pie chart. In our example, we have already marked the data in column B—cells B7 to B15. We confirm that these cells are correct when clicking on the Next > button.

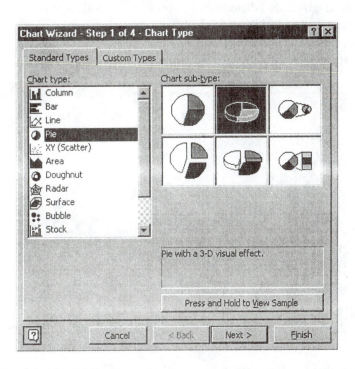

Fig. 9.10 The first of four dialog boxes presented by the chart wizard.

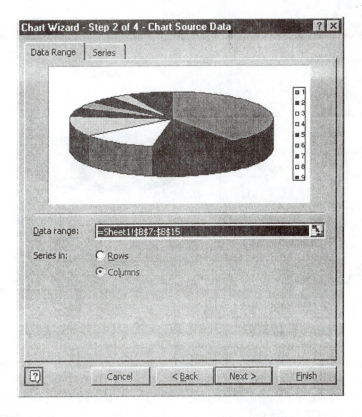

Fig. 9.11 The second dialog box in the chart wizard identifies the range of data to be employed.

The Chart Options dialog box for step 3, illustrated in Fig. 9.12, previews our pie chart. Since it looks perfect, we do not alter the default settings for the range of data. Also, this dialog box assists us in titling the graph and providing legends to identify the different slices of the pie. We click on the title tab to add a title. The legend tab yields a dialog box that provides several choices for the placement of the legend. However, in this example, we decide not to show the legend. Instead we click on the data labels tab and the dialog box shown in Fig 9.13 appears. We decide to identify each slice of the pie with a label and a percent by clicking on the appropriate dot.

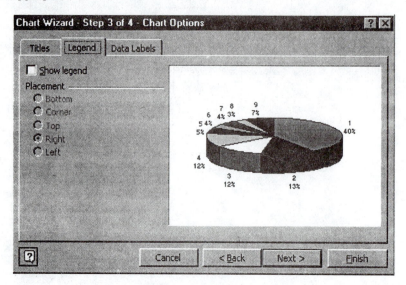

Fig. 9.12 Step three—placement of the legend, title, and data labels and a preview of the chart.

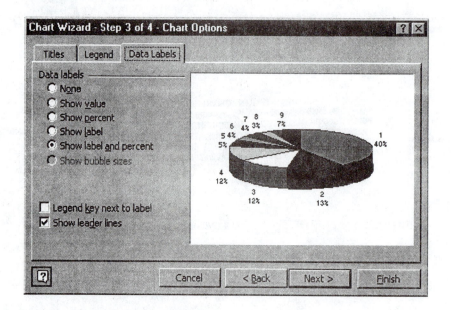

Fig. 9.13 Step three—placement of data labels.

Except for sizing the chart and locating it on the spreadsheet, we have completed the pie chart. We click the Next > button on the Step 3 dialog box and the final page from the chart wizard, presented in Fig. 9.14, provides a choice for the placement of the chart. We have decided to show the chart as an object on Sheet 1 of the spreadsheet. Click the finish button and examine the results on the monitor. If the chart is too small or too large, click and drag on a corner of the chart to adjust the size.

When you are satisfied with the appearance and location of the chart, click on File and select Print Preview. You may add gridlines and adjust margins on the Print Preview display. Print the spreadsheet and the pie chart and observe its professional appearance (see Fig. 9.9).

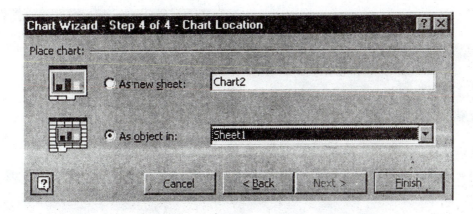

Fig. 9.14 The final step in using the chart wizard involves placing the chart.

9.5.2 Bar Charts

Bar charts are usually employed for showing comparisons. In the previous chapter 8 on Tables and Graphs, we introduced an example of a bar chart that compared the percentage of high school students graduating with three or more years of secondary mathematics and science in 1982 and 1994. It is a very simple comparison, but it indicates the power of visual graphics in carrying a message.

We begin by loading EXCEL, which opens to a new book. Again, we title this book by clicking on File, Save As and then name the file—BAR CHART. The title of the new spreadsheet is entered in the first five rows, and then the data for the graduating high school students is entered in the block A7 to C9, as illustrated in Table 9.5. In the spreadsheet, the column headings are in row 7, and the years used in the comparison are listed in column A. We begin to create our bar chart by marking the region on the spreadsheet where the numerical data exist—B8 to C9. We then click on the Chart Wizard button; the first dialog box shows illustrations for many different chart types. At the top of the list is an option for column charts and another for bar charts. They are nearly the same. The bar charts display comparison categories along the Y-axis and numerical values along the X-axis. The column charts display comparison categories along the X axis and numerical values along the Y-axis. We consider both the column and the bar chart representations in EXCEL to be suitable for presenting data comparing two quantities. Take your pick; we have selected the three dimensional bar chart. The first dialog box, presented in Fig. 9.15a, shows the available selections. Near the bottom of this dialog box is a long key—when pressed a preview of the bar chart is presented. It is advisable to use this feature to determine if the data range has been marked correctly. If the chart is not correct, cancel and revise your selection of the data to be included in the bar chart.

Table 9.5
Comparison of Skills of Graduating Seniors

BAR CHART						
COMPARISON OF GRADUATING HIGH SCHOOL STUDENTS						
COMPLETING THREE OR MORE YEARS OF MATH AND SCIENCE						
FROM 1982 TO 1994						
YEAR	MATH	SCIENCE				
1982	38	33				
1994	60	52				

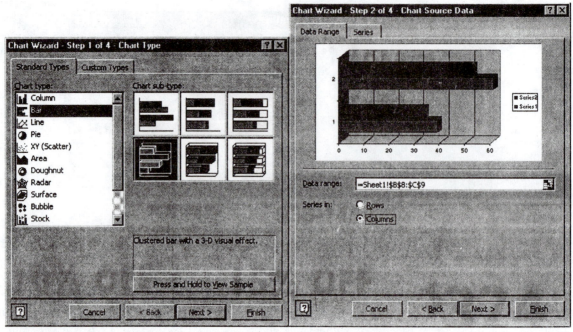

Fig. 9.15a   Fig. 9.15b

The next dialog box (Step 2) provides two tabs (two pages) for data entry—both are extremely important. The first is for the data range, which we marked prior to initiating the chart wizard. However, the series is in a column format and this choice should be made as indicated in Fig. 9.15b. Click on the series tab for the Source Data dialog box and identify the series as MATH and SCIENCE by pointing with the mouse to the headings of the respective columns (B7 and C7). The "category (X) axis labels" is confusing. In EXCEL the (X) axis is the ordinate on a bar chart. We identify the (X) axis labels by giving the cell addresses for the years 1982 and 1994 (A8 and A9). The series page for the source data dialog box is presented in Fig. 9.15c.

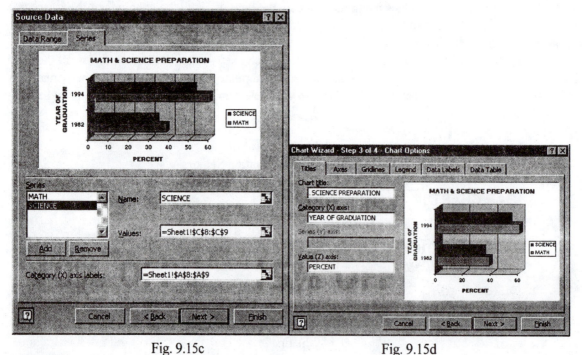

Fig. 9.15c   Fig. 9.15d

Fig. 9.15 Chart wizard dialog boxes showing data entries used to create the bar chart of Fig. 9.16.

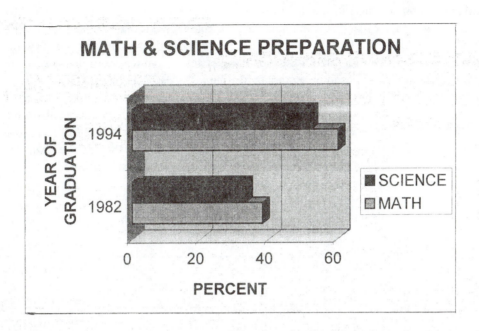

Fig. 9.16 Bar graph presentation of data comparing two quantities at two different times.

The dialog box associated with step three provides several tabs (pages) for entering information on the bar chart concerning the title, defining the axes, providing gridlines, showing legends, labeling data and showing the data table. A preview screen in the dialog box is very helpful in selecting axes and in deciding on the positioning of the gridlines. We show the page associated with the titles tab in Fig. 9.15d. Click on the various tabs and explore the various techniques for modifying the bar chart.

The final dialog box, Step 4, is identical to that shown in Fig. 9.14; it provides the opportunity for you to select the location of the bar chart. We clicked on the circle, which places the bar chart as an object on a new sheet. The screen of your monitor shows a large bar chart on a sheet labeled CHART 2. When you are satisfied with the appearance of the chart, click on File and Print Preview. Adjust the margins and then print the output. Try to duplicate the spreadsheet presented in Table 9.5 and the bar chart shown in Fig. 9.16.

9.5.3 X-Y Graphs

The X – Y graph is the type of visual representation most frequently employed by engineers. It is very effective in visually conveying trends indicated by numerical data. Is the trend increasing or decreasing? Is the quantity level, or oscillating with respect to time? The X-Y graph quickly and dramatically shows these trends. To demonstrate the method for producing X-Y graphs in EXCEL, we will consider the strain produced by a centrally-loaded, simply-supported beam as predicted by Eq. (9.3).

We have previously used Eq. (9.3) to demonstrate the technique for performing calculations using EXCEL. Let's save some work by recalling the spreadsheet presented in Table 9.3 that gives the results obtained in evaluating Eq. (9.3). These calculations provide the numerical data necessary for preparing our X-Y graph. Using the chart wizard, we develop an X-Y graph from the results as shown in Fig. 9.17.

Let's develop the step-by-step procedure for converting the numerical data on the spreadsheet to an X-Y graph. Start by marking the data block—A9 to H26. Note, we have included the headings for the seven columns of Y data that are given in columns B to H of the data block. Remember that the data for the scale and caption along the X-axis are listed in column A. EXCEL will plot the results for the strain automatically along the Y-axis. After the data block is marked (A9 — H26), click the chart wizard button and select the type of chart from the options listed in the first dialog box. **Do not select the line graph** from the available options because it often distorts the data along the X-axis. Instead, **select the X-Y scatter graph** as shown in Fig. 9.18a. Press and hold the long button on this dialog box to review the X-Y graph.

Click on the Next > button, and the Chart Source Data dialog box, associated with Step 2, appears as presented in Fig. 9.18b. The tab for the data range is active, and we check that the data range is correct. We also indicate that the series to be plotted is in columns. Also note that EXCEL has assigned the beam height h in the legend in the preview of the graph. Click on the Series tab, and verify the location of the series of numerical results to be plotted. In this example, a series of numerical results is available for each of the seven beam heights. The cell locations of the values plotted for X and for Y are given in a box for data entry. If these cell locations do not identify the correct parameter they can be changed by clicking on the button on the right side of this box. The correct series designations are indicated in Fig. 9.18c.

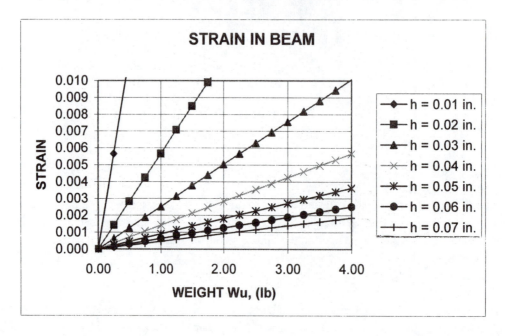

Fig. 9.17 X-Y graph showing the strain in beams of various heights as a function of applied load.

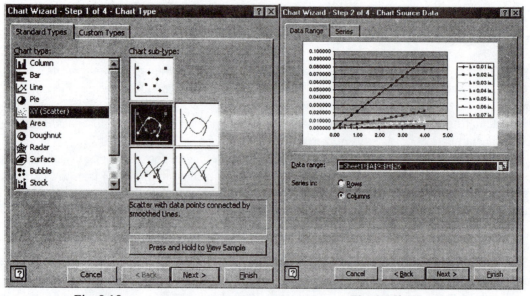

Fig. 9.18a   Fig. 9.18b
Fig. 9.18 Dialog boxes used in constructing X-Y (scatter) graphs.

The Chart Options dialog box, Step 3 in Fig. 9.18d, provides an opportunity to add the title and caption the X and Y-axes. We added the title **STRAIN IN BEAM**, and captioned the X and Y-axes as **WEIGHT, W$_u$ (LB)** and **STRAIN,** respectively. We may also add or remove grid lines and move the position of the legend by activating the appropriate tab. The draft of the graph is essentially completed when we click the Next > button.

In the fourth and final dialog box, we choose to show the graph on the same worksheet as the numerical data. We click on the graph to activate it, and then drag the corner and/or side markers of the graph to enlarge it. When the graph is the correct size it is positioned with the pointer to a suitable location on the spreadsheet. Before printing the graph, click on File then Print Preview, and check to determine if you have arranged the display so that it conveys the numerical data in tabular form on the spreadsheet as well as the X-Y graph.

The combination of the spreadsheet and the graph is an effective technique for presenting the results of theoretical or experimental data. However, in some instances we want to show the graph without including the spreadsheet. We can show the graph alone without the spreadsheet by displaying it on another worksheet. Click on the graph to activate it and then click on the Copy icon. Next, click on the tab for Sheet 2 to bring a new spreadsheet onto the screen. Finally, click on the Paste button. The graph from Sheet 1 is copied onto Sheet 2. Click on the graph to activate it, then drag its corners or sides to position and size it on the sheet. Examine the appearance of the graph with Print Review prior to printing. The result is shown in Fig. 9.17.

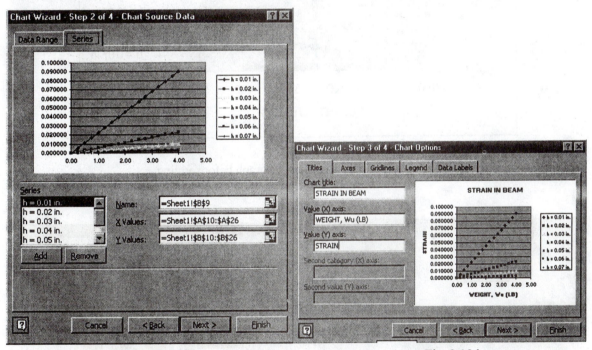

Fig. 9.18c   Fig. 9.18d
Fig. 9.18 Dialog boxes used in constructing X-Y (scatter) graphs.

A close examination of Fig. 9.17 shows that the graph is not exactly the same as that presented in the previews displayed in Fig. 9.18 or the view of the graph shown in Fig. 9.19. The scales on both the abscissa and the ordinate are different and the gridlines for the X-axis are missing in the preview graphs. The choice of scale for the X and Y-axes is very important. The very high values of strain shown in Fig. 9.19 tend to distort the graph. These strains are excessive and indicate that the beams with small heights would fail by plastic bending long before achieving a load of four pounds.

To change the scale on the graph in Fig. 9.19, point to the number 5.00 on the caption for the X-axis and double click the left mouse button. The Format Axis dialog box appears, as shown in Fig. 9.20a, with five different tabs. Select the tab for scale and enter the number 4.00 for the maximum value for the

X-axis. To change the scale on the Y-axis, point to the number 0.10 in the caption for the ordinate and repeat the process. Again, select the scale tab and enter the value of 0.01 for the maximum strain to be plotted along the Y-axis. When the new selections are made, the graph is rescaled so that it displays more meaningful values of strain as shown in Fig. 9.17.

A close examination of the graph in Fig. 9.19 indicates that the gridlines for the X-axis are missing. To correct this oversight, click on the graph to activate it; then click on the chart wizard icon. Proceed quickly through the dialog boxes, which are complete because the graph is active. When the third dialog box appears, click on the gridlines tab and select gridlines for the major markings along the X-axis. Click on the finish button to add the grid lines.

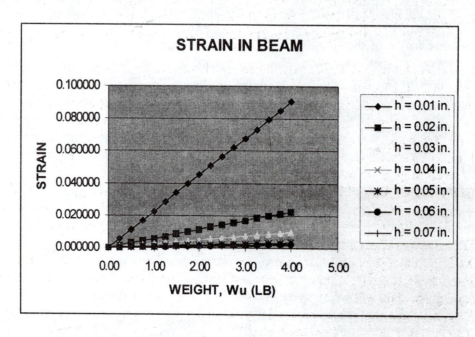

Fig. 9.19 Appearance of the X-Y graph before editing.

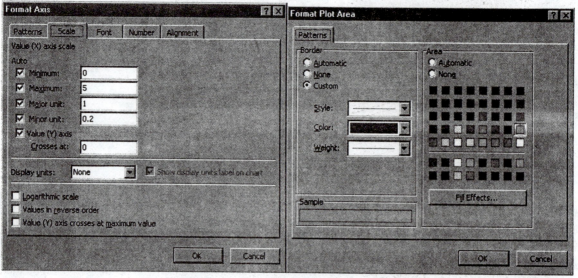

Fig. 9.20a   Fig. 9.20b
Fig. 9.20 Dialog boxes for editing graphs.

Finally, the background color (shading) of the chart area in Fig. 9.19 is too dark. To remove the shading double click on an open area of the graph. The Format Plot Area dialog box appears, as shown in Fig. 9.20b, with selections for the color of the border and the area. Under that part of the dialog box for the area, select white for the background color. Click on the OK button and the background shading vanishes.

9.6 SUMMARY

We have introduced EXCEL–2000 one of the most popular spreadsheet programs in use in the world. The introduction was limited because our purpose was to help you develop only entry-level skills. You should know how to interpret the screen, the menu, and the toolbars. You should also be able to navigate on the spreadsheet with the keyboard or the mouse, and to mark cells or blocks of cells by dragging the cursor with the mouse.

Three examples were described that will be useful in this course and in other courses that you will encounter later in the curriculum. First, we illustrated a technique for preparing a parts list. In this instance, the spreadsheet was utilized to organize a table containing both descriptive and numerical data.

The second example demonstrated techniques for performing calculations using EXCEL to evaluate simple formulas. We first illustrated methods for performing a single calculation in a cell. Then we copied this calculation and performed a line of calculations by varying one parameter. Finally, we performed a spreadsheet full of calculations when two parameters were varied.

The third example involved the preparation of three different charts or graphs. We demonstrated the techniques involved in preparing a pie chart, a bar chart, and an X-Y graph. We made extensive use of the chart wizard in EXCEL that leads you through the technique in a four-step process. We also described techniques for editing X – Y graphs.

EXCEL is a tool somewhat like a hammer. With time and practice, you learn to use a hammer—you hit the nail more squarely, with more force and become more effective. The same is true for any spreadsheet program. You will grow more proficient with time. The benefits provided with a spreadsheet program, throughout your tenure in college and in your professional and personal life, make the investment of time in learning the program well worthwhile.

REFERENCES

1. Gottfried, B. S., Spreadsheet Tools for Engineers, Mc Graw Hill, New York, NY, 1996.
2. O'Leary, T. J. and L. I. O'Leary, Microsoft EXCEL 7.0a for Windows® 95, Mc Graw Hill, New York, NY, 1996.
3. Anon, Discovering Microsoft Office 2000: Small Business and Standard, Microsoft Corporation, Redmond, WA, 1999.

EXERCISES

9.1   Explore the EXCEL screen and learn the purpose of the icon buttons.

9.2   Explore the EXCEL menu and learn the available options.

9.3   Prepare a parts list for the human-powered pump that your team is developing.

9.4   Suppose that you have a rich grandmother, who has provided an inheritance for you. Unfortunately the money is in a trust, and you cannot place your hands on the loot until you are a mature 50 years of age. If the amount of the inheritance was $ 7,000.00, and the trust yields 7% per annum compounded semi-annually, determine the value of the trust each year until your fiftieth birthday. Also prepare an X-Y graph showing the value of the trust with time. Assume the money is in a Roth account and not subject to either state or federal income taxes.

9.5   Duplicate the work involved in producing the spreadsheet determining the strain in the beam.

9.6   Prepare a bar chart comparing the percentage of men and women in your high school graduating class with the percentage in the freshman engineering class in 1996. Your high school yearbook will provide information on the gender of your graduating class, and the Office of the Dean of Engineering will provide data on the freshman class.

9.7   Suppose that you decide that the chart in Fig. 9.17 covers too large of a range of height of the beam. Prepare a new chart covering the range in height from 0.04 in. to 0.07 in.

9.8   Suppose that you decide that the chart in Fig. 9.17 covers too small of a range of weight. Prepare a new chart covering the range in weight from zero to 10 pounds.

9.9   Suppose that you decide that the chart in Fig. 9.17 covers too small of a range of height of the beam. Prepare a new chart covering the range in height from 0.01 in. to 0.12 in.

9.10   Suppose that you decide that the chart in Fig. 9.17 covers too large of a range of weight. Prepare a new chart covering the range in weight from zero to one pound.

9.11   Evaluate the relation $Y = a + bX$ when $a = 4$ and $b = 2$. Let X vary from zero to 15. Prepare an X – Y graph showing the results.

9.12   Evaluate the relation $Y = a + bX$ when $a = 1$ and $b = 5$. Let X vary from zero to 15. Prepare an X – Y graph showing the results.

9.13   Evaluate the relation $Y = a + bX$ when $a = 2$ and $b = 3$. Let X vary from zero to 15. Prepare an X – Y graph showing the results.

9.14   Evaluate the relation $Y = a + bX^2$ when $a = 2$ and $b = 3$. Let X vary from zero to 6. Prepare an X – Y graph showing the results.

9.15   Evaluate the relation $Y = a + bX^2$ when $a = 4$ and $b = 4$. Let X vary from zero to 6. Prepare an X – Y graph showing the results.

9.16   Evaluate the relation $Y = a + bX^2$ when $a = 2$ and $b = 5$. Let X vary from zero to 6. Prepare an X – Y graph showing the results.

9.17   Evaluate the relation $Y = a + bX^2 + c X^3$ when $a = 0$, $b = -4$ and $c = 2$. Let X vary from zero to 5. Prepare an X – Y graph showing the results.

9.18   Evaluate the relation $Y = a + bX^2 + c X^3$ when $a = 1$, $b = 4$ and $c = -2$. Let X vary from zero to 5. Prepare an X – Y graph showing the results

9.19   Evaluate the relation $Y = a + bX^2 + c X^3$ when $a = 4$, $b = -2$ and $c = 4$. Let X vary from zero to 5. Prepare an X – Y graph showing the results

9.20   Evaluate the relation $Y = ae^X$ when $a = 3$. Let X vary from zero to 6. Prepare an X – Y graph showing the results.

9.21   Evaluate the relation $Y = ae^X$ when $a = 4$. Let X vary from zero to 5. Prepare an X – Y graph showing the results.

9.22   Evaluate the relation $Z = a + b Y + c X$ when $a = 3$, $b = 2$ and $c = 4$. Let $X = 1, 2, 3$ and 4 and $Y = 1, 2, 3,$ and 4. Prepare a graph showing Z as a function of X and Y.

9.23   Evaluate the relation $Z = a + b Y + c X$ when $a = 2$, $b = -3$ and $c = 4$. Let $X = 1, 2, 3$ and 4 and $Y = 1, 2, 3$ and 4. Prepare a graph showing Z as a function of X and Y.

CHAPTER 10

MICROSOFT PowerPoint 2000

10.1 INTRODUCTION

During this course, your team will be responsible for two design reviews. You will be required to make a presentation before the class and the instructor. In the design briefing, you will describe the product specification, the important features that your team is incorporating in the design, and illustrate your concepts with engineering drawings. If you want to make a very good presentation, it is important that the audience closely follow your descriptions of the team's activities. To help in this regard, you should use graphics, which enable transmission of information through two senses—audio and visual. This chapter provides instruction for a graphics presentation program (GPP), which facilitates the preparation of the visual aids needed for your presentations.

Microsoft PowerPoint 2000 is a graphics presentation program that markedly reduces the time needed for you to prepare professional quality visual aids. It has a wide range of capabilities from simple overhead transparencies to sophisticated on-screen electronic displays. GPPs assist you in producing either black and white or colored transparencies, 35 mm slides, full screen projected electronic slides, and valuable support materials such as hard copy printouts, notes and outlines.

Before beginning to learn about PowerPoint, it is important to understand that preparing a presentation is a five-step process.

1. Plan the presentation to provide the information necessary for a design review. In planning, determine the amount of time that you have to speak, the size and layout of the room and the type of equipment available for visual aids.
2. Formulate the presentation by preparing an outline that identifies all of the topics that will be described.
3. Prepare the slides needed in your presentation by using a GPP such as PowerPoint to guide you. As you compose the slides, edit them and make certain that there are no misspellings. After you have completed the slides, review them and make the changes necessary for their enhancement. Can you add graphics or bullets to better capture the attention of the audience? Is the font size correct? Have you used the bold, underline or shadow options to the best advantage? If you are using electronic displays, significant latitude exists for enhancing the visuals and controlling the pace of the presentation.
4. Finally, you must rehearse the presentation. It is imperative that you completely understand the material on your slides. Nothing is more deadly than a speaker reading his or her material from the slide to the audience. Rehearse until the presentation is smooth and polished. You should feel confident about the message to be conveyed and your ability to handle the slides and the projector.

10.2 THE PowerPoint AUTO CONTENT WIZARD

Let's begin by loading PowerPoint. The window that opens, presented in Fig. 10.1, includes the PowerPoint dialog box with options for creating presentations. We have selected the Auto Content Wizard to help us develop a presentation.

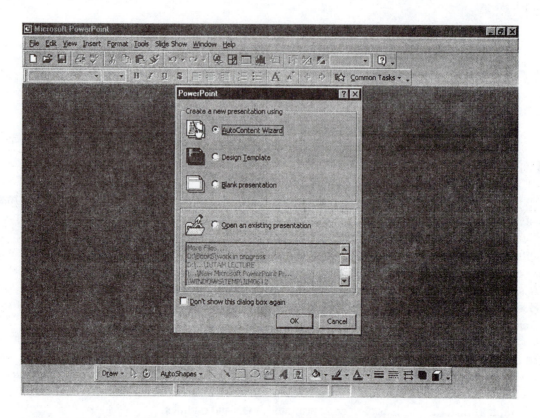

Fig. 10.1 The opening screen in PowerPoint that is used to create a new presentation.

When we click on the OK button at the bottom of the PowerPoint dialog box, a new dialog box for the Auto Content Wizard appears as shown in Fig. 10.2a. This dialog box describes the step-by-step process that PowerPoint follows in structuring the slides for your presentation. Click on the Next > button and the Presentation Type dialog box appears as shown in Fig. 10.2b.

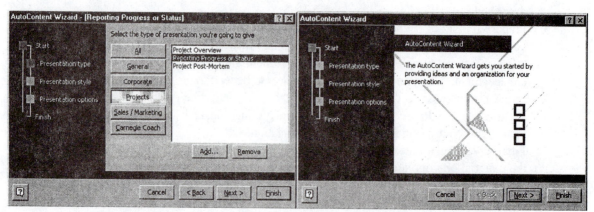

Fig. 10.2a   Fig. 10.2b

Fig. 10.2 The first two dialog boxes associated with the Auto Content Wizard.

From the Presentation Type dialog box, we click on the Projects button and note the three options that are available in PowerPoint:

- Project Overview
- Reporting Progress or Status
- Project Post-Mortem

We select Reporting Progress or Status, and click the Next > button. The Presentation Type dialog box presented in Fig. 10.3a appears that prompts us to select the type of visuals to be used. We have selected black and white overheads since this book is printed using only one color—black. There are much better choices that we will cover later. Click the next > button and the Presentation Options dialog box appears as illustrated in Fig. 10.3b.

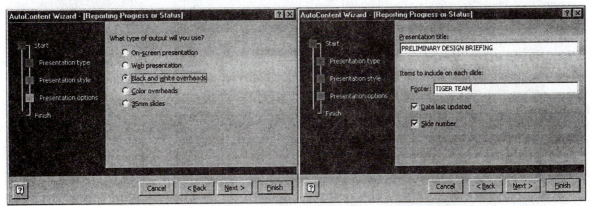

Fig. 10.3a   Fig. 10.3b

Fig. 10.3 Dialog boxes used for selecting presentation style and options.

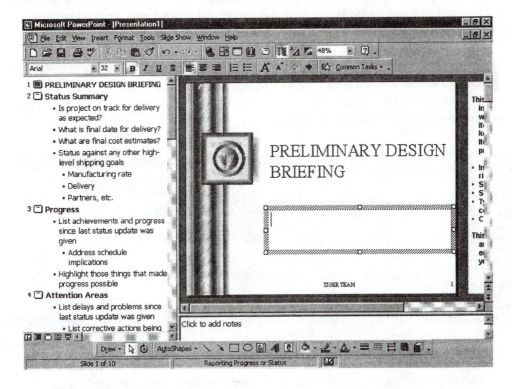

Fig. 10.4 The working screen for preparing a presentation reporting on project progress or status.

We use this dialog box to title the presentation—PRELIMINARY DESIGN BRIEFING—and also to show the team name. We also decide to date the overheads and to number them by clicking in the appropriate toggle squares. We click on the Next > button and the wizard indicates that we have finished the set up for the presentation. Click on the finish button to obtain the screen illustrated in Fig. 10.4.

10.3 THE PowerPoint WINDOW

The screen in Fig. 10.4 is extremely busy. It includes several toolbars and a working area divided into four different regions. Let's first consider the toolbars.

10.3.1 The Toolbars

There are four important rows or bars located above the body of the window shown in Fig. 10.4. The top row, called the title bar, displays the file name that you assign to the presentation when you first save your file. In Fig. 10.4, the title is Presentation 1—the default title assigned by PowerPoint prior to saving the file. At the far right side of the title bar, you will find three buttons: the left one sets the program aside, the middle button changes the size of the display, and the button marked with the × on the right closes the file.

The second row contains Microsoft's standard menu items. Click on one of these items and a pull down menu is displayed. Click and explore the pull down menu listings until you are comfortable with each item. You should develop a good understanding of what can be accomplished with each of the menu items. The Slide Show menu item is particularly interesting. We will describe some of its advantages later in this chapter.

The third row is the standard toolbar that displays 20 buttons and a zoom control symbol ▼ to change the size of the display. Point to any button and note that its purpose is defined in a small box attached to the pointer. The icon buttons on the left side of this toolbar are common to all of the Microsoft applications; however, those on the right side of the bar are unique to PowerPoint. We will introduce several of them later in an example presentation for a design review.

The fourth row down from the top is the formatting bar. Again, most of the icon buttons are common with other Microsoft applications. With the mouse, point to some of the icon buttons on the right side of the formatting bar and the purpose of each icon is displayed. The Common Tasks button at the far right of this toolbar is particularly important since it enables us to:

- Insert a new slide in the presentation.
- To change the slide layout.
- To select a different design template.

At the bottom of the window, we have three more rows. The top one in this group is located at the left side of the screen and contains the five buttons shown in Fig. 10.5.

Fig. 10.5 The buttons used to control the format of the display.

These five buttons control the presentation views. From left to right the buttons enable the following views:

1. The normal view with the workspace divided into four regions as shown in Fig. 10.4.
2. The outline view with more space provided to show the content of the slides.
3. The slide view enlarges the region for the slides and for action item notes.
4. The slide sorter enables us to view all of the slides and to rearrange them.

5. The slide show enlarges each slide to fill the screen and by clicking the mouse you may review the complete presentation.

These buttons will be described in more detail later when we describe the preparation of a presentation for a preliminary design review.

The next to the last row is used for the drawing toolbar. We may add drawings, arrows, clip art, etc. to any slide to enhance its visual impact. The drawing toolbar may be moved to a different location on the screen by clicking on it and dragging it to a preferred location.

The bottom row contains the status bar where the slide number and helpful messages are displayed.

10.3.2 The PowerPoint Workspace

The center area of the window is the workspace where you will prepare the slides. We usually work in the normal view, which is illustrated in Fig. 10.4. In the normal view the workspace is divided into four regions.

The region on the left side of the workspace shows the outline of the presentation. Each slide is numbered and its contents are shown. You can click at any location in the outline and edit the text in any slide. The scroll bar to the right of the outline region enables you to scroll forward or backward in the presentation to locate the slide of interest.

The large region to the right center of the workspace shows a view of the slide in sufficient detail to judge its appearance. Vertical and horizontal scroll bars are also evident to the right and below the slide region. The vertical scroll bar enables you to move either forward or backward in this group of slides. The presentation style and the design template of the slide presented in Fig. 10.4 may not be suitable. However, by selecting more suitable styles and design templates from the Common Tasks options, we may easily change the appearance of the slide. We will discuss changing the style and design templates later in this chapter.

The region to the right of the slide appears only when the first slide in the presentation is active. This region serves as a note pad to list the action items that occur as a result of the discussion during and following the presentation. Following the presentation, these notes are assembled in an action item slide. The technique followed in using the action item pad is shown in Fig. 10.6.

Fig. 10.6 Directions for using the action
item pad in PowerPoint.

This presentation will probably involve audience discussion, which will create action items. Use PowerPoint to keep track of these action items during your presentation

- In Slide Show, click on the right mouse button
- Select "Meeting Minder"
- Select the "Action Items" tab
- Type in action items as they come up
- Click OK to dismiss this box

This will automatically create an Action Item slide at the end of your presentation with your points entered.

The fourth region is located at the bottom of the workspace just below the slide. It is used to add notes as you prepare the slides. These notes are not visible on the slides projected during the presentation. While the pad for the notes in the normal view of the workspace is very small, it becomes much larger if you shift to the outline view.

Finally, the double delta buttons, on the right side of the screen below the vertical scroll bar, are used to move forward or backward in the sequencing of the slides.

10.4 PLANNING AND ORGANIZING WITH PowerPoint 2000

PowerPoint has built-in features to assist you in planning, organizing and timing your presentation. To show these features, we have copied the outline embedded in PowerPoint for "Reporting Progress or Status on Projects." This outline, presented in Table 10.1, provides an excellent guide for you to follow in preparing your design briefings. It includes ten slides that cover the important issues arising in most design projects. Indeed, it may contain many more questions than your team may wish to consider. If this is the case, it is easy to delete them while preparing the presentation.

Table 10.1
PowerPoint Outline for Reporting Progress or Status on Projects

1. Preliminary Design Briefing
2. Status Summary
   - Is the project on track for delivery as expected?
   - What is the final date for delivery?
   - What are the final cost estimates?
   - Status against any other high-level shipping goals.
     - Manufacturing rate
     - Delivery
     - Partners, etc.
3. Progress
   - List the achievements and progress since the last status update.
   - Address the schedule implications.
   - Highlight the key actions that made progress possible.
4. Attention Areas
   - List delays and problems since the last status update was presented.
   - List the corrective actions being taken.
   - Address the schedule implications.
   - Make sure you understand.
     - Issues that are causing the delays or impeding progress
     - Why the problem was not anticipated
     - If customer will want to discuss any issue with upper management
5. Schedule
   - List the top high-level dates.
   - Keep it simple so audience does not get distracted with details.
   - Distribute a more detailed schedule if appropriate.
     - Make sure you are familiar with the details of schedule so you can answer questions
6. Deliveries
   - List the main critical deliverables.
     - Yours to client.
     - Yours to outside services.
     - Outside services to you.
     - Other departments to you.
   - Understand your confidence rating to each deliverable.
     - Indicate this confidence level on slides if appropriate

7. Costs
   - List the new projections of the costs.
   - Include original estimates.
   - Understand the source of differences in these numbers -- be ready for questions.
   - Are there cost overruns?
     — Summarize why
     — List the corrective or preventative action you've taken
     — Set realistic expectations for future expenditures

8. Technology
   - List the technical problems that have been solved.
   - List the outstanding technical issues that need to be solved.
   - Summarize their impact on the project.
   - List any dubious technological dependencies for project.
   - Indicate the source of doubt.
     — Summarize the action being taken or the backup plan

9. Resources
   - Summarize the project resources.
     - Dedicated (full-time) resources.
     - Part-time resources.
   - If project is constrained by the lack of resources, suggest alternatives.
   - Understand that the customers may want to be assured that all possible resources are being used, but in such a way that the costs will be properly managed.

10. Goals for the Next Review
   - Date of next status update.
   - List the goals for next review.
   - Specific items that will be done.
   - Issues that will be resolved.
   - Make sure everyone involved in project understands the action plan.

Many of these topics for the slides are pertinent for your preliminary design briefing, although the answers to many of the questions posed in the outline may be incomplete. In some cases, such as deliveries, the topic may not be applicable. Delete this slide and add another that is more suitable. You may decide to add many slides to completely describe all of the mechanical details involved in the design of the system. However, the preliminary design review, or any presentation for that matter, has a time constraint (usually imposed by the instructor in this class). Your presentation should require 30 minutes or less. Many accomplished speakers believe that 18 minutes is the ideal length for a presentation. This amount of time is sufficient to include substance, but short enough to avoid boring the audience.

10.4.1 The Title Slide

Let's begin with the title slide. Note that you have already provided the title in responding to the questions raised by the Auto Content Wizard. You may want to provide additional information in the subtitle block. To add this material, use the normal view, click on the subtitle block to activate it and you are ready to enter additional material. Keep the title and the subtitle short. You will have the opportunity to expound later in the presentation. Typing information onto the slides is the same as typing in a word processing program. Click in the appropriate area and type. An example of a title slide appropriate for a design briefing is shown in Fig. 10.7. It gives the purpose of the presentation, the title, identifies the team and its members, and gives the date of the presentation.

Examination of Fig. 10.7 shows a marked difference from the view shown in Fig. 10.4. We have changed the design template by clicking on the Common Task button and selecting the Apply Design Template Option. The dialog box used in selecting a design template is illustrated in Fig. 10.8.

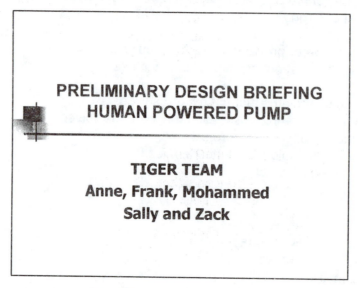

Fig. 10.7 An example of a title slide.

We have selected the Blends template because it has a light background color. There are many choices. Click on one, and examine the preview that appears in the window to the right of the template options. Note that many of the presentation templates have a dark background that impairs the visibility of the slide's content. The dark background is particularly detrimental when the light from the projector is limited or when the room in which the presentation is made cannot be darkened.

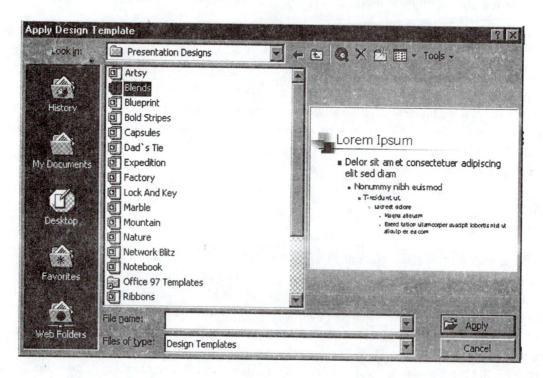

Fig. 10.8 The design template dialog box enables the selection of different presentation templates.

10.4.2 The Status Summary Slide

Let's continue with the preparation for the presentation by considering Slide 2. Shift to the outline view, and respond to the questions posed by PowerPoint for the Status slide. The answers to these questions, adapted for this design class, provide the current status of the project. As you make entries in the outline, examine the slide to check the layout of the text. You may wish to change to the normal view when inspecting the appearance of the slide because it enlarges the image of the slide. A typical Status Summary slide is presented in Fig. 10.9.

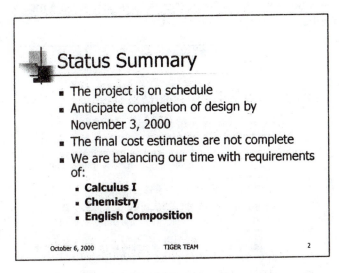

Fig. 10.9 Example of the Status slide.

10.4.3 The Progress-to-Date Slide

Next, consider Slide 3, which describes progress made to date in the development of the human powered pump. Again answer the questions posed by PowerPoint by editing the outline. The answers provided, as depicted in Fig. 10.10, are rather general. The various tasks that are complete or nearly complete are cited. Problems that the team has encountered are identified. Details pertaining to design problems, schedule, cost, etc. are described in slides presented later in the design briefing.

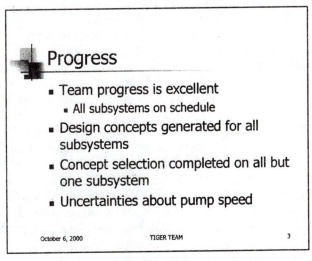

Fig. 10.10 Example of a Progress slide.

10.4.4 The Attention Areas Slide

After you have briefly reviewed the progress made in the development of the various subsystems, it is important to identify the problems that the team has encountered. We briefly identified the problems in the Progress slide. In the Attention Areas (slide 4), we describe the problems in more detail. This is your opportunity to raise any issues (questions) that the design team has discovered to date. Do not hide uncertainties; come forward and seek help. The design review is a formal venue for this purpose. The example shown in Fig. 10.11 indicates that the team is encountering problems in performing the analysis to determine the rotary speed for the centrifugal pump. This problem has resulted in a significant schedule delay. The team understands the criticality of the issue and has scheduled a meeting with the instructor to resolve the questions about the analysis.

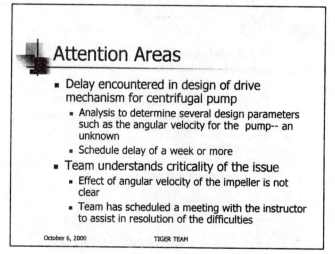

Fig. 10.11 The Attention Areas slide provides an opportunity to seek help in resolving problems.

10.4.5 Goals for the Next Review Slide

The final slide, illustrated in Fig. 10.12, describes your team's plans for action in the near future. Wizard titles this slide as Goals for the Next Review, which is appropriate. Wizard then prompts you to:

- List the goals to be accomplished in the near future.
- Specify detailed task that will be completed.
- Define any issues outstanding that must be resolved.
- Insure the instructor and your peers that all of the team members understand and agree to the action plan.
- Indicate the time of the next design review.

We have followed PowerPoint prompts in preparing the concluding slide shown in Fig. 10.12.

You are now ready to print the output. Click on File and select Print. The print page, presented in Fig. 10.13, is adapted for PowerPoint. It contains a selection box labeled—Print What. You have several choices including: slides, handouts with 2, 3 or 6 slides per page, note pages, and an outline view. You can prepare the overheads for your presentation and the handout for the audience. Print your presentation using these options and examine the output to decide which style is most suitable for the handout intended for the audience. The black and white paper versions of the slides can be copied to give the transparencies necessary for the presentation. If you have your own laser or ink jet printer, the transparencies can be printed directly.

Be certain you save the file before exiting the program.

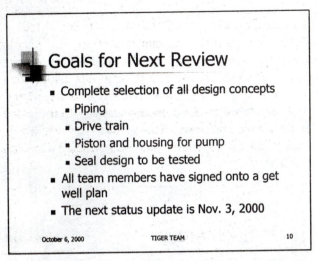

Fig. 10.12 The final slide outlines the actions the team plans to complete before the next briefing.

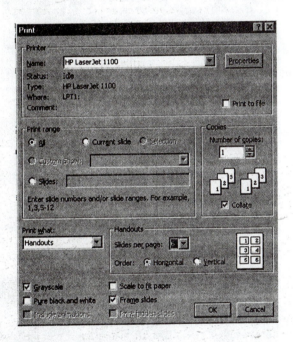

Fig. 10.13 Dialog box for printing the output from PowerPoint.

10.5 ENHANCING THE SLIDES

Suppose you decide to use color slides instead of black and white. It will cost more to use color if you take the job to a commercial copy center, but you may decide that it is worth it. How can you change the black and white slides to color slides? It is easy, and you do not have to retype all of the information into a new version of the presentation.

To change to color slides click on the Gray Scale Preview button on the right side of the PowerPoint toolbar. This button toggles the slide between color and gray scale. To change the design template Click on the Common Tasks button, and select the apply design Template option. The design template page displayed on your screen has over 30 different options that start with Artsy and end with Sumi Painting. With the mouse, select one of the templates and note that it is displayed for your review in a small preview box. Try several templates before selecting the one you believe is the most appropriate.

If you would like to change the color of the background, go to the menu bar and click on Format, and select Slide Color Scheme. The Color Scheme dialog box that is displayed gives you seven options for the colors of the text, titles and background employed. Click on one of the options and preview the results until you obtain the colors that you believe will be the most effective in your presentation. Be careful and stay with relatively light colors because they are more visible to the audience. Dark slides do not project well unless the room is extremely dark. Also, dark rooms are an invitation for your audience to sleep.

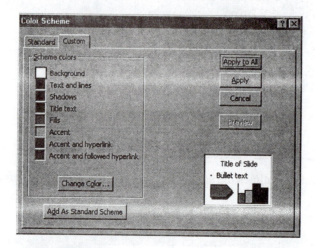

Fig. 10.14 The Color Scheme dialog box showing the colors for each feature on the slides.

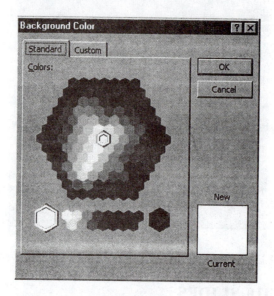

Fig. 10.15 The color hexagon shows all of the available options for customizing your slides.

Suppose that you like some of the colors used, but are unhappy with the color of the text and the bullets. Change them by clicking on Format and select Slide Color Scheme. The Color Scheme page has two tabs—standard and custom. The standard tab illustrates seven choices of color for the titles and bullets. If none of the colors are suitable, click on the custom tab and the Color Scheme dialog box shown in Fig. 10.14 appears on the screen. Prepare to become a painter. You have the opportunity to select the colors that you want to use in the background, shadows, fill, text and title and bullets. To begin changing the colors select a feature on your slide such as the background. Click the Change Color button and the multicolored hexagon, presented in Fig. 10.15, appears with many choices of different shades of the basic colors. Select the color and its shade that you wish to employ for each of the features on the slide. You can have a single design for all of your slides (recommended), or you can have a different color scheme for each slide. Have fun; PowerPoint gives you the means to make great slides.

Some presenters like to use either a header or footer on their slides. A footer is helpful if you give many presentations and want to retain the slides for your files. The information in the footer can include date, occasion for the presentation, location, etc. For example, you may decide to use a footer stating October 6, 2000—Preliminary Design Review, and slide number. How do we incorporate this footer on our PowerPoint slides? Click on View on the Menu bar and select Header and Footer. The Header and Footer dialog box is displayed with two tabs—one for the slide and the other for the notes and handouts. Let's select the slide tab and add a footer to our slides. We click on the Date and Time square, select Fixed, and type the date of the planned presentation. We mark the slide number because it will help us in timing the delivery of the presentation. We click on the footer, and type in a short message to appear in the center region of the footer. We also click on the box indicating that we do not want the footer to appear on the title slide. We delete the footer from the title slide because we want the audience to remain focused on our subject in the very beginning of the presentation. The Header and Footer page has a small preview window that permits us to review the input prior to printing. We illustrate the footer developed with this procedure in Fig. 10.7. The slide was in beautiful color on the monitor, but the figure is in black and white. Our apologies, but the use of color in printing in a low-volume, low-cost textbook is prohibitively expensive.

Still not happy with the impact that your slides will make? What about adding some clip art? Select a slide that you want to enhance with the addition of some photo or sketch that is available in the Microsoft Clip Art files. You don't know what's available? Not to worry; click on Insert on the menu bar, select Picture and then Clip Art. There is a pause while the clip art files are loaded into the active PowerPoint program. A dialog box entitled Insert Clip Art appears on the screen, as shown in Fig. 10.16, with 51 categories of illustrations. Select the Animal category and 965 illustrations of the selections of clip art available under this category appear in the preview window. To assist in the search we typed turtle into the line for search topics and hit the enter key. We then selected turtle (No. 4) from the eight images available, and clicked on the insert button.

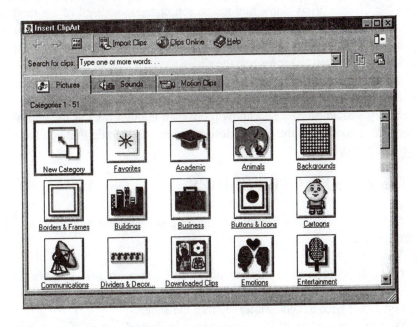

Fig. 10.16 Clip Art dialog box.

Now examine the slide you have selected for enhancement. We have a large picture of a turtle— smack in the middle of the slide. To move the picture and change its size, we drag it by the one of the little squares either on the corners or the middle of the sides to position the insert appropriately on the slide. We show the turtle in the lower right hand area of the Status Summary slide in Fig. 10.17. Note, if you push while dragging the object, it becomes smaller, but when the handles are pulled the object becomes larger. Try it, and you will soon manage to both move and size the turtle.

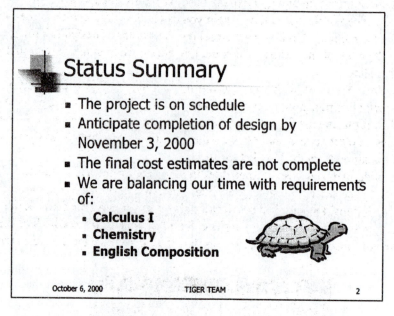

Fig. 10.17 The status slide shown with clip art enhancement.

10.6 REHEARSING

Rehearsing is an essential part of preparing for a design review. Fortunately, PowerPoint has a neat feature that helps in handling the visuals as you rehearse. Go to the first (title) slide and then click on the Slide Show on the menu bar. Select View Show and the screen changes with all of the toolbars disappearing. The slide occupies the entire screen of the monitor and the PowerPoint presentation simulates the screen in the classroom. You can practice your delivery moving from one slide to the next by using the directional arrow keys on the keyboard to forward or reverse the slides.

Timing is an important part of the delivery of a design review. Many presenters make the mistake of delivering material slowly in the beginning and hurrying at the end of the presentation to meet the time constraint. The delivery should be paced to allow more than enough time at the end of the presentation. The last slide is the most important in the overall design review. Be certain that you have sufficient time to present the conclusions properly.

10.7 COMPUTER PROJECTED SLIDES

The ability to project your slides directly from the computer gives you several significant advantages, which are available in PowerPoint.

1. You may store a large number of slides in memory. If questions arise during your presentation, you can quickly locate any slide necessary to improve your response.
2. You do not need a color printer to make the slides.
3. You can use special effects incorporated in PowerPoint to give your presentations some real class.

Let's first explore the special effect known as transitions where PowerPoint controls the manner in which we switch from one slide to the next. Go to the title slide in your presentation, click on Slide Show on the menu bar and select Slide Transition. The Slide Transition dialog box that appears is illustrated in

Fig. 10.18. Click on the arrow located beside this message and a list of many different types of transitions is given. Select one of them and watch the slide switch from one view to another in the small preview window. Select the type of transition that you like the best. We think the dissolve transition is an excellent choice. Next, select the speed of the transition. Again you can review your choice in the preview window. Finally, select the advance method. Until you are a real pro at delivery of design briefings, we suggest you advance the slides on the click of the mouse. Click the OK button when you have completed the set up of the transition of the title slide.

Click on the Slide Sorter View button and the first six slides of your presentation are displayed. The first slide has a small slide transition icon visible near its lower left corner. A new toolbar also appears at the top of the window replacing the formatting toolbar. You are now ready to select the type of transition for the remaining slides. Select slide 2, and click on the transition arrow button on the new toolbar. You can select the type of transition from the long list for slide 2.

Proceed to mark any slide that you wish to transition in the Slide Sorter View, and to select the type of transition that you wish to use when presenting that slide. After you have made your choice, it is confirmed by the presence of a small transition icon just below the slide when you have the Slide Sorter View active as shown in Fig. 10.19. If you employ the same transition for all of your slides, mark all of them (Edit/Select All), and select the type of transition that you believe is the most appropriate for the presentation.

Fig 10.18 Dialog box for Slide Transition in PowerPoint.

PowerPoint also permits you to build your slides progressively for your audience. It is a bit like watching the construction of a building from the foundation to the roof. For slides, we start with the title and add one line at a time to the text in the block below the title. To modify your slide to incorporate this feature, repeat the slide transition procedure described in the preceding paragraph. The new toolbar that is evident just above the window has a box that is titled No Build Effect. Click on the arrow beside that box to reveal a wide choice of techniques for building your slides. Pick an option; fly from the left is very common. When you select an option for building the slide, a second small icon shows under the view of that slide in the slide sorter view as shown in Fig. 10.10. You can proceed to build your slides, one by one, or you can use the same style on all of them. Select all of the slides on the slide sorter view by pressing the Control + A keys.

When you have completed the transitions and the building effects, you are ready to review the dynamic nature of your electronic presentation. Real high tech. Go to slide 1 and click on the Slide Show icon button on the horizontal scroll bar. As you advance from slide to slide with arrow keys on the keyboard or by clicking the mouse button, you will see the transitions and the lines of text for each bullet item being moved onto the screen.

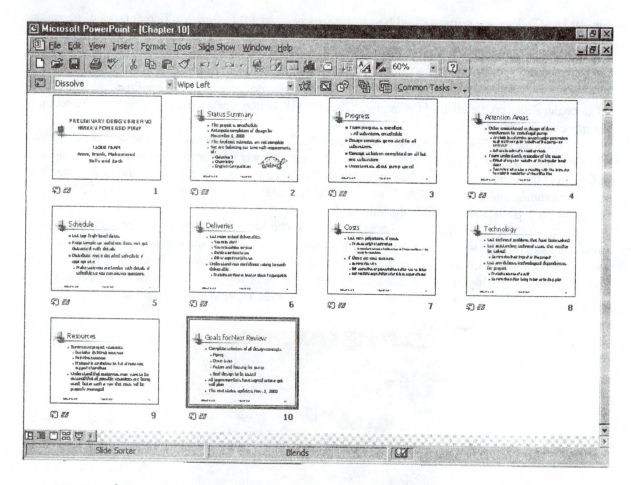

Fig. 10.19 Slide sorter view with transition and building effects icons below the individual slides.

If you have computer facilities available in the classroom, electronic presentations should be mandatory. Creating electronic slides and using them in a professional presentation are skills that are important for you to master. The transitions keep the audience on their toes and the building effects permits you to pace the flow of information. It takes less than an hour to learn how to prepare a very professional, dynamic presentation using PowerPoint. Take this opportunity to develop your computer graphic presentation skills.

10.8 SUMMARY

We have introduced you to PowerPoint, a graphics presentation program that is an extremely useful tool in preparing for design reviews or any other type of presentation. We strongly suggest that you take a few hours to acquire PowerPoint skills. At the very least you will master the skill necessary to produce professional quality slides or overheads. If appropriate equipment is available in the classroom, you will also develop the ability to make presentations with computer-projected slides.

The screen for PowerPoint has been briefly described. As with all Windows based programs produced by Microsoft, many of the buttons on the toolbars have a common purpose. The coverage given here is brief because learning is more rapid if you click, study, and understand. Hopefully, our limited coverage and examples will be helpful.

We suggest that you start learning to use PowerPoint by preparing your preliminary design review using the Auto Content Wizard. The wizard organizes your presentation, arranges the sequencing of your slides, and provides excellent suggestions for content in each slide.

After you have completed the standard presentation with the Auto Content Wizard, you should enhance the slides to improve their appearance. Procedures have been described for changing the design template, the colors of the background and the text, and by adding a footer and clip art. Clip art is particularly impressive and easy to implement.

Finally, we introduced techniques for preparing electronic presentations. An electronic presentation depends upon the availability of a computer-controlled projector for the classroom. If projection equipment is available, the use of a computer in your presentation has significant advantages. Employing transitions, as you move from one slide to the other, maintains the audience interest. Utilizing build effects, where the text is presented one line at a time, has the significant advantage of controlling the flow of information to the audience. The procedures for preparing and rehearsing a computer-aided presentation are described in this chapter.

EXERCISES

10.1 Prepare a title slide suitable for a final design briefing using PowerPoint.

10.2 Prepare a title slide suitable for a preliminary design briefing using PowerPoint.

10.3 Prepare a status slide that provides an overview for a preliminary design briefing using PowerPoint.

10.4 Prepare a progress slide using the slide outline format in PowerPoint. After completing the slide in the outline format, edit the slide, and print a copy of it.

10.5 Prepare a pair of slides describing your progress in the development of one of the subsystems for the human-powered pump.

10.6 Prepare a slide that describes the attention areas facing your team as you design the drive mechanism for the human-powered pump.

10.7 Prepare a slide showing your plans for action in the near future.

10.8 Using clip art, enhance a slide by inserting a caricature of an eagle in its lower right hand corner.

10.9 Prepare a transition from the title slide to the second slide. What type of a transition did you select?

10.10 Prepare a dissolve transition from the status summary slide to the progress slide.

10.11 Prepare a slide using the building effect where single lines of text are added to the slide one by one as you click the mouse button.

10.12 Add sound (chimes) to the transition from one slide to another during the sequencing of your slides.

PART IV

PRODUCT DEVELOPMENT PROCESSES

CHAPTER 11

DEVELOPMENT TEAMS

11.1 INTRODUCTION

The development of quality, leading edge, and high-performance products requires the coordinated efforts of many skillful individuals from different disciplines over an extended period of time. This group of individuals is organized within the corporation to act in an integrated manner to successfully complete the development. The organizational structure employed varies from one corporation to another, and it also depends to a large degree on the size of the company that engaged in the development. For relatively new firms in the entrepreneurial stage, team structure is usually not an issue. These firms are small with a single product and only ten to twenty individuals are involved in the development process. Everyone on the payroll is deeply committed to this product, and communication is usually accomplished around the lunch table. On the other hand for very large corporations, where the number of employees may exceed 100,000, the company is often organized into divisions or operating groups. These large companies have many products or services that are offered by each division, and in some instances one division competes to some degree with another. For example, General Motors with its five automotive divisions produces many different models of each of its lines that compete for the same market segment.

In a large corporation, the total number of new products that are introduced each year may exceed 100. Communication is often very difficult because hundreds or even thousands of miles sometimes separate personnel involved in the development of a given product. Sometimes different divisions are in different countries, and business is conducted in different languages. Clearly, in a large firm, organization of the product development team and the physical location of its team members becomes extremely important.

In the past decade, there have been many examples [1] demonstrating that a cross disciplinary team provides a very effective organizational structure for product development. The idea of forming a cross disciplinary team to design a new product appears simple until we examine the existing organizational structure of all but the smaller of our product oriented corporations. Larger corporations are usually organized along functional lines. The organization chart presented in Fig. 11.1 shows many of the functional departments that provide the personnel needed on a typical product development team. Each department has a functional manager. Also, several departments are often grouped to form a division, and of course, each division has a director. Titles for the division leaders vary from firm to firm, but Director or Vice President is commonly employed.

The functional organization with its leadership and its ongoing operations presents a management problem when forming product development teams. As a team member, do you report to the team leader (product manager), or to the functional manager? When important decisions must be made, are they made by the product manager, by one or more of the functional managers, or by the division directors? Clearly, the establishment of a product development team within a company that is organized along functional lines creates several operational problems for management. Functional organizations may also create problems for working engineers attempting to serve several different managers. In the next section, we address these problems, and describe a popular team structure employed in many corporations, which are organized along functional lines.

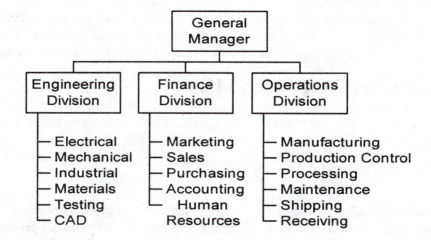

Fig. 11.1 Typical organization chart showing the disciplinary functions, three divisions and three levels of management.

11.2 A BALANCED TEAM STRUCTURE IN A FUNCTIONAL ORGANIZATION

There are several organizational arrangements that are frequently used in forming teams within companies that are organized along functional lines. They include:

- Functional teaming.
- Modified functional teaming.
- Balanced team structure.
- Independent team organization.

Each of these techniques for forming product development teams in a functional organization has advantages and disadvantages. A complete discussion of these four different team structures is given in reference [2]. To introduce you to the concept of forming teaming within a functional organization, we will briefly describe the balanced team structure.

The balanced team structure is represent by the organization chart presented in Fig. 11.2. The team is formed with members drawn from the functional departments; however, in most situations, the team members are located in close proximity in a separate room or a building wing reserved for the team. The team members are usually dedicated to a single project, and their responsibilities are coordinated and focused. The team members retain their reporting relationship to the functional manager, but the contact is less frequent than in functional teaming arrangements, and the relation is less intense. Members of a balanced team often consider the product manager to be more important than their functional manager. The product manager is frequently a senior technical administrator with more experience, status and rank than the typical functional manager. The status of the product manager is dependent on the product development costs and the size of the development team. The product manager reports to the general manager, and usually is equal in status to the division directors in the organization. The senior program manager with notable status and clout leads the team, and insures that resources will be available as required to meet schedules. The team members recognize the freedom from functional constraints that the balanced team structure implies. They assume ownership of the product specifications and commit fully to the success of the development activities. Communication is accomplished though the use of division and department coordinators and daily meetings. The team makes most decisions and only the higher-level decisions require the approval of the appropriate division directors.

The primary disadvantage of the balanced team structure is the dedication of the team members to a single project. Many talented team members are needed by the functional managers to support design activities across several product lines. With some of the functional resources dedicated to a single product development, other products may not receive adequate attention. The balanced team structure enhances the development capability of a given product at the cost of reducing the technical capability that can be applied across the company's entire product line. A second disadvantage is that some team members from a given department or division may not have sufficient expertise to perform adequately. With loose and distant functional supervision and review, this fact may not be apparent. The new design may not be at the leading edge of technology.

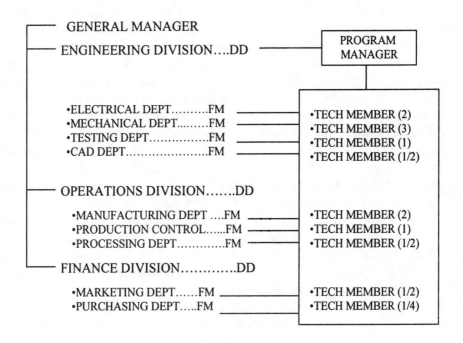

DD ....DIVISION DIRECTOR
FM.....FUNCTIONAL MANAGER
1/2......PART TIME (50%)

Fig. 11.2 Organization with a balanced team structure.

11.3 DEVELOPMENT PLANNING

As the development team becomes more independent and moves away from the functional departments, planning becomes more important. The team planning documents are often prepared by a small group of senior team members with talents in business, marketing, finance, engineering design and manufacturing. The planning documents incorporate detailed descriptions of the product, market analysis, business strategy for capturing market share, product performance specifications, development schedules and budgets, team staffing requirements, material selections, design strategies, technology prerequisites, manufacturing processes, tooling requirements, inspection procedures, test specifications, facility availability and distribution capabilities. An element of the plan covers techniques and personnel that are to be employed to insure quality and product reliability. Finally, the plan includes a section specifying the deliverables and another section on techniques to measure performance of the product and the productivity of the team.

The development planning documents are extremely important since they outline strategies for achieving the goals and objectives in developing a new product. Some of the documents may be used by management to monitor progress and to judge the team performance during periodic design reviews. The plan is essentially a well-defined contract between the team and the division management involved in the product development. The team members are expected to sign-on; to commit to the level of effort required to achieve the goals, meet the schedules and to produce the deliverables. Management is expected to fund the program, to provide adequate staffing and to provide the necessary facilities, tooling and equipment available in a timely manner.

11.4 STAFFING

Cross-disciplinary teams are necessary in staffing a product development team regardless of the structure employed by the organization. Usually, at least five disciplines or functions should be represented on the team including, marketing, finance, design engineering, operations or manufacturing, and quality control. The level of participation of each team member depends on the product being developed and the stage of the development. In some cases, team members work on more than one product and have responsibilities on two or more different projects. In other instances, several engineers from the same discipline are needed to complete long and complex tasks to meet the schedule for a single product. These members are usually dedicated to a single project.

Staffing on a product development team often changes with time over the duration of the development period. Initially the team is small and staffed with senior personnel with demonstrated skills and talent. This small cadre may stay with the project for its duration. As the development proceeds, additional design engineers are required as technical strategies are converted into design concepts, then to design proposals and finally into detailed engineering drawings. Manufacturing engineers are needed in larger numbers when the early prototypes are being produced. After release of the design by engineering, the number of designers is reduced, and the emphasis is shifted to production where staffing from operations (production control, quality control, purchasing and plant maintenance) increases sharply. Clearly, the staffing of the development team is dynamic. Usually only a relatively small proportion of the total team is committed to the project on a full time basis from its initiation to its completion. In some companies, individuals are committed to the development team for the duration of the project, but at varying levels of participation. This approach enhances ownership of the project by the individual team members.

11.5 TEAM LEADERS

The experience and leadership skills required of the product manager are strongly dependent on the team structure employed as well as the size and cost of the product development. Some products can be developed with a relatively small commitment of staff (say less than 10 members). In these circumstances, the team established is often structured along functional lines, and the product manager is usually one of the functional managers. The functional managers are experts in their respective disciplines, skilled in managing personnel and knowledgeable (but not expert) in the other disciplines. They are also familiar with the company's entire product line.

Other product developments require a larger commitment of personnel (say a team numbering 10 to 50), and the balanced team structure shown in Fig. 11.2 is probably the most suitable organization. The product manager is a seasoned staff member well-known for his or her expertise and respected for performing at a level that exceeds expectations. The appointment as a product manager (equal in rank and status to the functional department managers) often represents a promotion for the individual involved and a first experience in managing a technical group. The functional managers and the divisional directors provide close support through periodic reviews that monitor progress and reveal problems early in the development process.

As the size of development team becomes even larger, 100 or more members, team leadership assumes major importance. The product manager in this situation is a senior member of the company's executive management with a well-established record of successful accomplishments. The program manager is senior to the managers of the functional departments, and is at least equal in status to the division directors. The authority and responsibility given the product manager is significant, and his or her career depends strongly on the success of the product in the market and the productivity of the development team.

11.6 TEAM MEMBER RESPONSIBILITIES

Members of a product development team have two different sets of responsibilities—one set to his or her disciplinary function and the other to the team. A team member, from say the electrical engineering department, provides the technical expertise in his or her specialty and ensures that the product is correctly designed with leading edge, state-of-the-art technology. The team member acts to bring all of the important functional issues that affect product performance to the attention of the team. The team member also represents the functional department to ensure that the team in the design of the product maintains the functional principles, goals, and objectives.

The most important team activity for the individual member is to share responsibility. The team member shares responsibilities with the others to increase the team's effectiveness and to insure its success. Next, the team member must recognize and understand all of the product features, and fully participate in the methods and techniques employed to meet the design objectives. Individual team members must assess team progress and participate in improving team performance. The team member must cooperate in establishing all of the reporting relationships required to maintain the communication among the team and functional departments. In many of the team structures, an individual may report to three managers—the Division Director, the Department Manager and the Program Manager. If a Functional Coordinator is assigned to the team representing a certain discipline, it is necessary to keep him or her informed of your progress and plans. The matrix organization structure, inherent in forming development teams, creates complex reporting relations. Flexibility and cooperation on the part of the individual team members are required for the team structure to be effective.

Finally, the team members must be able to communicate in a clear concise manner in all three modes: **writing, speaking and graphics**.

11.7 TEAM MEMBER TRAITS

When a group of individuals work together on a team to achieve a common goal, they can be extremely effective [3]. The team interaction promotes productivity for several different reasons. First, meeting together is synergistic in that one's ideas freely expressed stimulate additional ideas by other members on the team. The net result is many more original ideas than would have been possible by the same group of individuals working independently. Another advantage is in the breadth of knowledge available in the team. The team is cross-disciplinary and the very wide range of skills necessary for the product development process should be included within the team. Each member of the team is different with some combination of strengths and weaknesses. Acting together the team can build on the strengths of each member and compensate for any weaknesses. The grouping of individuals provides social benefits that are also very important to the individuals and to the corporation. There is a bonding that occurs over time, and a support system develops which team members appreciate. Feelings of trust and mutual understanding, so important to everyone involved, develop over the course of the project. The team develops a sense of autonomy valued by each of its members. The team develops solutions, implements them and then monitors the results with little or no direction from management. Empowered teams assume ownership of the project. They commit to the product development process and consistently perform beyond expectations.

While cross disciplinary teams have been employed in many different companies for a decade or more, little has been done by our educational system to develop team skills. On the contrary, both the secondary school and college systems tend to encourage competitive attitudes and independence. Students compete for the few A grades that might be given in a course, and are discouraged from cooperating on the assignments. The system trains students to work independently, and team-building skills are not addressed. The industrial workplace is much different. Team members compete, but not within the team. They compete with a similar development team from a rival corporation. Team members bond, cooperate and consistently help each other by sharing assignments. Competition between team members is discouraged. Recognition and rewards go to the team as a whole much more often than to select individuals.

There is a set of characteristics that describe a good team member, and another set that depicts an individual who can destroy the efforts of a team.

The characteristics of a good team member are:

1. Treat every team member with respect, trust their judgment and value their friendship.
2. Maintain an inquiring attitude free of predetermined bias so that team members will all participate and share knowledge and opinions in a free and creative manner. Listen carefully to the other team members.
3. Pose questions to those team members who are hesitant to encourage them to share their knowledge and experience more fully. Share your experiences and opinions in a casual, easy-going manner. Help other members to relax and enjoy the interactive process involved in team cooperation. Participate but do not dominate.
4. Observe the body language of the other team members because it may indicate lack of interest, defensive-attitudes, hostility, etc. Act with the team leader to defuse hostility and to stimulate interest. Disagree if it is important, but with good reason and in good taste.
5. Emotional responses will occur when issues are elevated to a personal level. It is important to accept an emotional response even when you are opposed. It is also important to control your emotions and to think and speak objectively. You should be self confident in your discussions, but not dogmatic.
6. Seating arrangements once established tend to become permanent. It is productive to break these patterns, change the seating pattern and to mix team members. The result, over time, is to enhance bonding and to stimulate new ideas generated by different combinations of members.
7. Allocate time for self-assessment so the team can determine if it is performing up to expectations. Make suggestions during these assessment periods to build team skills.
8. Be comfortable with your disciplinary skill level and continue to study to improve your skills. Communicate effectively by speaking clearly in good English, writing well and in using modern methods in graphics.

The characteristics of a destructive team member are:

1. No member of the team should ever participate in a conversation that is derogatory about a person on the team or in the corporation. If you cannot make complimentary remarks about a person, keep your thoughts to yourself. Respect, trust and friendship are vital elements in the founding of a successful product development team. Derogatory remarks destroy the foundation necessary for respect, trust and friendship.

2. Arguments among team members are to be avoided. You are encouraged to introduce a different opinion, and participate in a discussion with a different viewpoint, but the discussion should never degenerate into an argument. Every member on the team is responsible for quickly resolving arguments that arise among the team members.

3. All members of the team are responsible for their attendance at the meeting and their timely arrival. If a member is absent, the leader should know the reason beforehand and explain it to the team. We must be certain that all members appreciate and respect the reason for everyone's role on the team and the importance of every member attending every meeting.

4. Team progress is hindered when one member dominates the meeting. These people are often overly critical, intimidating and stimulate confrontations. If you have these characteristics, work very hard to suppress them.

5. The team leader must be extremely careful about intervention. If the team is moving and working effectively, the leader should be quiet because intervention in this instance is counter productive. If the team is having difficulty, intervention may be necessary depending on the problem. If a team member becomes hostile, intervention should be quick, and the disagreement producing the hostility should be dealt with immediately. If the team is seeking consensus, the process may require time and the leader should wait 5 to 10 minutes before intervening.

6. No member of the team can be opinionated including the leader. He or she is not a judge determining the correct solution. The leader seeks to facilitate so that the team can reach a consensus on the correct solution. It is only when the team accepts the solution that implementation can begin.

As we compare our own traits with those listed above, we will note characteristics from both the good and bad lists. To be effective team members, we must work to enhance the favorable traits and to suppress those that are destructive to the efforts of the team.

11.8 EVOLUTION OF A DEVELOPMENT TEAM

If a number of workers or students are assigned to a new team, it is possible to observe several behavioral phases as the team melds and matures [4]. In the very beginning (Phase 1), the team members usually exhibit both excitement and concern. They are excited about working on a new project with a new group of talented people. It should be an opportunity to learn, make new friends, advance in position and/or stature and have fun in the process. At the same time, the team members exhibit signs of concern. They are worried about meeting and understanding the other team members. The tasks assigned to the team are probably extremely vague at this stage of the development, and vagueness leads to uncertainty. The skill levels and areas of expertise of fellow team members have yet to be determined. The personality of the team leader is unclear. There is worry over the new reporting relations that the new team membership implies. In this orientation phase, the team members are searching for their role in the team and evaluating their possibilities for success or failure. It is a tense time.

Dissatisfaction is the second phase in the evolution of the team. We have met our fellow team members, and the news is not good. We recognize the differences in personalities and in work habits. A few of the team members do not understand how to tell time and never arrive on time when attending team meetings. Our schedules make it very difficult for the team to meet as a whole. Some team members lack minimal social graces. Some members are inexperienced and unable to cope with their assignments. We have met the team leader and he or she is very difficult (stressed, demanding, exacting, impatient and abrupt). The schedule calls for us to leap over mountains on a daily basis. The budget for the project is totally inadequate. Surely we are in a lose—lose situation. Our thoughts are dominated by schemes to transfer to a better team. During this phase, the progress of the team toward scheduled milestones is very slow.

With time progress is made, and the team members start learning to work together. This is the resolution phase. We agree to attend meetings and arrive on time. Personality conflicts are smoothed over. We may never be true friends, but we manage to get along and show mutual respect. We have come to terms with the schedule, and have committed to the extra effort that it requires. The manager has mellowed, or we have learned how to handle his impatient demands. Solutions for reducing development costs have been found. There appears to be a good possibility of meeting the product development plan. The team is beginning to perform well, and more members are committed.

The team has melded into an efficient unit. Conflicts rarely arise. Lasting friendships begin to be formed. The strengths of each team member are fully utilized. The manager recognizes the talents of the team. Tasks are completed ahead of schedule and under budget. Upper level managers recognize the accomplishments of the team. You now find yourself in a win—win situation. This is the production phase of the development team.

The project has been completed and the ribbons have been cut. The team prepares to disband. In this termination phase, we reflect on our experiences, good and bad, over the duration of the project. We examine our individual performance and evaluate various factors that improved the effectiveness of the team. We meet with the manager for a performance evaluation. The team meets to share a sense of accomplishment in achieving the goals and objectives of the development.

11.9 A TEAM CONTRACT

Teams are effective because they meet together to focus their wide range of disciplinary skills and natural talents toward the solution of a well-formulated problem. The team meeting is the format for the synergistic efforts of the team. Unfortunately, not all teams are effective because some members are so disruptive they destroy the cohesiveness of the team. Bonding of team members is vital if the team is to successfully solve the multitude of problems that arise in the product development process.

Teams fail because of three main reasons. First, they deviate from the goals and objectives of the development, and cannot meet the milestones on the development schedule. Second, the team members become alienated and the bonding, trust and understanding, critical to the success of a team, never develops. Third, when things begin to go wrong and adversity occurs, "finger pointing" starts and fixing blame on someone replaces creative actions.

Clearly, many problems may arise to impair the progress of a development team. We recommend the team establish a set of simple rules for working together. We have provided some suggestions for your team to consider. Team members should meet and modify or add more rules as necessary to govern your team behavior.

- The information discussed by our team members will remain confidential.
- We will acknowledge problems and deal with them.
- We will be supportive rather than judgmental.
- We will respect differences.
- We will provide responses directly and openly in a timely fashion.
- We will provide information that is specific and focused on the task and not on personalities.
- We will be open, but will respect the right of privacy.
- We will not discount the ideas of others.
- We are each responsible for the success of the team experience.
- The team has all of the resources needed to solve the problems that arise during the development process.
- Every member of the team will contribute to its success.
- We will try to know and understand our fellow team members so as to identify ways for enhancing everyone's professional development.

- We recognize the importance of the team meeting to the success of the group, and accordingly we agree to:

  - Always attend the scheduled team meetings.
  - When attendance is impossible, notify the team leader and as many other members as possible in advance.
  - Use meeting time wisely
  - Start on time.
  - Limit our breaks and return from them on time.
  - Keep focused on our goals.
  - Avoid side issues, personality conflicts and hidden agenda.
  - Share responsibility for briefing members when they miss a meeting.
  - Avoid making phone calls or engaging in side conversations that interrupt the team.

Writing a team contract has been included in this chapter in Exercise 11.19. We recommend that each team prepare a contract, and each team member sign and date it.

11.10 EFFECTIVE TEAM MEETINGS

Leadership is important in ensuring that the team remains focused on the overall goals and objectives of the project and on the agenda at each meeting. Creating the correct team environment or atmosphere is essential so that the team members cooperate, share ideas and support each other to achieve solutions [5]. It is important for the team leader to assess the performance of the team as a whole at each meeting. When the team fails to make progress, it is critical for the team leader or possibly an individual member to make changes as necessary to enhance the team's effectiveness. There are several measures of team effectiveness that can be exercised by everyone on the team during the course of a typical team meeting. Corrective actions to improve team performance are not difficult to implement. We will identify some of the difficulties encountered during team meetings and suggest actions to be taken by members for enhancing its productivity.

One common problem, impairing team productivity, is when the goals and objectives of the meeting are not clear or they appear to be changing. While in many cases team meetings are planned well in advance and are supposed to follow an agenda, we observe that new topics are introduced and that the meeting drifts from one item to another. This team behavior indicates that either some of the team members do not understand the goals, or they do not accept them. Instead, they are marching to their own drummer and trying to take the team with them. Unless corrective actions are taken to focus the team on the original goals, this team meeting will fail and time and effort will be lost. A team that is focused stays on the agenda, the team members accept the goals and their discussions deal with all of the issues important in generating a wide set of solutions to the problem being considered. The discussion continues until the team members reach a consensus. This meeting is successful because a problem has been solved and all of the team members accept the solution. The team can move forward and organize its talents to address the next problem.

Leadership in the team is an essential element for its effectiveness. The most effective teams are usually democratic with a large amount of shared leadership. While there is a recognized leader, with a considerable degree of responsibility and authority, the team leader is supported by each member of the team. The team member with the appropriate expertise (relative to the problem being considered) will usually lead the discussion and effectively act as leader of the team during this period. However, in some instances, the team leader maintains strict control of the meeting and does not share the leadership role with any member of the team. Some team members resent this style of leadership and may not participate as fully as possible. The result is that complete utilization of the resources of the team does not occur. The "tightly controlled" team leader will almost always insist on sitting at the head of the table.

Attitude of the team members is an important element in the success of the team. Are the team members committed? This commitment is evidenced in their attitude. If they come to the meeting table exhibiting interest and willingness to participate, they are committed and will make a positive contribution. However, if they are bored and indifferent, they will not become engaged in a meaningful way. In fact, if they are sufficiently disinterested, they may initiate side conversations or arguments and destroy the atmosphere of trust necessary for an effective meeting. It is important to evaluate each of the team members, and to secure their commitment to the goals and objectives of the project and to the agenda for each meeting.

As the meeting progresses, it is important to assess the discussions that are occurring. For a meeting to be successful, the discussion should include everyone on the team. The discussion should stay focused on the agenda items with few if any deviations to unrelated topics. The team members must listen carefully to one another. All of the ideas presented are given a serious hearing. No one is intimidated or is made to appear foolish for making suggestions that may appear to be too radical. There is an informal and relaxed attitude exhibited by the participants. Teams tend to accomplish less when the discussion is dominated by only a few of the participants. They talk well, but they do not bother to listen to others. They may introduce a topic that is not on the agenda, and ignore suggestions for getting back on track. These members tend to intimidate others in their efforts to control and dominate the team. The result is disastrous because they essentially eliminate the contributions of the other team members.

Teams do not operate with total agreement on every issue. There must be accommodation for disagreements and for criticism. All of the ideas or suggestions made by every team member at any meeting will not be outstanding. Indeed, some of them may be ridiculous; criticism of these ideas must occur. However, the criticism should be frank and without hostility. Personal attacks must not be a part of the critique of an idea. If the criticism is to improve an idea or to eliminate a false concept, then it is of benefit to the progress of the team. The criticism should be phrased so that the team member advancing the flawed idea is not embarrassed. When disagreements occur, they should not be suppressed. Suppressed disagreements breed hostility and distrust. It is much better for the team to deal with the disagreements when they arise. The root causes of the disagreement should be ascertained, and the team should take the actions necessary to resolve them. On some occasions voting is a mechanism used to resolve conflict. This procedure must be used with care particularly if the vote indicates the team is split almost evenly in their opinion. A better practice is to discuss the issues involved and attempt to reach a consensus. For a team to reach a consensus, may require more time than a simple vote, but the results are worth the effort. Consensus implies that all members of the team are in general agreement and willing to accept the decision of the team. When the team votes, a simple majority is sufficient to resolve an issue. However, the minority members may become resentful if they are always on the short end of the voting. In a very short time, they will not accept the outcome of the voting, and they will not commit to the actions necessary to implement the decision.

The team meetings must be open. The agenda should be available in advance of the meetings and subject to change during the new business agenda item. A sample agenda is shown in Table 11.1. Team members should believe that they are important and have the authority to bring new issues before the team. They should feel free to discuss procedures used in the team's operation. Hidden agenda items detract from the harmony and trust developed through open and fair operation of the team. Secret meetings of an inside group should be carefully avoided. When the fact leaks that secret meetings are being held and that issues are prejudged by a select few, the effectiveness of the team is destroyed.

Table 11.1
Sample Agenda
Weekly Meeting of the Tiger Team
September 19, 2000 7:59 am

1. Weekly status report   Team Leader
2. Review of outstanding action items
   - Action item #12   Member responsible
   - Action item #13   Member responsible
   - Action item #N   Member responsible
3. Report on progress
   - Subsystem –Power   Member responsible
   - Subsystem –Electrical   Member responsible
   - Subsystem – Mechanical   Member responsible
   - Subsystem –Control   Member responsible
   - Market study results   Marketing representative
4. Identify new problems   All members participate
5. Assignment of action items   Team leader and volunteers
6. New business   All members participate
7. Summary   Team leader
8. Adjourn at 8:59 am

At least once during a meeting it is useful to evaluate the progress of the team. Are we still following the agenda? Is someone dominating the meeting? Is the discussion to the point and free of hostility? Is everyone properly prepared to address his or her agenda items? Are the team members attentive, or are they bored and indifferent? Is the team leader leading or is he or she pushing? If a problem in the operation of the team is identified during the pause for self-appraisal, it should be resolved immediately through open discussion.

Table 11.2
Action Items
Tiger Team—Meeting of September 19, 2000

Decisions made	Follow-up required	Responsible member	Date complete
1.	1.	1.	1.
2.	2.	2.	2.
3.	3.	3.	3.
4.	4.	4.	4.
Action at next meeting	**Preparation required**	**Responsible member**	**Date complete**
1.	1.	1.	1.
2.	2.	2.	2.
3.	3.	3.	3.
4.	4.	4.	4.

Finally, the team must act on the issues that are resolved and the problems that are solved. To discuss an issue and to reach a consensus is part of the process, but not closure. Implementation is the closure, and implementation requires a plan for action. When decisions are made, team members are assigned action items. These action items are tasks to be performed by the responsible individual;

only when the tasks are completed is the problem considered solved. It is important that the action items are clearly defined, and the role of each team member in completing their respective tasks is evident. A realistic date should be set for the completion of each action item. The individual responsible should be clearly identified; he or she must accept the assignment without objection or qualification. A checking system must be employed to follow up on each action item to insure timely completion. If there is a delay, the schedule for the entire product development cycle may be at risk. It is important to deal with delays immediately and for the team to participate in the development of plans to eliminate the cause of the delay. We have included a form in Table 11.2 to facilitate the assignment of action items and the follow-up on each item that the team may employ.

11.11 PREPARING FOR MEETINGS

Effective team meetings do not just happen. It is necessary to prepare for the meeting and to execute post-meeting activities to insure success. The preparation usually involves selecting, arranging and equipping the meeting room, scheduling the meeting so that the necessary personnel are in attendance, preparing and distributing an agenda in advance of the meeting and conducting the meeting following a set of rules accepted by the team. Equally important, are actions taken by the team members following the meeting to implement the decisions that have been reached.

The meeting room and its equipment are also important to the progress made by the team. The room should be sized to accommodate the team and any visitors that have been invited. Rooms that are too large permit the members to scatter and the distance between some members becomes too long for effective and easy communication. The chairs should be comfortable, but not so soft as to promote napping. The seating arrangement should be around a table so that everyone can observe each other's face and body language. It is very difficult to engage anyone in a meaningful conversation if they have their back to you. Water, coffee, tea or soft drinks should be available if the meeting duration exceeds an hour. Smoking is strictly prohibited. The equipment that will be needed for presentations must be available, and its operation should be checked prior to the start of the meeting. Flip charts and white boards that are useful for recording the key ideas generated during the meeting are essential. Tape, pens, markers and other supplies necessary for preparing and displaying charts must also be available. A computer with projection capability is of growing importance in a well-equipped meeting room. The availability of a computer during the meeting permits one to draw from a large database and to modify the presentation in real time. Also results of analyses or experiments can be displayed in graphical form for the entire team to review. Finally, adjust the room temperature until it is comfortable. Check to ascertain that the light intensity can be controlled from low for projecting overheads to high for round table discussions.

Prior to the scheduled meeting, it is important to make careful preparations. Minutes from the previous meeting must be distributed with sufficient time for review before the next scheduled meeting. A detailed agenda, similar to the one shown in Table 11.1, is distributed which informs the team members of the topics and/or problems that will be addressed in the next meeting. Individual team members responsible for specific agenda items are identified. The details are clear to all and responsibility shared by individual members is defined. In some instances, information will be needed from corporate employees, suppliers or visitors that are not members of the team. In these cases, arrangements must be made to invite these people to the meeting so that they can provide the necessary expertise and respond to questions from all of the disciplines represented on the team.

In conducting the meeting, it is important to start on time. It is very annoying to the majority of the members to wait five or ten minutes for a straggler or two. The team leader must make certain that everyone involved understands that 8:00 am means 8:00 am and not 8:08 or 8:12 am. The objectives of the meeting should be displayed and the time scheduled for each item should be estimated. A team member should be assigned as the timekeeper, and another should act as a secretary to record notes and to prepare the minutes of the meeting. If team meetings are frequent and they are held over a long period of time (several months), the duties of the timekeeper and the secretary should be shared with others on the team. The meeting should follow a set of rules that govern the behavior of the individual members. We have assigned the preparation of a list of rules, which should be followed in conducting an effective meeting as Exercise 11.17. As the meeting draws to a conclusion, the team leader should take a few minutes to summarize the outcome of the discussions. This summary gives an ideal opportunity to insure that assignments are understood, responsibility accepted and that completion dates are established. The meeting must be completed on time and everyone's schedule should be respected. If the meeting is not periodic (e.g. every Tuesday at 8:00 am), then it is important that the time and place for the next meeting be scheduled.

After the meeting, it is productive for the team leader to check to make certain that the minutes have been prepared, that they are complete and that they have been distributed. If any member was not able to attend the meeting, the team leader should brief that person on the happenings. Finally, the team leader should follow up on each action item to ensure that progress is being made, and that new or unanticipated problems have not developed. It is clear that effective meetings do not happen by accident. Many members of the team work intelligently before, during and after the meeting to make certain that the goals and objectives are clearly defined, that the issues and problems are thoroughly discussed and that the solutions developed result in assignments that are executed with dispatch.

11.12 POSITIVE AND NEGATIVE TEAM BEHAVIOR

All the members on a team and particularly the team leader must behave in a positive manner for the team to function effectively. Negative behavior, that occurs frequently, is detrimental and inhibits the progress that can be made by the team and delays the product development cycle. Group behavior is often divided into three categories including [6]:

- Leading
- Supporting
- Hindering

The leading and supporting roles are consistent with team progress and positive contributions to the development project. However, the hindering role is destructive as it impedes progress and causes conflict and concern to many members of the team.

11.12.1 Leading Roles

Any team member can play a leading role at any time in a team meeting. Of course, we expect the team leader to provide direction, but leadership is shared in a high-performance development team. Leading tasks include:

1. **Initiate:** Suggest ideas or concepts for the design. Propose objectives or goals. Define a problem and give an approach for its solution. Raise issues that are important to the progress of the team. Recommend procedures for completing tasks.
2. **Provide:** Give relevant information pertaining to issues before the team. Offer suggestions, ideas and theories. Furnish facts and background knowledge. Present opinions with supporting material.
3. **Seek:** Ask for suggestions and ideas. Solicit options. Request facts relative to a team problem. Look for information leading to resolution of team concerns.
4. **Clarify:** Indicate alternatives and issues for the team. Provide examples to support another's statement. Interpret ideas and suggestions. Define issues and terms. Offer illustrative conclusions.

5. **Test:** Check to ascertain the degree of agreement within the team. Inquire to determine if the team is prepared to reach a consensus.
6. **Summarize:** Offer conclusions or decisions at the termination of a team discussion. Gather together related ideas offered by several team members. Restate suggestions after team discussions.

11.12.2 Supporting Roles

Many team members are not always comfortable with the leading roles. Instead they can make significant contributions in a supporting role. Team members providing support engage in the following activities:

1. **Monitor:** Assist in maintaining open communication among team members. Recommend procedures for sharing ideas. Promote the participation of less active team members.
2. **Encourage:** Respect other team members by providing opportunities for their recognition. Appear friendly, warm and responsive to all members of the team. Accept others and recognize their contributions.
3. **Compromise:** When team conflicts arise, offer a compromise approach. Admit errors readily. Modify ideas or concepts in the interest of team cohesion and growth.
4. **Sense:** Share your feelings with the team. Identify feelings, moods and relationships within the team.
5. **Harmonize:** Act to defuse conflict. Reconcile disagreements. Reduce tension and stress. Encourage team members to explore their differences.
6. **Testing:** Ascertain if the team is satisfied with the procedures in place. Suggest new procedures or guidelines to improve the operation of the team.

11.12.3 Hindering Roles

Unfortunately, we sometimes find lemons in the group. These are team members who do more harm than good. Behavior indicating a member impeding the team progress includes:

1. **Uncooperative:** Stubborn resistance of ideas of others for personal motives. Disagrees for no apparent reason. Employs a hidden agenda to deliberately impede the advancement of another's idea.
2. **Degrading:** Jokes in a sarcastic manner. Hostile criticism of ideas and suggestions of others. Comments harmful to the esteem of other team members.
3. **Dominating:** Interrupts the contribution of other team members. Asserts authority to manipulate one or more team members. Controls some team members by manipulation.
4. **Avoiding:** Absent or late to team meetings on many occasions without valid justification. Frequent changing of the topic under discussion. Uncomfortable with conflict.
5. **Withdrawing:** Avoids participation when possible in team activities. Offers few if any suggestions, ideas or recommendations. Responds to questions with abrupt and brief answers.
6. **Distracting:** Ignores the team meeting and begins side conversations with other team members.

Most of us are capable of playing all three roles at any time. We can help or hinder the progress of the team. It is important that we pause occasionally toward the end of a team meeting to assess the roles played by each team member. We have provided an assessment form in Exercise 11.20 to facilitate collecting the data from your assessment.

11.13 SUMMARY

We have presented arguments for the importance of development teams in industry. Experience has shown that effective development teams are vital if a corporation is to develop highly successful products and introduce them in time to win a major share of the market.

We have described functional organizations that exist within most corporations, and have given reasons why this organizational structure often interferes with rapid low-cost product development. The balanced team structure commonly employed in establishing development teams within large functionally organized corporations has been described.

We have not discussed the importance of location of team members in either enhancing or inhibiting communication, but you should be aware of its importance. The rule is simple. To enhance communication, minimize the distance between those that are to communicate with each other. The ideal situation is to place the entire development team in the same room.

Development planning and staffing of the development team has been covered. The success of the team depends strongly on generating a comprehensive plan. The plan is used initially to justify the development budget and its schedule. During the course of the development, the plan provides guidance for monitoring the progress and for judging both the team and product performances. Team staffing is cross disciplinary with adequate representation from both the engineering and finance divisions. Engineers interact with the business personnel from marketing, sales and purchasing. Responsibility is shared between disciplines and functions.

We briefly discussed team leadership in industry, and related the organizational structure of the team with the level of management provided by the team leader. This is important because of the authority that goes with the level of management. Higher level managers carry more authority, many decisions can be made more rapidly, communication is more effective and team progress is often enhanced if the team leader has status and clout associated with executive management.

Next, team member responsibilities and team member traits were discussed in considerable detail. Team members work in a matrix organization and often report to two or more managers. This is sometimes difficult particularly when you get mixed signals from different managers. It is important to be flexible and always remain cooperative with the program manager and the functional managers. Team member traits are very important to your career. Please evaluate your own traits in an honest self assessment. This is a critical first step in building team skills.

Most design teams evolve over the duration of a development project. We have described five phases usually experienced by the team in this evolution. Not all of the phases provide pleasant experiences. Early in the evolution, a team encounters the dissatisfaction phase with very slow progress, low productivity and member gloom. To minimize the time in this phase, we suggest the team prepare and execute a contract that provides guidelines for member behavior.

Finally, we have emphasized team meetings because the progress of the team is very dependent on effective meetings. Also, you will spend more time than you can imagine in meetings. From the outset, learn the techniques for a successful meeting. We provide reasons for the failure of some meetings and the success of others. The role of the team leader and the behavior of the team members are described. The meeting room is more important that most engineers imagine. We described features and furniture in a room conducive to productive meetings. We also emphasized positive and negative behavior on the part of the individual team members and gave explicit examples of both types of behaviors.

REFERENCES

1. Smith, P. G., D. G. Reinertsen, <u>Developing Products in Half the Time</u>, Van Nostrand Reinhold, New York, 1991.
2. Cunniff, P. F., Dally, J. W., Herrmann, J. W., Schmidt, L. C. and Zhang, G., <u>Product Engineering and Manufacturing</u>, College House Enterprises, Knoxville, 1998.
3. Barczak, G. and Wilemon, D., "Leadership Differences in New Product Development Teams," Journal of Product Innovation Management, Vol. 6, 1989, pp. 259-267.
4. Lacoursiere, R. B. <u>The Life Cycle of Groups: Group Development State Theory</u>, Human Service Press, New York, 1980.
5. Barra, R. <u>Putting Quality Circles to Work</u>, McGraw Hill, New York, NY, 1983.
6. Foxworth, V., "How to Work Effectively in Teams," Class Notes ENME 371, University of Maryland, College Park, September 5, 1997.
7. Clark, K. and Fujimoto, T., <u>Product Development Performance</u>, Harvard Business School Press, Boston, 1991.
8. Wheelwright, S. C. and Clark, K. B., <u>Revolutionizing Product Development</u>, Free Press, New York, 1992.

EXERCISES

11.1   Write an engineering brief describing why team structure is so important in the product realization process that takes place in a large corporation. Add a second paragraph indicating why it is much less important in a very small company.

11.2   Prepare an organization chart, like the one shown in Fig. 11.1, for the University or College that you are attending. Describe the logic, as you see it, upon which the organization is based.

11.3   List the advantages and disadvantages of the balanced team structure.

11.4   Outline a development plan that you would prepare if you were a senior team member representing the engineering function on a newly formed development team. The product to be developed is a new trail bike.

11.5   Write an engineering brief describing staffing required for the development team for the trail bike. Give the reasons for changing the staff as the bike evolves during the development process.

11.6   Describe the experience and status of the product manager for the balanced team structure.

11.7   Describe a matrix organization structure. Explain why this organization impacts an engineer assigned to a product development team with a balanced structure. If you were this engineer, how would you handle your functional manager?

11.8   Write a letter to a prospective employer explaining why you are a good team member.

11.9   Write a letter, with the best intentions, to a friend explaining why you think he or she is not an effective team member.

11.10 You have been a member of a functional team for four months. The team is not doing well and is beginning to miss milestones in the schedule for the development plan. You are to meet with your functional supervisor for your formal six month performance review. Write a script for a two-person play, with your supervisor's questions and comments, and your answers and comments covering what you expect to occur during your review.

11.11 You (a male/female figure) are attending weekly team meetings and are always seated next to Sally/Bill who is young and very attractive. Sally/Bill is apparently interested in you because she/he frequently involves you in side conversations that are not related to the ongoing team discussions. Write a plan describing all of the actions you will take to handle this situation.

11.12 Brad, the team leader, is a great guy who believes in leading his team in a very democratic manner. He encourages open discussion of the issue under consideration for a defined time period, usually 5 or 10 minutes. At the end of the time period, he intervenes and calls on the team to vote to resolve the issue. Write a critique of this style of leadership.

11.13 You together with Zack and Mary are senior members on a development team with 14 other members. The three of you are really very knowledgeable and experienced. You get along very well and have developed a habit of gathering together at the local watering hole the evening before the scheduled team meeting. During the evening you discuss the issues on the agenda and reach some decisions between the three of you. Write a brief describing the consequences of continuing this behavior.

11.14 Sue has a wonderful personality, and as a team leader is skillful in promoting free and open discussion by all of the team members. After the team reaches a consensus she calls for volunteers to implement the required action. When two or more people volunteer, she assigns them as a group to handle the action item. When no one volunteers, she assigns the least active person on the team to the action item and moves promptly to the next item on the agenda. Write a critique of this leadership style.

11.15 Efficient Ed prepares a meeting agenda that includes 16 major items. He schedules each item on the agenda with a 15-minute discussion period without indicating which team member will lead the discussion on each topic. The meeting is scheduled to begin at 8:00 am and to adjourn at 11:59 am. Write a description of what you imagine would happen during the meeting. If you were the team leader, how would you handle the 16 issues that had to be resolved in a short time frame?

11.16 You are the leader of a team of eight members meeting for the first time to develop a new hair dryer. During this initial meeting the following events occur.

- The temperature in the meeting room is 88 °F.
- The room is set up for a seminar speaker.
- The meeting is scheduled to begin at 9:30 am and at 9:35 am only five of the eight members have arrived.
- During the meeting Horrible Harry begins to verbally abuse Shy Sue.
- At 9:50 am Talkative Tom begins to describe his detailed positions on the world situation and is still going at 10:10 am.
- The team breaks for coffee at 10:15 am and has not returned at 10:30 am.
- Sleepy Steve begins to snore.
- Procrastinating Peter refuses to accept an action item after leading the discussion on the issue for 15 minutes.

Describe the approach that you would follow in dealing with each of these problems.

11.17 Prepare a list of ground rules that should be followed by all the members of a team in conducting an effective meeting.

11.18 You are a member of a team with an ineffective leader. What can you do to improve the productivity of the team? In responding to this question consider the irresponsible actions on the part of the team that are listed in Exercise 11.16. Remember in your response that you are not the leader and the most that you can do is to share the leadership role from time to time.

11.19 Meet with your team and discuss the list of rules that will be used in guiding the conduct of the team during the semester. Prepare this list of rules in the form of a contract similar to that shown in Section 11.9. As a final step, each member of the team is to sign the contract indicating his or her commitment to the rules of conduct.

11.20 Record your assessment of the leading, supporting and hindering roles that are played by each member of your team. We have provided you with a form shown below to facilitate this process. Use a score from 1 to 10 in each cell with 1 indicating very poor team skills and 10 superb team skills. Place the name of each team member in the column headings

TEAM MEMBERS' NAMES						
SCORES	1 -10	1 -10	1 -10	1 -10	1 -10	1 -10
LEADING ROLES						
1. Initiate						
2. Provide						
3. Seek						
4. Clarify						
5. Test						
6. Summarize						
SUPPORTING ROLES						
1. Monitor						
2. Encourage						
3. Compromise						
4. Sense						
5. Harmonize						
6. Test						
HINDERING ROLES						
1. Uncooperative						
2. Degrading						
3. Dominating						
4. Avoiding						
5. Withdrawing						
6. Distracting						
TOTAL SCORE						

CHAPTER 12

A PRODUCT DEVELOPMENT PROCESS

12.1 INTRODUCTION

The design and development of new products and services are essential for the welfare of most industries operating in the world. The only industries that can operate today, disregarding the marketplace and continuing to ignore the consumers' needs, are those protected by government regulations (i.e. the U. S. Post Office). For private enterprise to be successful, their product line must be competitive in every category. The products must be attractive, functional, efficient, durable, reliable, affordable, and most importantly they must satisfy the needs of the customer. How do we systematically design products that will be a success in the marketplace while generating profits for the company? Over the years most companies have evolved a product development process that is intended to provide a steady stream of new successful products. During the past decade, there have been many changes, and new approaches to product development have been advanced by several of the more successful companies. A general outline of a nine-phase product development process, which is followed by many companies in the U. S. today, is listed below:

1. Identify the customer needs.
2. Establish the product specification.
3. Define alternative concepts for a design that meets the specification.
4. Select the most suitable concept.
5. Design the subsystems and integrate them.
6. Build and test a prototype and then improve it with modifications.
7. Design and build the tooling for production.
8. Produce and distribute the product.
9. Track the product after release developing an awareness of its strengths and weaknesses.

A graphic describing the product realization process, derived from references [1, 2], is depicted in Fig. 12.1. The idea stage of the development involves interpreting the information gleaned from customer interviews and benchmarking. Then the product is defined in the product specification. The product specification establishes objectives for all of the important parameters controlling the performance of the product.

In the second stage of the process (concept generation), the development team considers many different ways to design each subsystem in the product. The ideas flow and different design concepts are explored. The better ideas are differentiated from inferior concepts. Models may be constructed and tested to determine initial feasibility of the newer concepts. The number of design ideas initially increases with time during the concept generation phase, and then decrease during the concept evaluation process as the weaker concepts are eliminated.

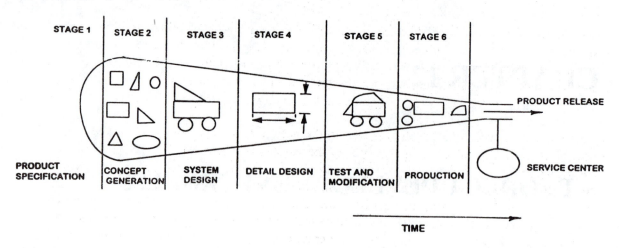

Fig. 12.1 Stages in a product development process showing design activities occurring from the initiation of a development to release of a product to market. After [1, 2].

The third and fourth stages, in the pictorial of Fig. 12.1, deal with design. System design encompasses the entire product line. Functions of the product are considered that lead to the division of the product into various subsystems. Interfaces between the subsystems are defined and the subsystems are integrated. Only after the subsystems have been integrated, is the detail design initiated. Engineering drawings are made of the component parts, and final selections of materials are made. Manufacturing and process control personnel begin to design tooling and to plan the production sequences. Purchasing agents begin to negotiate price with qualified suppliers. Quality assurance procedures are developed and delivery schedules for parts to be procured are established.

Stage 5 in the pictorial in Fig. 12.1, entails building one or more prototypes. Building these prototypes insures that the parts fit together properly and that the unit can be assembled without difficulty. Testing the prototypes verifies performance and gives a basis for design modifications leading to product improvements or refinements.

Stage 6 of the development process, includes procuring the final tooling, establishing the production procedures, and writing the production routings. The product is manufactured at increasing production rates until the inventory is sufficient; then the product is released to the market.

After product release, service centers are contacted to investigate the reasons for early product returns. Failures of the product are diagnosed; corrections to the design are made. Product modifications are implemented at the earliest opportunity.

The illustration in Fig. 12.1 implies that product development is a sequential series of steps. However, it is important to recognize that we must attempt to conduct many stages of development concurrently. The phrase **"concurrent engineering"** has been defined to convince managers that the various stages of development are conducted in parallel and not in series. By overlapping the phases, beginning certain aspects of stage 2 and possibly 3 before completing stage 1, we can significantly reduce the development time and reduce the time to market.

In this chapter, we will describe many aspects of the product development process. Hopefully, you will gain an understanding of the activities that take place within a typical corporation as a new product is developed for the marketplace.

12.2 THE CUSTOMER

We develop a product so that it may be sold at a profit. If a product is to be sold, someone must like it well enough to spend his or her money to purchase it. How do we design a product that the customer will like? The answer is to talk to the customer. Yes, design engineers go into the field and talk with the customers one on one. These talks (interviews) are conducted at the customer's location for 30 minutes to an hour. Another approach for learning about the customer's needs is to quietly watch the customer use the product over an extended period (several hours). By observation, we determine all of the ways the product is used and record the customers' reactions in each application. Interviews conducted by two members of the design team with a user of the product and observation periods with the customer represent very cost effective methods of establishing what improvements the customer prefers in a new product. Studies [3] indicate that 30 to 40 interviews usually reveal the needs of the customer sufficiently well to provide direction for developing a new improved product.

Focus groups, similar to those used in the news programs on television to determine political preferences, are also effective in determining the customers' needs. The focus group (eight to ten customers) is assembled at an off-site location to discuss the product. A moderator with skills in group dynamics leads the discussion. The focus session, which is about two hours in duration, is video taped. The development team reviews and analyzes the tape. They gain insight regarding the needs of the customer and the value they place on features that may be included in the development of a new product. Focus groups have the advantage that they encourage cross communication among the customers, which often reveals information that is not discovered in the individual interviews. However, the disadvantage of this approach is higher costs because of the need to hire a moderator, tape the session and rent the meeting room in a hotel.

Who is the customer? Clearly the person buying the product in the marketplace is a customer. However, usually there are others involved with the product that we need to consider as stakeholders if not customers. To illustrate this point, let's consider Black & Decker® (B&D), a company that designs, manufactures, and sells moderately priced power tools. Clearly, Harry and Harriet, happy homeowners, are customers because they go to the store and buy power tools. B&D will certainly interview Harry and Harriet to obtain their input, but they will also interview the store managers at retail outlets like Home Depot, Walmart, Lowe's, etc. These store managers are stakeholders because they share in the success of the product. They are concerned with attractive packaging, the size of the package, the lead-time for the delivery, the market introduction date, price, etc. The retail store manager's needs are as important as the needs of the individuals buying the product. They control the entry to the marketplace because they place the purchase orders with the company's distributors.

The service centers are another rich source of data that must be carefully considered before beginning to write the product specification. The service centers receive the products that are returned for a variety of different reasons. They listen to the complaints of the users, and they repair products that have failed in service. They know in great detail the strengths and the weaknesses of the product line. They know the importance of each flaw to the customers because they are exposed to the remarks made when a faulty product is returned for either refund or repair.

Suppose you are assigned to a development team, and your first task is to interview six previous customers all located in the same city. You schedule a 30-minute interview period with each, leaving a 30-minute period for travel between their locations. What questions do you ask? We suggest that you try the following questions:

- What are the features about the product that you like the most?
- What are the characteristics about the product that you do not like?
- Has the product ever failed?
- If so, what component failed?
- Do you like the appearance of the product?
- How do we compare to the competitor?
- How would you improve the product?
- What features would you incorporate in a new product?
- What are the important product characteristics that you consider before purchasing?
- Show us how you use the product in your applications.

Your job in the interview is to listen. Talk only enough to encourage the customer or stakeholder to relate his or her experiences with your product or one of your competitors' products. Keep the flow of information moving from the customer to you. Ask the customer to demonstrate the use of the product as a way to trigger his or her memory. Keep your own ideas for product improvement to yourself. You want to learn about the customer's ideas, and should not try to sell your ideas or to influence his or her thoughts.

It is a good idea to make notes or to use an audiotape to record the conversation. If you use an audiotape, make certain that the customer is aware that you are taping the session and agrees, on record, to the use of the tape recorder. If you interview six customers in the course of a day, notes and the audio record will help immensely in developing a complete listing of their significant comments at the end of the day. The audiotape has an advantage because it provides, by the emotions evident in the voice of the customer—a record of the importance of the issue being discussed.

12.3 INTERPRETING THE CUSTOMER NEEDS

If the development teams interviews 30 to 40 different customers, they probably will have generated several hundred significant comments, favorable and unfavorable, about the features and characteristics of the product. Some comments will indicate satisfaction, others will be critical, and still others will be superfluous. What are we to do with all of this information?

In examining the listings of the customer's comments, you will attempt to identify their needs. What does the customer need from the product? At this stage of the process, do not worry if changes suggested by the customers are feasible or about the changes that you already intend to make in redesigning the product. Keep the listing of the needs positive by **avoiding** wording like—it should not—. Arrange the listing of the comments and the needs in two columns as shown in Table 12.1.

Table 12.1
Listing of customers' comments converted to customers' needs.

Customers' comments	Customers' needs
I hate it when the tool breaks when I drop it.	Impact resistant tool.
I can't quit working even in the rain.	Waterproof tool.
I keep losing the key to the Jacob's chuck.	Keyless chuck.
I want to use the tool all day	Continuous power available.

Only four of the customers' comments are illustrated in Table 12.1. Well organized and properly conducted interviews with about 40 customers will usually generate a hundred or more significant comments that the team considers in establishing the customers' needs. It will be necessary for the development team to place priorities on these comments by listing the most important first and the least important last. Judgments by the team are necessary in prioritizing the comment list. Three rules often followed in ranking the importance of the customers' comments are:

1. Frequency—if many customers describe the same problem.
2. Emotion—if some of the customers show enthusiasm (positive or negative) when describing a feature of the tool.
3. Price—if the customer demonstrates a willingness to pay more for a new feature.

The final step in identifying the customers' needs is to expand the table shown above by expressing the need as a product feature. We accomplish this by adding a third column to our listing as demonstrated in Table 12.2. Note the three step process moves from the customers' comments, which are often vague and ill defined, to a need defined in engineering terms and then to the product feature involved in the suggestion for product improvement.

Table 12.2
Adding the product feature required to meet the customers' needs.

Customer's Comments	Customer's Needs	Product Feature
I hate it when the tool breaks when it's dropped.	Impact resistant tool.	Toughened case. Shock proof components.
I can't quit working even in the rain.	Waterproof tool.	Waterproof switch and motor.
I keep losing the key to the Jacob's chuck.	Keyless chuck.	High gear ratio chuck.
I want to use the tool all day without worrying about the cord.	Continuous power available.	High capacity batteries. Interchangeable batteries.

12.4 THE PRODUCT SPECIFICATION

Once the customers' and stakeholders' needs are established and prioritized, we must convert the vague statements gleaned from the customer interviews into meaningful engineering terms. A meaningful engineering term is a number or a statement that gives a well-defined design objective for each product feature or attribute. If the customer states that the tool vibrates too much, we must conduct experiments and measure the vibration level. Only then can we set a realistic vibration level in the product specification that will satisfy the customer. We need metrics (measurable quantities) describing each product feature that can be employed as targets to include in the product specification. We establish these metrics by testing not only our own products but also those of the competitors. In this extensive testing program (called benchmarking in industry), we establish the market norms for the product. If our new product is to be successful, we must improve on these norms to produce a superior product.

Benchmark testing provides the basis for writing the product specification. In preparing the product specification, we attempt to provide metrics associated with each feature or subsystem in the product. We establish a target value for a certain feature in the specification with a lower limit that is marginally acceptable and a higher limit that is essentially the best that we hope to achieve. Design is a compromise particularly when several metrics are to be satisfied simultaneously. We design to meet all of the subsystem targets; however, with some features, it may not be possible to achieve the highest level of performance while maintaining target values for cost and development time.

There is a systematic procedure for generating a product specification, where the vague statements made by the customer are converted to well-defined metrics that are then included as design targets in the product specification. This procedure called **quality functional deployment (QFD)** is a set of planning and communication tools that are used by a development team to design and produce products that closely meet the customers' needs. The products designed also conform to the capabilities of the company to manufacture, procure and assemble the product on time and at a predictable cost. Quality functional deployment is outside the scope of this book, but the interested reader may learn more about this important concept by reading the excellent book by Dr. Don Clausing [4].

12.5 DEFINING ALTERNATIVE DESIGN CONCEPTS

Let's suppose we begin an assignment on a development team the day after review of the product specification has just been completed. Our next task is to develop design concepts. A design concept (an idea) determines the approach taken to develop a specific feature or subsystem in the product. For example in the development of an electronic system to measure flow, we must generate an electrical signal that is uniquely related to the height of the water flowing in a defined channel such as a V-notch weir. Clearly, it is necessary to utilize a sensor of some type to convert the height of the water level in the weir to an electrical signal. Which of the sensors listed below is the most suitable?

- Resistive
  - Potentiometer
    - Rotary
    - Linear
  - Strain gage
    - Metal foil
    - Semi-conductor
- Capacitive
  - Flat plate
  - Cylindrical
- Inductive
  - Linear variable differential transformer
  - Eddy current
- Electro-optical
  - Photo cell, cadmium sulfide type
  - Pin photo diode
  - Phototransistors

We have presented many different design approaches (concepts) for converting the height of the water level to an electrical signal. Some of these ideas may be well known and others may be new but wild. Hopefully, one of the design concepts will be new and feasible. In the initial stages of generating design alternatives, we consider all ideas acceptable (the good, the bad and the very bad). We will evaluate all of them and select the better ideas later in the concept development process.

While several different types of flow measuring instruments are already available in the market, suppose your team is to develop an improved version. In this phase of the product development process, it is important to generate as many design ideas as possible to consider in designing each product feature or subsystem. Do not worry about their feasibility in this first effort to generate ideas. We will sort the good ideas from the crazy ones later in the development process.

Is there an organized method that we can follow to generate design concepts? **Yes!** There are so many methods available that complete books have been written to adequately describe all of them. We include three of these books as references [5, 6, and 7]. In this textbook, we will describe a few of the more popular techniques for generating design concepts including functional decomposition and both creative and rational techniques. We will use the pumping system together with the flow rate transducer as a product we are designing to provide examples of each method where appropriate.

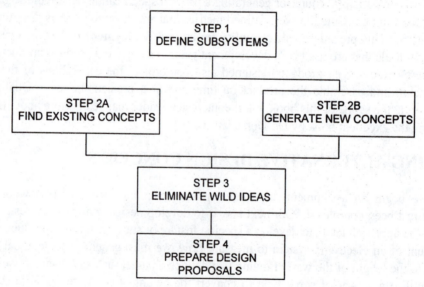

Fig. 12.2 A block diagram showing an approach for concept generation and subsequent evaluation.

A systematic approach to the concept generation phase of the development process is shown in Fig. 12.2. The first step in this approach is to define the subsystems involved in the operation of the product. In our case, the design of a pumping system with flow rate instrumentation will depend on the overall approach. If we decide to develop a system with a piston pump and an electronic flow rate instrument, we will have several subsystems including:

1. The inlet piping with a foot valve.
2. The inlet and outlet directional control valves.
3. The piston pump.
4. The reciprocating drive.
5. The delivery piping.
6. The flow control channel including the weir.
7. The height tracking mechanism and sensor.
8. The power supply and signal conditioning circuits.
9. The voltmeter.

We have examined our concept of a pumping system with flow rate instrumentation and have subdivided it into nine different subsystems. This is not a unique list. In fact, your team may decide to divide the many functions of the system differently. The uniqueness of the list is not important providing it is complete. We divide the product into its major parts to simplify the design problem. It is easier to think about several smaller design tasks involving a more limited number of concepts than it is to handle a giant task involving a very large number of concepts.

The second step is to generate design concepts for each of the subsystems identified above. We approach the concept generation in two different ways. First, look to others for ideas. You are not the first design team in the world to design a pumping system or any other product. Someone else was first. If you go to the library and search the literature, you will find scores of papers or books describing many different types of pumps and flow rate instruments that have been developed over the past century or even earlier in history. A careful review of this literature, and a search of the patents will enable you to develop a listing of design concepts that already exist for each of the nine subsystems listed above. Don't worry about using another person's idea. We do it all the time. We call it by a special name—"reverse engineering". We perform a detailed examination of the competitor's product, test it, adopt the good features, and reengineer the features that we believe to be inferior. We also try to improve on the good features by making them better. Providing we do not violate a patent, there is no requirement that the design be original. However, the new product must be superior to the competing products—feature by feature in a rigorous comparison.

Okay! We have a complete listing of the existing design concepts used in previous designs of our pumping system. Let's next generate some new ideas from within the design team. There are two approaches to follow. First, we suggest that each team member think about one or more of the subsystems. It is important to understand their purpose, function, the way that they work and the interface among the various subsystems. Then after developing a thorough understanding, the individual team members should list as many design ideas as possible for each subsystem. The independent thinking by team members is important early in our attempts to generate alternative design concepts. Write your ideas on index cards, with one idea on each card, and then give the cards to the team leader. Keep a copy of the card for yourself because you will use it in the next part of the new concept generation task.

The second approach in generating alternative design concepts involves a collective effort of the team. One of most widely used methods for generating ideas is brainstorming. The development team, five or six people, is a good number of participants in a brainstorming group. The team leader moderates the group keeping it on task and avoiding any tendency to begin a round table discussion of any idea. The team leader must formulate the questions posed to the team that initiates its response. A typical question is:

- What ideas do you have for the design of the check valves to be used to control the flow direction in and out of the pump?

The questions should be focused on a subsystem, but the inquires should not be so narrow as to limit the response from the team.

The team spends the next few minutes quietly thinking about the question. If ideas occur to you during this period for quiet thought, write them on an index card. You may also use the index cards that you prepared when you considered the design concepts independently. The team leader goes around the table in sequence, and each team member briefly presents one idea to the team. After presenting the idea, he or she hands the index card to the team leader. The brainstorming method is very effective when a few simple rules are followed:

1. No criticism is allowed during the session. The critique comes later.
2. The team responses should be short and to the point. Detail is deferred until later.
3. Wild ideas are okay because our goal is to generate as many ideas as possible.
4. Improvements and combinations of ideas are sought.

Brainstorming should be fun with everyone participating. Begin with prepared ideas from your index cards, and then add spontaneous ideas as they occur to you from the synergism of the group. You should, if possible, try to embellish the ideas of other team members with improvements. Humor helps during the brainstorming process because it breaks down barriers and improves the bonding of the team members. As the session moves from discussion based on the index cards to the generation of spontaneous ideas, the team needs a recorder to write the ideas on the blackboard or a flip chart (a very large pad of paper visible to the entire team). The visual display of the emerging ideas keeps the team on track and facilitates more thoughts on improving the ideas already identified.

There are many questions that the team leader can put to the brainstorming team to trigger ideas. Let's consider several of these questions. The questions are underlined and the descriptions that follow are to provide an example that relates to the pumping system:

1. Adapt
   - Can we copy a previous design of the piston pump?
   - Can we emulate the appearance of the pump of a competitor?
   - Are there other fundamental methods for displaying the flow rate?
   - Is the design of the lever drive mechanism that was abandoned five years ago suitable?

2. Substitute
   - What other material can be used for the piping?
   - What other means can be used to control the flow direction?
   - What new additive can be used with the lubricant to reduce friction?
   - Is there an improved process for sealing the piston within the housing?

3. Modify
   - Can we change the color of the housing to make the product more attractive?
   - Can the shape of the lever be changed to improve the pump's efficiency?
   - Can we change the motion of the lever to reduce the effort required?
   - Can we change the size of the piston to reduce the force required for pumping?

4. Rearrange
   - Can we change the location of the flow rate sensor?
   - Can we change the sequence of development in the schedule?
   - Can we interchange components for the inlet and outlet valves to reduce inventory?
   - Can we introduce pilot prototypes to the distributors earlier?

5. Combine:
   - Can we combine the cylinder head with the body of the pump to eliminate a part number?
   - Can we combine the lock washer with the screw to reduce assembly time?
   - Can we blend glass fibers in the plastic to improve the toughness of the housing?
   - Can we combine your idea with mine to reduce the size of the drive mechanism?

6. Eliminate
   - What feature can we eliminate to reduce the cost of the product?
   - What process can we eliminate in manufacturing to reduce cycle time?
   - What practice can we eliminate to reduce inventory?
   - What fasteners can we eliminate to reduce assembly time?

7. Reverse
   - Can we reverse roles in the interview process to improve communications?
   - Can we reverse the position of the opening in the flow channel to improve the accuracy of the flow rate measurement?
   - Can we reverse the color of the printing and the background of the transparencies for the design briefing?

8. Magnify
   - Can we make the housing stronger?
   - Can we make the outlet pipe longer?
   - Can we add more time for the tooling in our schedule?
   - Can we provide more value with an added feature?

9. Reduce
   - Can we omit the padding on the handle of the lever drive?
   - Can we reduce the diameter of the piston?
   - Can we reduce the weight of the total system?
   - Can we reduce the number of parts required to produce the pump?

The questions listed above are not exhaustive. You can certainly think of many more. They should be suggestive of the types of questions that can be raised at a brainstorming session. Do not worry that different questions may lead to the same answer. The idea is to learn answers to as many questions as you can imagine. Also, we want to share those answers with all of the team members.

A brainstorming session should run for 20 to 30 minutes, or it should be concluded when the team has exhausted all of their ideas. The team leader takes the index cards and recorder's notes, and is responsible for sorting and classifying the team's contributions. These team-generated ideas are added to the list of concepts that were determined from our literature and patent searches. Together they give us relatively complete lists of design ideas for each subsystem. At this point in the process, the lists—one for each function or subsystem—are composed of many ideas with very brief descriptions.

A follow-up meeting of the team is conducted to consolidate the lists. Wild ideas were encouraged in the brainstorming session to create a fun-like, open and creative environment. The following session is a return to realism. Its purpose is to remove the crazy ideas from our list. The team reduces the list so that it includes only three or four of the most feasible ideas for each subsystem. Reducing the lists means that someone's idea must be eliminated. It is important that the team reach consensus in rejecting ideas. The procedure to cull ideas should not generate ill feelings. Good reasons for excluding an idea should be given in good humor. Considerable time is needed for thorough discussion. No one should feel belittled or the butt of a joke by the idea elimination process.

When the lists of the design concepts have been reduced so that they include only feasible ideas, we expand them. Short written statements on index cards or notes without significant detail represent the ideas. Discussion of the concepts occurred in the meeting to reduce the number of alternatives, but we have not prepared detailed written descriptions of the concepts. To proceed with the evaluation of the remaining concepts, it is necessary to develop brief design proposals that include more detailed information and drawings or sketches showing the key features of each idea. Individual team members should develop a design proposal for his or her concept. The proposal should describe the idea in a few paragraphs, present a sketch if required, a list of advantages and disadvantages, and, where appropriate a design analysis. The proposal should contain much of the information necessary for the team to select the superior idea from among the few remaining design concepts.

12.6 SELECTING THE SUPERIOR CONCEPT

We have reduced the number of design concepts originally generated in a team meeting that weeded out the obviously weak and deficient design ideas. We have also expanded the ideas in the design proposals, providing the team with sketches, descriptions and perhaps even analyses. But, hopefully, we still have too many good ideas. We must now select the best from the good, and that is not always easy to do. Fortunately, we have the technique developed by Stuart Pugh [8] to help us in the selection of the best concept. The Pugh selection technique is a team activity that involves the preparation of an evaluation matrix. Keep calm, it's not a math matrix, and it is very easy to evaluate. Let's consider the nine steps in the Pugh selection technique that are:

1. Select the alternative design concepts to be evaluated.
2. Define criteria upon which the evaluation is based.
3. Form the Pugh matrix.
4. Choose a datum or a reference for concept comparisons.
5. Score the concepts.
6. Evaluate the ratings.
7. Discuss the good and bad elements of each concept.
8. Select a new datum and repeat the Pugh matrix if necessary.
9. Revise select concepts as needed and evolve a superior concept.

Step 1 concerns the selection of the design concepts to be evaluated; however, we have already selected the most feasible three or four concepts based on previous team discussion. The essential element in successfully completing this step is that all of the team members understand each concept in considerable detail. If all of the team members are to participate effectively in the Pugh selection process, then every one must understand each concept in great detail. For example, consider that we have generated four concepts for the sensor to be employed in an electronic flow rate measurement instrument, which include:

A. A cylindrical potentiometer.
B. A metal foil strain gage.
C. A linear variable differential transformer.
D. A cadmium sulfide photocell.

Do you understand how each of these four concepts can be implemented in the design? The sensor is a critical part of the transducer that converts the height of the water level in the flow channel to an electrical output signal. Can the concept be implemented in the limited time allotted for prototype evaluation in this course?

Step 2 defines the criteria that will be used in the selection process. These criteria may change in evaluating concepts for different features or subsystems, but the criteria is fixed for any one feature or subsystem. All of the concepts being considered must be judged against the same measure. What are the criteria that should be used to evaluate the four concepts listed above for sensing the water level in a weir? In the instructions provided to you in Chapter 2 for the design of the pumping system, we specified the performance expected from the pumping system and listed several design criteria. Finally, the importance of safety and avoiding all hazards is vitally important. Clearly, the criteria used to judge the concepts must take into account all of the requirements imposed by the course instructors as well as any additional criteria that your team imposes. Let's list some items that you may wish to include in the concept selection criteria:

1. Cost of materials
2. Availability of materials or components
3. Simplicity
4. Ease of manufacturing
5. Ease of assembly
6. Ease of disassembly
7. Impact on the environment
8. Weight
9. Power requirements
10. Friction effects on accuracy of flow rate measurement

Your team may want to consider additional items to include in this list of criteria. For instance, do you believe that stability of the electrical signal output or the ease of maintenance should be included as a criterion?

Step 3 involves the development of the Pugh matrix as shown in Fig. 12.3. The matrix is a simple table with rows and cells. You could use a spreadsheet program to prepare the matrix or a sheet of engineering paper as illustrated, in Fig. 12.3. The concepts A, B, C are placed in the column headings, and the criteria 1, 2, 3…10 are placed in the row headings.

CRITERIA	CONCEPTS			
	B	C	D	A
1				
2				
3				DATUM
4				
5				
6				
7				
8				
9				
10				

Fig. 12.3 Format for the Pugh chart used for concept selection.

In step 4, you choose the datum or reference against which your comparisons will be made. The datum is one of the concepts that you are evaluating. We select what is initially believed to be one of the better concepts, and this design approach serves as our reference. Suppose that we consider Concept A (the cylindrical potentiometer) to be the datum. We then compare the remaining three concepts B, C and D to the cylindrical potentiometer to reach a decision regarding the selection of the sensor.

In Step 5, you score the concepts and enter the results in the Pugh matrix. The scoring is simple. We consider each criterion separately and compare a selected concept (say B) against the datum concept (say A). The evaluation is a team activity. If the team reaches a consensus and decides that concept B is superior to concept A for criterion 1, then we score the intersecting cell in the Pugh matrix with a plus sign (+). On the other hand the team may decide that concept A is superior to B, and mark the intersecting cell with a minus sign (−). If the team decides that there is not a significant advantage of one concept over the other, the intersecting cell is marked with the letter S indicating that the concepts are essentially the same. It is important to recognize that the team discussion leading to the scoring is of great benefit. The team members interchange information in reaching a consensus. This discussion heightens everyone's awareness and understanding of the design issues involved in implementing the concept. The scoring continues with each concept compared to the datum for every criterion. An example of a completed Pugh matrix is shown in Fig. 12.4.

Let's use the matrix presented in Fig. 12.4 as an example to illustrate the scoring technique. Note, this scoring was performed by one of the authors without the significant benefit of team discussion. In your team's evaluation, items may be scored differently because you are considering different information or the concept may be modified during the discussion. The matrix is not unique since it is totally dependent on the information available to the team at the time of the evaluation.

CRITERIA	CONCEPTS			
	B	C	D	A
1	S	−	S	
2	S	−	S	
3	−	−	S	DATUM
4	+	+	+	
5	S	S	S	
6	S	S	S	
7	S	S	S	
8	+	−	+	
9	S	S	S	
10	+	+	+	

Fig. 12.4 Example of a completed Pugh chart where four concepts have been compared.

Let's consider the author's scoring for the 1st criterion in Fig. 12.4. Concepts B, C and D are compared with concept A on the basis of cost. First, recall the concepts:

A. A cylindrical potentiometer.
B. A metal foil strain gage.
C. A linear variable differential transformer (LVDT).
D. A cadmium sulfide photocell.

Second, recall the criteria:

1. Cost of materials
2. Availability of materials or components
3. Simplicity
4. Ease of manufacturing
5. Ease of assembly
6. Ease of disassembly
7. Impact on the environment
8. Weight
9. Power requirements
10. Friction effects on accuracy

To begin, consider the first criterion—cost. The author estimated the cost of all four options. A catalog for electronic components showed a wide variety of potentiometers both cylindrical and linear. Prices for cylindrical potentiometers ranged from a few dollars to more than $40. A standard "stock pot" is priced at $3.18. A metal foil strain gage suitable for the sensor beam is priced at about $25 for a package containing five gages or $5 per gage. The price of a linear variable differential transformer is difficult to determine because they are not available from electronic supply houses. The LVDT is a displacement transducer sold by a few firms supplying specialty instruments and transducers to engineering companies. One of the few companies producing LVDTs is Schaevitz Engineering. You may contact them on the Internet at www.schaevitz.com and request prices by e-mail. The quote I received by e-mail for a LVDT suitable for our flow rate sensor was $35. Cadmium sulfide photocells are available from many electronic supply houses. The cost of these semiconductor devices ranges from a couple of dollars per unit to about $10 depending on size, sensitivity and range. Other light sensitive transducers fabricated from silicon are in the same price range.

Now that the approximate cost for implementing each concept has been determined, we proceed to score the matrix. There is a relatively small difference in the cost of the sensors for concepts A, B and D; therefore, we mark cell B1, and D1 with the letter S. The LVDT, concept C, is relatively expensive; thus, it is scored with a (–).

We repeat the scoring for each criterion, and mark each cell in the matrix with +, –, or S. Each score represents a team consensus, and the discussion leading to that agreement gives each member of the team an opportunity to share information in depth about every concept as it is judged against every criterion. The matrix should not be considered fixed. If a criterion is not effective in the selection process, eliminate it from the matrix. Concepts may be modified during the rating process, or elements from one concept can be incorporated in another. The Pugh matrix is not a strict scoring device. Instead, it is a technique that leads to the interchange of information between team members, which gives additional opportunities to modify and improve the design concepts.

Step 6 evaluates the ratings, by examining the scores for concepts B, C and D when compared to concept A. In Fig. 12. 4—concept B (strain gages), we have three plus signs, one negative sign and six Ss. Do not cancel the negative with one of the positives because the negatives will be examined later in the Pugh process. For concept C (LVDT), we have two plus signs, four negative signs and four Ss. For concept D (photocell), we have three positive signs, no negative signs and seven Ss. Concept C (LVDT) is clearly inferior to concept A (potentiometer), but concepts B (strain gage) and D (photocell) appear to have positive ratings relative to the datum concept A (potentiometer). We have evaluated the ratings and summarized the pluses and minuses, but we defer the final selection.

Step 7 is inserted in the Pugh selection process to formalize the discussion of the advantages and disadvantages of each concept. Let's examine concept B (strain gage) and focus on the minus sign in cell B3. That minus score is due to the fact that the low signal output from the strain gage signal conditioning circuit will increase the complexity of the design because of the need for an amplifier to increase the signal output. Can we retain the same concept but reduce the complexity? The amplifier cannot be eliminated, but it is possible to reduce the complexity by using a modular design where an instrument amplifier is a single integrated component. The minus sign in the Pugh matrix permits the team to focus its attention on a weakness in a concept and improve the design by eliminating the weakness.

Sometimes minus signs lead us to abandon a concept. Let's consider cells B10, C10 and D10, all of which contain a plus sign indicating that we anticipate friction problems if we implement concept A (the potentiometer). The friction due to the wiper moving over the resistor will detract from the accuracy of the measurement. If the friction force is large, it will prevent us from achieving the specified accuracy requirements. Is it possible to change the concept and eliminate friction effects while retaining all of the advantages of the potentiometer? If we cannot determine a technique for eliminating friction effects, concept A will not be feasible and will have to be abandoned.

Another minus sign for cell C1 indicates that the cost of the LVDT is prohibitive. It is approximately ten times more expensive that the other three types of sensors. Is it possible to reduce the cost of this sensor? Suppose Shirley Sharp suggests that the sensor is made of three coils of small diameter magnet wire. Magnet wire is not expensive and winding a few coils is easy. If you design and manufacture the LVDT, the costs become comparable to the other sensors. This approach eliminates the cost disadvantage effects, but increases the time required for the team to procure a sensor. As a consequence, this new approach will change the scoring on the Pugh chart for some of the other criteria. If we make modifications to improve a concept relative to one criterion, it probably will be necessary to repeat the scoring of the Pugh chart.

Step 8 selects a new datum and repeats the Pugh procedure. We recommend this additional step if there are many concepts that remain after the first evaluation which represent uniquely different approaches. For the example given here, a second round of the Pugh procedure is recommended. The strain gage (concept B) and photocell (concept D) sensors are nearly identical in scoring the Pugh matrix. With only two concepts to consider, the criteria could be improved and more detail regarding the characteristics of the sensors could be brought to bear in the comparisons.

Step 9 revises the selected concept to evolve to a superior concept. Suppose that we have selected concept B (strain gages) for the sensor after eliminating the negative mark for simplicity. However, we can consider improvements by working to eliminate the reasons for the negative scores or even the scores of S. Keep asking questions about the factors detracting from the merits of an idea. New approaches emerge— negative scores or scores of S can change to positive scores. Answers to your questions often lead to design modifications that eventually provide the superior concept. When we finally identify superior concepts for every feature or subsystem, we proceed with the detailed design.

12.7 DESIGN FOR X, Y AND Z

No, we have not made a typo in the section heading. When we design, so many different aspects must be considered that the list has been replaced in the section title with X, Y and Z. For example, in the design of any system, subsystem, or component, we must consider the following major factors:

- Performance
- Appearance
- Manufacturing
- Assembly

- Maintenance
- Environment and sustainability
- Safety

12.7.1 Detail Design

Let's begin our discussion of design with the four Fs, namely form, fit, function and finish. The four Fs refer to the detail design of piece parts involved in some system. This is the easiest part of the design, which is the reason for considering it first. Form refers to shape. Is the part shaped correctly to perform its function? If it is too thin, it may fail by breaking or by deforming excessively. If it is too large or too thick, it will weigh too much and cost too much. If it has a sharp external corner, it represents a safety hazard because sharp corners are cutting edges. If it has a sharp internal corner, high stresses, which may lead to premature failure, will develop.

Fit is involved when we assemble two or more pieces to form a subsystem. Will the pieces fit together without gaps or discontinuities? If holes are drilled in a given component to accommodate a bolt pattern, are they in alignment so that the bolts fit without interference? Fit becomes critical when we deal with rotating shafts with bearings. Tolerances required for several of the diameters in a bearing housing are very tight and require precise machining of the component parts if the bearing is to fit properly. If the fit is too tight, the bearing will overheat and seize. If the fit is too loose, vibration will occur which detracts from performance and shortens the life of the bearing and other components.

Function refers to the ability of the system to perform satisfactorily with its components. If a part fails in service, then the system cannot function. Suppose you design a crankshaft for an auto that resonates at an engine speed of 3800 RPM. Your crankshaft is great. It has the right form, fits properly and has the correct surface finishes on the bearing journals—but the crankshaft vibrates at or near highway cruising speed. The design fails because the performance of the auto has been severely compromised by the vibrations induced in the engine by the crankshaft. The crankshaft has not functioned correctly.

Finish refers to the surface finish of the part. Is the surface rough, smooth or polished, or does it matter? Some surfaces are important because of appearance. The sheet metal on our auto is very smooth so that the paint that is applied will produce a high gloss finish. Other surfaces are not important. The aluminum block on an auto engine is produced by die-casting, and its outside surface reflects the finish of the die. Since one rarely looks at the engine until it dies, a relatively rough surface finish is acceptable. Some surfaces are polished or ground to enhance their performance in bearing applications. When designing a part, we must know how it will be used in service. With this understanding, we can specify the appropriate surface finishes on the detail drawing for that part.

12.7.2 Design for Manufacturing

In designing a piece part or a subsystem, it is essential that you consider how the parts are to be fabricated. Manufacturing methods are probably as important as function in the design of parts that are to be produced in large quantities. For a component that is mass produced, we must carefully consider manufacturing methods such as die casting, injection molding, forging, casting, drawing, machining, stamping etc. in the detail design of the parts. We have good and bad design rules to follow depending on the production methods to be employed in manufacturing [9].

In the development of the prototype, manufacturing methods remain important, but they are less critical. For the design of the prototype pumping system that we are developing in this course, you probably will be much more limited in your manufacturing capabilities. Hopefully you will have a model shop with power saws and safety qualified operators to cut wood and plastic according to your drawing. For sheet metal, you may have a shear and brake for cutting, bending and folding the thin sheets of metal employed in the design. You may also have hand tools such as hammers, saws, chisels, wrenches, etc. available for your use. Small power tools such as the drill, bayonet saw and belt sander, illustrated in Fig.

12.5, may also be available in the student workshop. Any component that you design must be fabricated with the tools available to you unless the components are procured from a retail outlet in finished form.

As an example, let's consider the piston for pump that must fit into a cylindrical housing with just enough clearance for a seal. We will assume that your piston is a flat, cylindrical plate, 8 inches in diameter fabricated from a piece of ¾ inch thick plywood. The circle for the piston is carefully drawn using a suitable compass and the precise location of its center is marked. The circumference of the piston is carefully cut from the plywood with a scroll or small saber saw. A hole for the attachment of the plunger rod is drilled at the center mark. If the piston circumference is not perfectly circular, it is possible to improve its shape. Insert a short bolt with a tight fit through the center hole with flat washers on each side of the plywood. Clamp the piston by tightening a nut onto the bolt. Next place the bolt into the chuck of a ½ in drill, and then clamp the drill body in a vise. The clamping should not be extremely tight because too much pressure from the vise will damage the drill motor. When the drill motor is activated, the piston will rotate and you can use a file, sandpaper, or lathe tool to produce a piston with circumference that is concentric with the center hole and very close to a perfect circle.

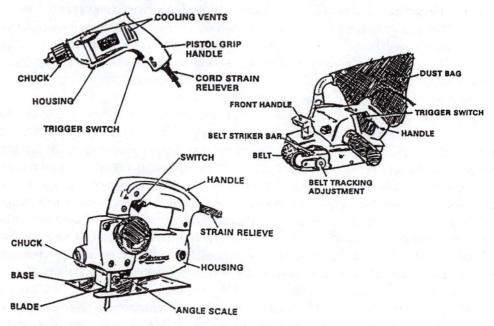

Fig. 12.5 Power hand tools available in the student workshop.

We can cut the piping necessary for the inlet and exhaust from plastic pipe to the appropriate lengths with a saw. Pipefittings, as required in the design of the directional flow control valves, may be procured at local retail outlets such as Lowe's or Home Depot. Adhesives are effective in joining the plastic pipefittings to the plastic pipe. If wood is used for select components, we can nail, screw, or glue pieces together to form strong joints. It appears that you will not encounter any problems manufacturing the pieces needed for the pumping system. Will you encounter any problems when trying to assemble all of these pieces in a short period of time?

12.7.3 Design for Assembly

Products are systems containing several subsystems with each subsystem fabricated from several components. In the design or selection of each component, we must consider the methods and procedures used to assemble them in the construction of the subsystems. We also must be concerned with the assembly of the various subsystems to form the complete system (product). There are several levels of consideration in design for assembly. The first and most important is whether or not the components can be assembled. Don't laugh. The author has seen designs that were impossible to assemble and had to be scrapped.

Second, the assembly operation should be easy and not require special operator skills. Finally, the time required for the assembly should be minimized.

Assembly time is very critical to the success of very complex products that are mass-produced. For example, Ford Motor Company leads the big three auto producers in the U. S. in assembly. They can assemble an auto with 24.72 people-hours compared to 26.32 and 27.76 people-hours for Chrysler and General Motors respectively. The Japanese managed company Nissan requires only 17.84 people-hours at its plant in Smyrna, Tennessee. It is estimated that General Motors could save about $3.5 billion each year if it could reduce the time required for assembly to that currently achieved by Nissan. Clearly, savings of billions of dollars can be made by employing design for assembly in the development of mass produced products.

One approach to design for assembly is to minimize the number of parts used in the product. As designers, we tend to use many more parts than are needed based on assembly theory. Boothroyd et al [9] have developed three criteria that are used to justify employing any part in the design of a product.

1. The part must move relative to other parts in the operation of the product.
2. The part is fabricated from a different material or must for some other reason be separated from adjacent parts.
3. The part must be separate from adjacent parts to permit either assembly or disassembly of the product.

Using these three criteria, we can develop a theoretical part count for any given product. With very rare exceptions, we greatly exceed the theoretical minimum part count when designing. Nevertheless, these rules are valuable because they stimulate the development team to consider combining parts and to design unique parts to serve dual functions. The number of parts required for a product is often reduced by 50% or more by applying the design for assembly criteria.

How does this affect you in the design of your pumping system? After all you are designing a one of a kind prototype to learn about the product development process. Well, you are stuck with an assembly problem, because you have a very limited time to assemble, test and disassemble your pumping system. If you need more than 45 minutes to assemble and 15 minutes to disassemble, then you will have to use test time for assembly, and you may not have enough time to accurately calibrate and adjust your flow rate instrumentation.

There is a lot of work involved in the assembly of the prototype. Can we shorten the time by conducting several tasks at the same time? After all we have five or six team members, and they should all be involved in the assembly operation. The team should divide the assembly operation into several assembly tasks that can be conducted in parallel. Many of the subsystems can be assembled simultaneously, and working tasks in parallel rather than series can shorten the overall time required to complete the assembly. You are not expected to work like a pit crew at the Indianapolis raceway, but every one should stay productively busy.

There are many aspects to design for assembly, and this treatment is only a brief introduction to a very important topic. For a more complete coverage see the excellent book identified in reference [9].

12.7.4 Design for Maintenance

Have you ever changed the oil filter on your auto? On some cars you have to raise the body, crawl under the car, and reach up to unscrew a horizontally oriented filter. When the filter is loosened, it leaks; oil runs down the side of the engine and onto you. The oil spill is difficult if not impossible to contain. Clearly, the design of this engine, and the placement and orientation of the oil filter did not accommodate the need for periodic and scheduled maintenance. Other cars have the oil filter on the other (front) side of the engine. The filter is replaced from the front without going under the car. The designers have also oriented the filter nearly vertically so that the oil does not spill when the filter is loosened and the seal is broken. This second filter is an example of design for maintenance. The designer recognized the need for periodic maintenance and developed the product to allow easy access, and rapid, inexpensive replacement of the required parts.

In the design of the pumping system, you should not have to perform any maintenance. However, you should recognize that many commercial products require periodic maintenance, and that it is often a very real design criterion in product development.

12.7.5 Design for the Environment

Customers and society at large are concerned with preserving the environment. Society enjoys new products and enhanced performance, but they clearly want to minimize the impact on the environment. In some countries in Europe, it is necessary for a company to show the methods to be utilized in recycling the product and disposing of any waste. In the U. S., most companies are sensitive to environment issues. Emissions are minimized, wastewater is cleaned before being discharged into streams, toxins are eliminated from processes, recycled paper is used in packaging, and recycled plastic and metal is used in many components.

We are in the early stages of design for the environment. As world population grows and third world countries become more affluent, preserving the environment and conserving resources will become even more important. Today many products are designed to be partially recyclable. Today about 75% of the weight of an auto is comprised of components that may be recycled. These percentages will grow with time as disposal becomes more difficult and concern over the environment assumes major importance[1].

A very new approach to conservation is remanufacturing. Remanufacturing involves rebuilding an older model of a product rather than building a completely new model. In the remanufacturing process critical components are replaced and some components are updated to enhance performance. Personal computers and copy machines are examples of products that can be remanufactured. Of course, auto components have been remanufactured for several decades because it costs much less to rebuild an internal combustion engine than to build a new one.

12.8 PROTOTYPE BUILD AND EVALUATION

There are several important reasons for building a prototype. The first is to determine if all of the component parts fit together. Frequently we find errors, and some of the parts do not fit properly when we attempt to assemble the subsystems. It is much easier to change the drawings and to fix the fit problems when we build the first prototype. Imagine trying to assemble 10,000 units of a product using parts that are manufactured with dimensional errors. The prototype, often called the first article build, is vital to insuring parts that fit perfectly. The assembly of the first article also reveals difficulties that may arise at a later date on the production line when the product is mass-produced. The idea is to find the assembly and/or performance problems and fix them immediately while the development is in the prototype stage. Finding problems later is much more expensive. Also, a problem discovered after the product has been introduced to the market requires a recall and harms the reputation of the company for producing a high quality product.

The prototype also enables engineers to test the product. These tests permit us to measure the performance parameters and to verify the analysis used in the design phase to predict performance. If any deficiencies in performance are revealed in the test program, design modifications are made to the prototype and the tests are repeated. The prototype is a vehicle for design change. It is instrumental in enhancing the performance of the product. In some instances, prototype development is nearly a continuous process with one prototype following another. At select times in this process, the company will freeze a design and introduce a new product based on these prototype experiences.

---

[1] A much more complete description of the issues involved in sustainable engineering is provided in Chapter 18.

You are building a prototype of a pumping system. As you build the first article, you will probably find some mistakes. Do not be depressed. Drawing errors and other analytical mistakes are common. Find the errors and fix them immediately. Make the prototype work as early as possible in the semester. When you fix a problem, find the mistake and learn from it. You should make certain that you revise the drawing to reflect the revisions made in correcting errors. An early test of your prototype, prior to the last week of class is strongly encouraged. The workshops and testing facilities will be ready for early evaluations of the prototype.

12.9 TOOLING AND PRODUCTION

In many instances we design and develop products for a very large market with sales of 100,000 units or more each year. Also, the life cycle of the product may range from five to ten years, so that the total production over the product life is huge. The production of large quantities requires extra care in the design phase. It is widely accepted that 70% of the final product costs are determined by the design [10]. Decisions made in the design affect the cost of producing the product over its entire life. If one million units are produced over the life cycle of the product, selecting a fastener with a cost premium of only 2 cents adds $20,000.00 to the production costs. Designing a component that requires an extra machining operation could easily add more than several hundred thousand dollars to the life cycle costs for high volume products. Clearly, when designing components intended for use in high volume products, we need an excellent understanding of costs, manufacturing process design rules, and assembly theory.

We trust that the procurement of the materials and the components for your pumping system will provide a small lesson in cost control. For lessons in manufacturing processes and assembly theory, we defer to later courses in the curriculum or independent study on your part. The references at the end of the chapter will help you begin the study.

12.10 MANAGING THE PROJECT

When a development team works on a project, they are charged with the responsibility for completing the project according to a schedule and a budget. The team leader should manage time and costs. Development costs can be extremely large. Chrysler spent about one billion dollars to develop the Concorde auto. The project took 3.5 years and involved a total of about 2250 people [2]. As you move up in a corporation, your position will likely involve some degree of management. You will need to learn about scheduling and project costs relatively early in your career. Let's start now by considering scheduling.

You already know something about scheduling because the University imposes a schedule on you from day one. Classes start on September xx or January xx and end on December yy or May yy. The final exam schedule is fixed before the semester starts. You have a schedule of classes—M, W and F at 9, etc. The instructor has imposed some deadlines for design reviews that you must meet in this course. You will need to incorporate these deadline dates into the Gantt schedule that we will develop later in this section.

Scheduling begins with listing the tasks that must be completed to develop the prototype pumping system. The list of tasks should be in reasonable detail to enable you to realistically assess the time to allocate to each task. Let's begin by setting up this list, presented in Table 12.3 that shows all of the tasks that must be completed during the development of a piston pump with a flow rate transducer. Note, the listing is in sequence according to time except for the design reviews.

Your team should review the list and the time estimates and make those modifications necessary to conform to your design. The time estimates are particularly important. We have scheduled 4 weeks for the team to generate, evaluate and select design concepts. That is a significant block of time (five or six people-weeks). Do you need that much time, is it the correct amount of effort, or will it prove to be insufficient? Carefully estimate the time required to complete a task, and then check to see if your estimates are correct. Estimating time for task completion is very difficult. Errors in underestimating time leads to delays in design releases and product introductions. These delays are costly to the corporation, and they are serious detriments to career advancement. Many of us even after many years of experience still underestimate the time needed to properly complete a task. Take this opportunity to start developing your time estimating skills.

TABLE 12.3
Task List for Developing a Pumping System

TASKS	TIME (Weeks)
CONCEPT DEVELOPMENT	
Receive and understand specification	1
Concept generation	2
Concept evaluation and selection	2
DETAIL DESIGN	
Piston and pump housing	1/3
Inlet and outlet piping	1/3
Direction flow control valves (3)	1/3
Lever drive mechanism	1/3
Flow rate measurement system	1/3
Drawing reviews	1/3
DESIGN REVIEWS	
Preliminary review	1
Final review and design release	1
MANUFACTURING	
Procure materials and components	1
Manufacture piece parts	1
Inspect piece parts and rework as required	½
Prepare assembly kit	½
ASSEMBLE AND TEST	
Pre-assembly	1
Final assembly and test	1
TOTAL	14

After completing the listing of tasks, we construct a Gantt chart as shown in Fig. 12.6. The Gantt chart shows a listing of the tasks in the rows down the left side. The time line, in this case expressed in terms of calendar months during the fall semester, is displayed along the X-axis. We draw bars to represent the time required to complete a specific task. The left edge of the bar corresponds to the time when the task is scheduled to begin. The right edge of the bar indicates the completion date. Of course, the length of the bar shows the number of weeks scheduled to complete the task. When the tasks are completed the bars are darkened. In the example presented in Fig. 12.6, the bars for the concept development tasks are darkened indicating that these tasks are complete. A vertical line is shown in Fig. 12.6 to indicate a current date—in this example it is sometime in mid October. The vertical line intersects the horizontal darkened bar for concept evaluation and selection. This configuration indicates that the team is ahead of schedule because the bar is darkened for a week ahead of the current date.

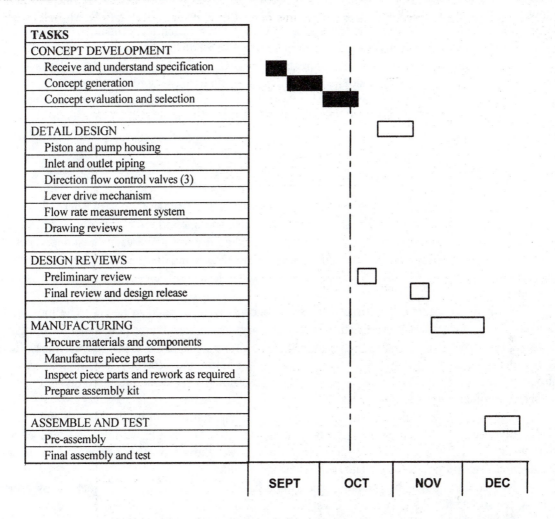

Fig. 12.6 Example of a Gantt chart for the development of a pumping system during a semester.

The placement of the bars in Fig. 12.6 is typical of what is called series scheduling. We schedule one task after the completion of the previous task. However, it is possible to overlap the bars along the time scale. Overlapping the bars (parallel scheduling) implies that we will start a new task before the current task is completed. In many instances, we can effectively pursue two tasks simultaneously, and important timesavings are achieved whenever parallel scheduling is possible. In procuring materials, we often encounter delays of weeks or even months in delivery of components necessary to build the prototype. In procuring these long-lead-time components, parallel scheduling is mandatory because it is necessary to place the purchase order long before the design is complete.

In Fig. 12.6, we have provided your team with an example Gantt chart for developing a pumping system with a flow rate transducer over a 15-week semester. You should consider your team's schedule and change the beginning and completion dates for each task to correspond with your team's plan. Prepare the Gantt chart early in the semester, and revise it periodically to reflect changes to the plan as the design of the pumping system evolves. Be sure to add the deadline dates imposed by your instructor as milestones on the Gantt chart.

The physical location of the team members is extremely important in the time required to develop a new product. Communication, so essential to the entire development process, is markedly affected by the distance. If two people are located only 20 meters apart, the probability of their communicating with each other once a week drops to only 10% [11]. To facilitate communication during the development process it is imperative that team members be located close together. Black & Decker, the largest manufacturing company in the State of Maryland, places the entire development team for a related group of products in a single room called a fusion cell. The room is large with desks for each team member located around its perimeter. Shoulder high screens to each side of the desks give a little privacy—but not much. The idea of the room arrangement and the office furniture is to facilitate communications and to eliminate private conversations. A conference table, situated in the center of the room, is used for all meetings concerning the program. The concept is effective because every team member is close at hand and within eyesight. You do not need e-mail, a phone, or a FAX machine to communicate when everyone is located in close proximity in the same room.

12.11 SUMMARY

A complete product development process has been outlined in this chapter. We very briefly described the beginning of the process, where the customers are identified, and interviewed to determine their needs and the amount that they would be willing to pay for new product features. The vague customers' comments are converted into meaningful engineering targets. The targets are incorporated into an engineering specification that defines the goals and objectives, which the designers are expected to meet in developing the product.

Methods for generating alternative design concepts were covered in detail. The most essential element of this process is the synergism of the team members in generating new ideas for design. The approach is to divide the product into subsystems and then generate as many design ideas for each subsystem as possible. From these many concepts, we select the best one and then improve it by introducing modifications that eliminate its disadvantages. A selection technique based on the Pugh matrix has been described in considerable detail with an appropriate example.

The four Fs of design were introduced to show the importance of form, fit, function and finish. We then briefly discuss several important aspects of design for manufacturing, assembly, maintenance and the environment. These are all extremely important subjects and the information presented here is only an introduction. Several references are cited for independent study.

Reasons are given for the importance of building the prototype. The prototype is critical to insuring the success of a development. It verifies that all of the parts fit, that they can be assembled with ease, and that the system performs in accordance with the product specification. The prototype is a vehicle for design change. It affords a critical opportunity to correct the design errors prior to the introduction of the product to the market.

Finally, we introduce a method for managing the project. The method involves a listing of all of the tasks that must be accomplished to complete the prototype development and a careful and intelligent estimate of the time necessary to complete each task. The data from the listing is then used to construct a Gantt chart that enables the team to track the completion of each task relative to a predefined time schedule.

REFERENCES

1. Wheelwright, S. C. and K. B. Clark, Revolutionizing Product Development: Quantum Leaps in Speed, Efficiency and Quality, The Free Press, New York, 1992.
2. Ulrich, K. T. and S. D. Eppinger, Product Design and Development, McGraw-Hill, New York, NY, 1995.
3. Griffin, A. and J. R. Hauser, "The Voice of the Customer," Marketing Science, Vol. 12, No. 1, 1993.
4. Clausing, D. Total Quality Development: A Step-by-Step Guide to World Class Concurrent Engineering, ASME Press, New York, NY 1994.

5. Cross, N., <u>Engineering Design Methods: Strategies for Product Design</u>, 2nd ed., Wiley, New York, NY, 1994.
6. Dym, C. L., <u>Engineering Design: A Synthesis of Views</u>, Cambridge University Press, New York, 1994.
7. Pahl, G. and W. Beitz, <u>Engineering Design: A Systematic Approach</u>, Ed. K. Wallace, Springer-Verlag, New York, 1977.
8. Pugh, S., "Concept Selection---A Method that Works," <u>Proceedings of the International Conference on Engineering Design (ICED)</u>, Rome, March 1981, pp. 479-506.
9. Boothroyd, G, Dewhurst, P. and W. Knight, <u>Product Design for Manufacture and Assembly</u>, Marcel Dekker, New York, NY 1994.
10. Anon, Company Literature, Munro and Associates, Inc., 911 West Big Beaver Road, Troy, MI 48984.
11. Allen, T. J. <u>Managing the Flow of Technology: Technology Transfer and the Dissemination of Technological Information within the R & D Organization</u>, MIT Press, Cambridge, MA, 1977.
12. Kalpakjian, S., <u>Manufacturing Processes for Engineering Materials</u>, Addison Wesley, Reading, MA, 1984.
13. Carter, D. E. and B. S. Baker, <u>Concurrent Engineering: The Product Development Environment for the 1990s</u>, Addison-Wesley, Reading MA, 1992.

EXERCISES

12.1 Consider a household refrigerator as a product. List the customers that you should consider in any redesign. Also give reasons for including each person or organization as a customer.

12.2 Interview a family member, Mom, Dad, Sister or Brother about some product that they own and use frequently. Try to list all of the favorable comments as well as their complaints. Are they willing to pay for product improvements? If so, how much?

12.3 Working together with your team divide the pumping system that you are designing into subsystems. Briefly describe each subsystem and define all of the interfaces that exist between subsystems.

12.4 Working with your team conduct a brainstorming session to generate design concepts for each of the subsystems defined in Exercise 12.3. List all of the concepts generated, the good and the bad, for each subsystem.

12.5 Define the criteria that you plan to use to evaluate one of your subsystems in the Pugh selection process.

12.6 Construct a Pugh matrix and evaluate several design concepts using the criteria defined in Exercise 12.5.

12.7 When the Pugh analysis of Exercise 12.6 is complete, discuss the implications of the positive and negative scores that appear in the Pugh matrix.

12.8 Write an engineering brief describing design for manufacturing.

12.9 Write an engineering brief describing design for assembly.

12.10 Write an engineering brief describing design for maintenance.

12.11 Write an engineering brief describing design for sustainability.

12.12 Why do we invest funds and time in building prototypes before we release a product to the market?

12.13 What was your reaction when your auto or a friend's auto was recalled to replace a design defect?

12.14 Discuss the implications of mass production on the development process.

12.15 Prepare a task list for your development of the prototype of a pumping system.

12.16 Prepare a Gantt chart for the pumping system development.

PART V

COMMUNICATIONS

CHAPTER 13

TECHNICAL REPORTS

13.1 INTRODUCTION

We hope that you enjoy writing, because engineers often have to prepare several hundred pages of reports, theoretical analyses, memos, technical briefs, and letters during a typical year on the job. We trust that you will learn how to effectively communicate—by writing, speaking, listening, and employing superb graphics. Communication, particularly good writing, is extremely important. Advancement in your career will depend on your ability to write well. We recognize that you will be taking several courses in the English Department and other departments in Social Sciences, Arts and the Humanities, which will require many writing assignments. The courses should help you immensely with the structure of your composition and the development of good writing skills. Most of the assignments will be essays, term papers, or studies of selected works of literature. However, there are several differences between writing for an engineering company and writing to satisfy the requirements of courses such as English 101 or History 102.

In school, you write for a single reader, namely the instructor. He or she must read the paper to grade it, subsequently determining how well you are doing in class. In industry, many people, within and outside the company, and with different backgrounds and experience, may read your report. In class, the teacher is the expert, but in industry, the writer of the report is supposed to be the one with the knowledge. Many of those working in industry do not want to read your report because they are busy: the phone is ringing continuously, meetings are scheduled back to back, and many important tasks need to be completed. They read—if not skim—the report only because they must be aware of the information that it contains. They want to know the key issues, why these issues are important, and who is going to take the actions necessary to resolve them. In writing reports in industry, we cut to the chase. There is no sense in writing a 200-word essay when the facts can be provided in a 40 to 50-word paragraph that is brief and cogent. An elegant writing style, so valued by our colleagues in the arts and humanities, is usually avoided in engineering documentation. In writing an engineering document, do not be subtle; instead be obvious, direct and factual.

In the following sections, we will cover some of the key elements of technical writing, including an overall approach, report organization, audience awareness, and objective writing techniques. Then a process for technical writing that includes four phases: composing, revising, editing and proofreading is described.

13.2 APPROACH AND ORGANIZATION

The first step in writing a technical report is to be humble. Realize that only a very few of the many people who may receive a copy of your report will read it in its entirety. Busy managers and even your peers read selectively. To adapt to this attitude, organize your report into short, stand-alone sections that attract the selective reader. Three very important sections—the title, summary and introduction are located at the front of the report so they are easy to find. At the front of your report, they attract attention and are more likely to be read. In college, you call the page summarizing your essay an abstract, but in industry, it is often called an executive summary. After all, if it is prepared for an executive, perhaps a manager will consider it sufficiently important to take time to read it. Follow the executive summary with the introduction and then the body of the report. A common outline to follow in organizing your report is provided below:

- Title page
- Executive summary
- Table of contents, list of figures and tables
- Nomenclature—only if necessary
- Introduction
- Technical issue sections
- Conclusions
- Appendices

The title page, the executive summary and the introduction are the most widely read parts of your report. Allow more time polishing these three parts, as they offer the best opportunity to convey the most important results of your investigation.

NASA
Technical Memorandum 107204

Army Research Laboratory
Memorandum Report ARL–MR–307

ALLSPD-3D
Version 1.0a

K.-H. Chen
Ohio Aerospace Institute
Brook Park, Ohio

B. Duncan
NYMA, Inc.
Brook Park, Ohio

D. Fricker
Vehicle Propulsion Directorate
U.S. Army Research Laboratory
Lewis Research Center
Cleveland, Ohio

J. Lee and A. Quealy
NYMA, Inc.
Brook Park, Ohio

April 1996

NASA
National Aeronautics and
Space Administration

U.S. ARMY
ARL
RESEARCH LABORATORY

Fig. 13.1 Illustration of the title page of a professional report.

13.2.1 Title Page

The title page provides a concise title of the report. Keep the title short—usually less than ten words; you will have an opportunity to give detail in the body of the report. The authors provide their names, affiliations, addresses, and often their telephone and fax numbers and e-mail address. The reader, who may be anywhere in the world, should be provided with a means to contact the authors to ask questions. For large firms or government agencies, a report number is formally assigned and is listed with the date of release of the report on the title page. A title page of a report on a computer code for designing combustors is shown in Fig. 13.1. This report was prepared by several authors from different organizations and was sponsored by NASA (National Aeronautics and Space Administration) and ARL (Army Research Laboratory).

13.2.2 Executive Summary

The executive summary should, with rare exceptions, be less than a page in length (about 200 words). As the name implies, the executive summary provides, in a concise and cogent style, all the information that the busy executive needs to know about the content in the report. What does the big boss want to know? Three paragraphs usually satisfy the chief. First, briefly describe the objective of the study and the problems or issues that it addresses. Do not include the history leading to these issues. Although there is always a history, the introduction is a much more suitable section for historical development and other background information related to the problems. In the second paragraph, describe your solution and/or the resolution of the issues. Give a very brief statement of the approach that you employed, but reserve the details for the body of the report. If actions are to be taken by others to resolve the issues, list the actions, those responsible, and the dates of implementation. The final paragraph indicates the importance of the study to the business. Cost savings can be cited, as can improvements in the quality of the product, gains in the market share, enhanced reliability, etc. You want to convince the executive that your work was worth its cost, valuable to the corporation and that you performed admirably.

13.2.3 Introduction

In the introduction, you restate the problem or issues. Although the problem was already defined in the executive summary, redundancy **is** permitted in technical reports. Recognize that we have many readers, so important information, such as a clear definition of the problem, is placed in different sections of the report. In the introduction, the problem statement can be much more complete by expanding the amount of information provided for the problem. A paragraph or two on the history of the problem is in order. People reading the introduction are more interested than those reading only the executive summary; therefore, more detail is appropriate.

Following the problem statement, establish the importance of the problem to both the company and the industry. Again, this is a repeat of what was included in the executive summary, but your arguments are expanded. You can cite statistics, briefly describe the analyses that lead to the alleged cost savings, give a sketch of customer comments indicating improved quality, etc. In the executive summary, you simply stated why the study was important. In the introduction, however, arguments are presented to convince the reader that the investigation was important and worth the time, effort and cost.

In the next segment of the introduction, you briefly describe the approach followed in addressing the problem. You begin with a literature search, followed by interviews with select customers, a study of the products that failed in service, interviews with manufacturing engineers, and the design of a modification to decrease the failure rate and improve the reliability of the product. In other words, tell the reader what you did to solve the problem.

The final paragraph in the introduction outlines the remaining sections of the report. True, that information is already covered in the table of contents, but again, we repeat to accommodate the selective reader. Also, placing the report contents in the introduction gives us the opportunity to provide a one or two sentence description about each section. Perhaps we can attract some reader interest in one or more sections by convincing a few people to study the report more completely.

13.2.4 Organization

An organization for the report was suggested by the bullet list shown at the beginning of this section. However, you should recognize that an organization exists for every section and every paragraph in the report. In writing the opening paragraph of the section, you must convey the reason for including this section in the report and the importance of its content. You certainly have a reason for including this section in the report, or you would not have wasted the time required to write it. By sharing this reason with your readers, you convince them that the section is important.

A paragraph is written to convey an idea, a thought or a concept. The first sentence in the paragraph must convey that idea, clearly and concisely. The second sentence should describe the significance or importance of the idea. The following sentences in the paragraph support the idea, expand its scope, and give cogent arguments for its importance. State the idea in the first sentence and then add more substance in the remaining sentences. If the report is written properly, a speed-reader should be able to read the first two sentences of every paragraph and glean 80 to 90% of the important information conveyed in the report.

As you add detail to a section or a paragraph, develop a pattern of presentation that the reader will find easy to read and understand. First, tell the reader what will be told, then provide the details of the story, and finally summarize with the action to be taken to implement the solution. Do not try this approach in the theme that you have to write for English class, because the instructor will clean your clock for being redundant. Technical writing is not the same as theme writing. In technical writing, we reiterate to the reader to convey our message. Often the readers must take corrective actions for the benefit of the company and the industry.

We frequently include analysis sections in engineering reports. In these sections, the problem is stated and then the solution is described. After the problem statement and the solution, we introduce the details. The reader is better prepared to follow the details (the tough part of the analysis) after they know the solution to the problem.

13.3 KNOW YOUR READERS AND OBJECTIVE

Another important aspect to technical writing is to know something about the readers of your report. The language that you use in writing the report depends on the knowledge of the audience. For example, we are writing this book primarily for engineering students. We know your math and verbal skills from the SAT scores required for admission. We recognize that you have very good language skills, and even better math skills. You are bright and articulate. We believe that most of the chapters in this book are in line with your current abilities. However, there are a few sections in the book that deal with topics usually found in the business school. Profits and costs are important in product development, and although they may not be as interesting as a computer program, they need to be clearly understood.

As you write a report for a class assignment, you understand that the instructor is the only reader with which you need to communicate. You also know that he or she is knowledgeable. In that sense, the writing assignments in a typical class are not realistic unless they are modified. Suppose we asked you to write an assembly routing (a step by step set of instructions for the assembly of a certain product) for the production of your new product in a typical factory in the U. S. What language would you use as you write this routing? If the plant is in Florida or the Southwest for example, Spanish may be the prevalent language. You should also be aware that a significant fraction of the factory workers in the U. S. are functional illiterates, and many others detest reading. In this case, it might be a good approach to use fewer words and many cartoons, photographs and illustrations.

Know the audience **before** you begin to write, and adapt your language to their characteristics. Consider four different categories of readers:

1. Specialists with language skills comparable to yours.
2. Technical readers with mixed disciplines.
3. Skilled readers, but not technically oriented.
4. Poorly prepared readers who read with difficulty.

Category 1 is the audience that is the least difficult to address in your writings; the audience in the last category is the most difficult. Indeed, with poorly prepared readers, it is probably better to address them with visual presentations conveyed with television monitors and videotapes.

A final topic, to be considered in planning, is the classification of the technical document that you intend to write. What is the purpose of the task? Several classifications of technical writing are listed below:

•Reports: trip, progress, design, research, status, etc.
•Instructions: assembly manuals, training manuals, and safety procedures.
•Proposals: new equipment, research funding, construction, and development budget.
•Documents: engineering specifications, test procedures, and laboratory data.

Each classification of writing has a different objective and requires a different writing style. If you are writing a proposal for development funding, you need persuasive arguments to justify the costs of the proposed program. On the other hand, if you are writing an instruction manual to assemble a product, arguments and reasons for funding are not an issue. Instead, you would prepare complete and simple descriptions that precisely explain how to accomplish a sequence of tasks involved in the assembly.

13.4 THE TECHNICAL WRITING PROCESS

Whether you are writing a report as part of your engineering responsibilities in industry or as a student in college, you will face a deadline. In college, the deadlines are imposed well in advance and you have a reasonable amount of time to prepare your report. In industry, you will have less time, and an allowance must be made for the delay in obtaining approvals from management before you can release the report. In both situations, you have some limited period of time to write the report. The idea is to start as soon as possible. Waiting until the day before the deadline is a recipe for disaster.

Many professionals do not like to write; consequently, they suffer from writer's block. They sit at their keyboard hoping for ideas to occur. Clearly, you do not want to join this group, and there is no need for you to do so. A technique is described for you to follow that helps to avoid writer's block.

13.4.1 Define Your Task

Start your report by initially following the procedures described in the previous section. Understand the task at hand, and classify the type of technical writing that you have to produce. Define your audience, to establish before starting to write the sophistication of the detail and the language that you may employ. Prepare a skeletal outline of the report. Start with the outline presented previously, and add the section headings for the body of the report. This outline is very brief, but it provides the structure for the report. This structure is important because it has divided the big task (writing the entire report) into several smaller tasks (writing the report a section at a time).

13.4.2 Gather Information

It is impossible to write a report without information. You must generate the information going into the report. You can use a variety of sources: interviews with peers, instructors, or other knowledgeable persons who are willing to help; complete literature searches at the library; read the most suitable references; and conduct additional Internet searches for information. Take notes as you read, gleaning the information applicable for your report. Be careful, however, not to plagiarize. While you can use material from published works, you cannot copy the exact wording; the statements must be in your own words. If the report has an analysis, go to work and prepare a statement of the problem, execute its solution, and make notes about an interpretation of the solution.

13.4.3 Organize Data

When you have collected most of the information that is to be discussed in your report, organize your notes into different topics that correspond to the section headings. Then incorporate your topics in an initial outline, which will grow from a fraction of a page to several pages as you continue to incorporate notes in an organized format.

13.4.4 Compose Document

You are now ready to sit down and write. Writing is a tough task that requires a great deal of discipline and concentration. We suggest that you schedule several blocks of time and reserve these exclusively for writing. The number of hours that should be scheduled will depend on the rate at which you compose, revise, edit and proofread. Some authors can compose about a page an hour, but most students need more time.

In scheduling a block of time for composing, you should be aware of your productive interval. Most writers take about a half an hour to come up to speed, and then they compose well for an hour or two before their attention and/or concentration begins to deteriorate. The quality of the composition begins to suffer at this time, and it is advisable to discontinue writing if you want to maintain the quality of your text. This fact alone should convince you to start your writing assignments well before the deadline date.

While you are writing, avoid distractions. Since writing requires deep concentration, you must remember your message, the supporting arguments, the paragraph and sentence structure, grammar, vocabulary, and spelling. Find a quiet, comfortable place, and focus your entire concentration on the message and the manner in which you will present it in your report.

Naturally, some sections of a report are easier to write than others. The easiest are the appendices, because they carry factual details that are nearly effortless to report. The interior sections of the body of the report carry the technical details, which are also easy to prepare, because describing detail is less concise and less cogent. Do not become careless on these sections because they are important; however, each sentence does not have to carry a knockout punch.

The most difficult sections are the introduction and the executive summary. It is advisable to write these two sections after the remainder of the first draft of the report is completed. We suggest that the introduction be written before the executive summary. Although the introduction contains much of the same information as the executive summary, it is more expansive. The introduction can be used later as a guide in preparing the executive summary.

13.5 REVISING, EDITING AND PROOFREADING

Writing is a difficult assignment. Do not expect to be perfect in the beginning. Practice will help, but for most of us, it takes a very long time to improve because writing is such a complex task. Expect to prepare several drafts of a paper or report, before it is ready to be released. In industry, several drafts are essential, since you often will seek peer review and receive mandatory manager reviews. In college, you have fewer formal requirements for multiple drafts, but preparing several drafts is a good idea if you want to improve the paper or report and your grade.

13.5.1 First and Subsequent Drafts

The first draft is focused on composition, and the second draft is devoted to revising the initial composition. Devote the first draft to getting your ideas down in reasonable form and in the correct sections of the report. In the second draft, focus on revising the composition and defer the editing process. Concentrate on the ideas and their organization. Make sure the message is in the report and that it is clear to all of the readers. Polish the message later in the process.

Several hours should elapse between the first and second drafts of a given section of the report. If we read a section over and over again, we soon lose our ability to judge its quality. We need a fresh, rested brain for a critical review. In preparing this textbook, the author composes on one day and revises on the next day. Revising is always scheduled for the early morning block of time, when the brain is rested and concentration is usually at its highest level.

Let's make a clear distinction between composing, revising, editing and proofreading. Composition is writing the first draft where we formulate our ideas and organize the report into sections, subsections and paragraphs. Unless you are a super talented writer, your first draft is far from perfect. The second draft is for revisions where you focus on improving the composition. The third draft is for editing where errors in grammar, spelling, style, and usage are corrected. The fourth draft is for proofreading where errors are eliminated.

13.5.2 Revising

As you revise your initial composition, be concerned with the ideas and the organization of the report. Is the report organized so that the reader will quickly ascertain the principle conclusions? Are the section headings descriptive? Sections and their headings are helpful to the reader because they help the reader to organize his or her thoughts. Additionally, they aid the writer in subdividing the task and keeping the subject of the section in focus. Are the sections the correct length? Sections that are too long tend to be ignored, or the reader gets tired and loses concentration before completing them.

Question the premise of every paragraph. Are the key ideas presented together with their importance, before the details are included? Is enough detail given, or have we included too many trivial items? Does the paragraph contain a single idea or have we tried to include two or even three ideas in the same paragraph? It is better to use a paragraph for every idea even if the paragraphs are short.

Have you added transition sentences or transitional phrases? The transition sentences, usually placed at the end of a paragraph, are designed to lead the reader from one idea to the next. Transitional phrases, embedded in the paragraph, are to help the reader place the supporting facts in proper perspective. You contrast one fact with another, using words like **however** and **although**. You indicate additional facts with words such as **also** and **moreover**.

13.5.3 Editing

When the ideas flow smoothly and you are convinced that the reader will follow your ideas and agree with your arguments, begin to edit. Run the spell checker, and eliminate most of the typos and the misspelled words in the report. Search for additional misspelled words, because the spell checker does not detect the difference between certain words like **grate** or **great**, or say **like** and **lime** and between **from** and **form**.

Look for excessively long sentences. When sentences become 30 to 40 words in length, they begin to tax the reader. It is better to use shorter sentences where the subject and the verb are close together. Make sure that the sentences are actually sentences with a subject and a verb of the same tense. Have you used any comma splices (attaching two sentences together with a comma)? Examine each sentence and eliminate unnecessary words or phrases. Find the subject and the verb, and attempt to strengthen them. Look for redundant words in a sentence, and substitute different words with similar meaning to eliminate redundancy. Be certain to employ the grammar check incorporated in the word processing program. It is not perfect, but it does indicate a significant number of grammatical errors.

13.5.4 Proofreading

The final step is proofreading the paper to eliminate errors. Start by running the spell checker for the final time. Then print out a clean, hard copy to use for proofreading. Check all the numbers and equations in the text, tables and figures for accuracy. Then read the text for correctness. Most of us have trouble reading for accuracy because we read for content. We have been trained since first grade to read for content, but we rarely read for correctness.

To proofread, read each word separately. You are not trying to glean the idea from the sentence, so do not read the sentence as a whole. Instead, read the words as individual entities. If it is possible, arrange for some help from a friend—one person reading aloud to the other with both having a copy of the manuscript. The listener concentrates on the appearance of each word and then checks the text against the spoken word to verify its correctness.

13.6 WORD PROCESSING SOFTWARE

Word processing software is a great tool to use in preparing your technical documents. The significant advantage of using word processing software is the ability to revise, edit and make the necessary changes without excessive retyping. You can mark, delete, and copy or move text. You can easily insert words, phrases, sentences and/or paragraphs.

While working with word processing software, it is difficult to revise, edit, and proofread on the screen because only a portion of the page is visible. Better results are obtained if you print a hard copy and can view the entire page. With the entire page, you can see several paragraphs at once and can check that they are in the correct sequence. Use double spacing when printing the first few drafts to give adequate space between the lines for your modifications.

After you have completed the revisions on the hard copy, make the required modifications to the text using word processing software and save the results on your own floppy disk. We recommend that you keep only the most recent version (draft) of the document. If you save several versions of the document, it is necessary to keep a logbook of the changes to each version. If the writing takes place over several weeks, it is easy to lose track of what changes you made and which version of the hard copy goes with which electronic copy. We find it easier to keep one electronic file (on your floppy disk) of the most recent draft with a hard (paper) copy to serve as a back up.

Word processing software has several features that are helpful in editing. The spellchecker finds most of our typos and provides suggestions to correct many misspelled words. The search or find command permits you to systematically examine the entire document so you may replace a specific word with a better substitute. A thesaurus is available to help you with word selection, but you must be careful when using it. Make sure you understand the meaning of the word that you select; do not try to impress the reader by using long or unusual words. Short words that are easily understood by the reader are preferred in technical writing.

Formatting the report is another significant advantage of word processing software. You can easily produce a document with a professional appearance. The formatting bar permits you to select the type of font, the point size (the height of the characters) and emphasis such as bold, underline or italic.

You can also format the page with four different types of line justification, which are commonly available. We are using word processing software to prepare this textbook, and applied "justify" alignment for both the right and left margins. The word processing software automatically added the spaces in each line and aligned both margins. In designing your page, use generous margins. One-inch margins all around are standard unless the document is to be bound; then the left margin is usually increased from 1.00 to 1.25 inch.

Tables and graphs can be inserted in the text. Take advantage of the ability to introduce clip art into the text or to transfer spreadsheets and drawings produced in other software programs into your report. Position the tables and figures as close as possible to the location in the text where they are introduced. Try to avoid splitting a table between two pages. If the table is longer than a page or two, consider placing it in an appendix. Identify figures with a suitable caption. The caption, placed directly beneath the figure, is to embellish the message conveyed by the illustration.

13.7 SEARCHING FOR INFORMATION

13.7.1 In the Traditional Library

In design, you do not have to reinvent the wheel in order to use it. The same is true for information. You do not have to go into the laboratory and recreate knowledge. It is acceptable practice to use the knowledge generated by others. In fact, you do not have a choice in the matter. Demands are placed on you to generate information for your technical reports and design briefings. This information may exist in your brain, but more likely it does not. The specific information that you need exists in a library[1]. The difficulty encountered in either the traditional library or the electronic library is finding the information that you need since we have information overload. There is so much information available that we need to be experts to sort among the excess to find the most applicable facts for the report.

To survive the information overload, we need to develop search procedures. These search procedures differ considerably depending upon whether we are searching in the traditional library or the electronic library. Let's begin by developing search routines for the traditional library and then consider the electronic library later.

The volume of scientific and engineering literature is huge. Today we have nearly 100,000 technical journals, which periodically publish new papers describing recent research and development. Add to these, new books, reports from government and private companies, proceedings of technical meetings and patents, and it is little wonder that it is so difficult to find narrowly defined facts in the information haystack. Where do we begin?

13.7.1.1 Encyclopedias

If you are at a complete loss, begin with an encyclopedia. The encyclopedias give an overview, a brief description of the topic, some history, and a few references useful for further inquiry. A list of useful encyclopedias is:

- Encyclopedia Britannica
- Encyclopedia Americana
- Harper's Encyclopedia of Science
- McGraw-Hill Encyclopedia of Science and Technology
- Van Nostrand's Scientific Encyclopedia

This list of encyclopedias offers information on general and scientific topics. It is not a complete list since there are many encyclopedias devoted to more specific areas of science such as astronomy, biology, chemistry, physics, mathematics, etc.

13.7.1.2 Library of Congress

If you are knowledgeable about the subject being investigated, then begin a more focused search. Card catalogs have given way to the computer, which does an electronic search based on author, subject or keywords. The electronic search is very quick, and if the library has books on your topic, call letters and numbers will be provided to help locate the book in the stacks. The call letters assigned to engineering subjects by the Library of Congress are:

---

[1] We will make reference to two different types of libraries in this section. The first, is a traditional library with its stacks of books and periodicals. The second is the new electronic library with information compiled at many web sites and available to you at your computer through the Internet and the World Wide Web.

- TA General engineering including civil engineering.
- TB Hydraulic engineering.
- TD Sanitary and municipal engineering.
- TE Highway engineering.
- TF Railroad engineering.
- TG Bridge engineering.
- TH Building construction.
- TJ Mechanical engineering.
- TK Electrical engineering.
- TL Motor vehicles. Aeronautics and Astronautics.
- TN Mining engineering. Mineral industries. Metallurgy.
- TP Chemical technology.
- TR Photography
- TS Manufacturers
- TT Handicrafts. Arts and crafts.
- TX Home economics.

The call letters and numbers provide a guide for locating books in the stacks. Browse through the books on adjacent shelves. If the subject is only a few years old, there is a good chance that you will find a book or two that provides excellent coverage on your specific topic. If the subject is older (a decade or more), you will probably find several books available to you. Browse through these books and note the different treatments provided by the various authors. Select the book that covers your topic in sufficient detail and provides the depth of coverage suitable for your purposes.

13.7.1.3 Textbooks

Books are one of the best sources of information for your technical reports or design briefings. The authors have studied the literature and have prepared a section or chapter that organizes and describes the topic with clarity. There is a great temptation to copy sections and/or paragraphs verbatim (word for word), and insert them into your report. Avoid the temptation— it is plagiarism. It is okay to use the **material** from any book or paper, but the **wording** must be yours. To avoid problems with plagiarism, read the reference; make notes outlining the organization, the ideas, and the supporting arguments. Then compose your text from your notes. Do not refer back to the book as you compose.

Most textbooks contain an extensive list of references of the material covered in each chapter. Sometimes these references are listed at the end of the chapter or listed at the end of the book. If you need more detail on a given topic, these references provide an excellent starting point for your search. They will identify the book or periodical and authors who developed the information you seek. A difficulty with references from books is that they are usually several years old. The information in these older books will be useful, but it will not be up-to-date. Therefore, you will need to fill in the gap to determine the current state of the art. To accomplish this objective, stay with the same periodical and even the same group of authors. Libraries bind the periodicals annually, and usually the December issue of a given journal has an author and a subject index. Look in this index for the author's names that are experts on your topic. Chances are they have continued to research the same subject. Find one of their most recent papers and examine their references. You will find additional authors and more periodicals that are current. Continue looking up additional papers, finding more authors and more periodicals. Soon you will have an abundance of information covering the past two decades showing a complete development of the topic of interest.

13.7.1.4 Science Citation Index

A common problem in performing an information search for a college assignment is that you find **too much** material and become overwhelmed. You have to weed out the leading material from the average material; otherwise, a section of your report will turn into a chapter. How do you judge the merits of one paper against another? One approach is to use a citation index. The **Science Citation Index (SCI)** lists the number of times a specific paper by a known author has been cited in the references of papers written by other authors. If one of the papers you have found has been cited frequently in the citation index, then it is well recognized in the field as being a significant reference. On the other hand, if one of your references has not been cited in the citation index, the research community has largely ignored it. The SCI keeps score as the research community judges the merits of all writers' contributions to the literature. The SCI is available in the library, on CD-ROM or on the World Wide Web (WWW).

You have performed a computer search, found several books in the stacks, updated your sources with select papers from the current periodicals, and found the well recognized sources from the SCI. It is now time to organize your materials, outline your sections, and begin to compose.

13.7.2 Electronic Libraries on the World Wide Web

During the past decade or so, an electronic library has evolved that differs in many ways from the traditional library. There are no books on any shelf and no stacks with periodicals on the fourth floor. The information resides in digital format in countless computer memories (web sites) distributed worldwide. You access this information from a computer that is connected to the Internet. The Internet is a network of computers that are connected together by wires, fiber optic cables, or microwaves. It is similar to a system of roads leading to the information, which resides at web sites. The roads, in this analogy, connect cities; whereas, the wires and fiber optic cables connect computers. The advantage of the Internet is that your personal computer provides access to the world's greatest libraries. All that is necessary is a computer, a modem to connect to a phone line, an Internet Service Provider (ISP) and Internet software. For readers still in college, the college computer center provides Internet access and software. All you need to access a college computer, is an account number and a password.

13.7.2.1 Basic Internet Service

The Internet, a system of many nets, provides the means of transporting the information from one site to another. The information resides on the World Wide Web (WWW) at individual web sites. At each web site, you will find one or more pages of information. There are a vast number of web sites[2], so there is no shortage of sources.

Each page shows information in the form of text, pictures, video clips, and audio clips. Some of this information may be useful, but much of it will be worthless for your application. Interesting are the linkages that exist between different pages on different web sites. You can obtain facts from several different web sites, by clicking on keywords or phrases (called "hot links") and your browser (to be described below) transfers to another information source at a new web site. These web sites may reside in any country, and the information may be accessed 24 hours a day, 365 days of the year. There are no "hours of operation" on the Web—the information is always available for you.

Another advantage of the links embedded in web sites is that they permit you to quickly expand the information search. The Web, with the appropriate links, connects pieces of information from different sources, viewpoints and regions of the world. Geography and time constraints are removed. The information exists, and the browser will quickly find it for you. Your problem is in organizing the information and presenting it in a form suitable for your audience.

When seeking information on the Web, you will need to use a browser, which is software that locates the web pages and then displays them on the screen of your monitor. Two of the most common browsers are Netscape and Internet Explorer. The screen for Internet Explorer is shown in Fig. 13.2.

The top row shows the title of the Web page. The second row gives the menu items such as File, Edit, View, Go, Bookmarks, etc. Click on the menu titles, and explore the options available to you on the

---

[2]The number of web sites is difficult to estimate because of the rapid addition of new sites. It is estimated that the number is doubling every two to three months. By the time this book reaches publication, the number will exceed several million.

drop-down menus. The third row shows a string of buttons that you click to move forward or backward, print, search or stop your search. The address of the web site that you intend to visit is typed into the box occupying the fourth row. Such an address is known as the Uniform Resource Locator (URL) and often looks like **http://www.title.com**. Type the address of the web page into the address box, press the Enter key, and your search is initiated. Review the web page that appears on your screen for the information, which you seek. Also, look for the "hot links" in the text of the web page. Underlined blue text or blue borders on photos indicate links to other web pages that may contain more useful information. Pointing at them with the cursor will change the arrow to a small hand. Click on the link, and the browser will connect you to another source of information. There is, however, no guarantee that the link will be helpful.

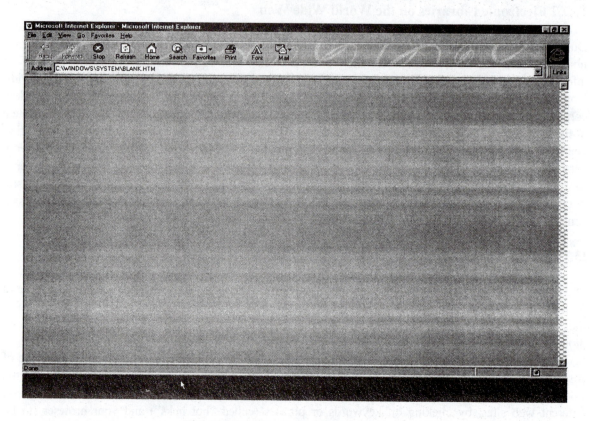

Fig. 13.2 Display for Internet Explorer.

For those readers who are curious, the "http" that is used in all of the addresses stands for HyperText Transfer Protocol. The use of http permits Web users to receive hypertext material that is linked to hypertext material on other web pages.

13.7.2.2 Advanced Search

Sounds cool! But what if you haven't developed an Internet address book and cannot locate the first web site? Fortunately, several software programs, called search engines, are available to help you search. Two of the most popular programs are **Yahoo!** and **AltaVista**. (Yes, the exclamation mark belongs after the word Yahoo. Again for the curious, the word Yahoo stands for Yet Another Hierarchically Organized Oracle.

The choice between these two programs depends entirely upon whether you prefer to work from a directory or an index in your search. **Yahoo!** uses a directory based search routine, similar to using the table of contents in a book to search for a topic. There are 370,000 sites in Yahoo's directories. **AltaVista** uses an index based search routine that employs keywords to locate the information. There are 30 million sites in AltaVista's index. The best way to learn to search is to begin using one search engine until you have found the web site of interest. Start with **Yahoo!** by typing the following address into the address block of either Netscape or Internet Explorer:

http://www.yahoo.com

Click on Reference and then Libraries to see the amount of information available to you. You may also be interested in clicking on Education, then Universities and explore the information available.

Next try **AltaVista** as a search routine by typing the following address into Netscape or Internet Explorer:

http://www.altavista.digital.com

A web "scooter" that continuously searches all of the web sites, cataloging keywords and addresses on every page, establishes the database for AltaVista. This data is transmitted to the index in **AltaVista** keeping it up to date as new web pages are added to the Internet. When you employ **AltaVista** for a search, type in keywords that describe the topic. Use lower case for most words, except for proper names, which begin with a capital letter, like Bill Clinton. The difficulty with this type of search is the large number of matches found. Often you will find thousands of sites relating to the keywords describing your topic. Look over the first few pages of the listings to determine if you were successful in locating the information you were seeking. If not, try using additional keywords, in an attempt to narrow the search.

Some other useful addresses for searching are:

http://www.lycos.com	An automated index.
http://www.webcrawler.com	An automated index.
http://www.infoseek.com	An index with a limited directory.
http://www.excite.com	An index with concept search and reviews.
http://www.hotbot.com	An index with 54 million sites.
http://www.inktomi.com	An index with 110 million sites.
http://www.about.com	A well-organized index with relevant listings.
http://www.savvysearch.com	Searches through 200 search engines.

13.7.2.3 Specific On-line Libraries

Okay, the search engines are running and able to seek information using either indices or directories. Let's return to the massive electronic library. We previously mentioned the library available to you if you call Yahoo!; however, there are many other libraries. Since you have a wide selection, let's review a few of the more popular offerings to give you a start.

The WWW Virtual Library is one of the oldest libraries. It maintains a collection organized into about 100 categories. A volunteer maintains each category, and its organization resembles a collection of reference journals published by different societies and companies. Its address is:

http://www.w3.org/hypertext/DataSources/bySubject/Overview.html

EINet Galaxy is another library that is maintained by a for-profit company. The organization of Galaxy is like a library with its card catalog sorted by subjects. Engineering and Technology is a main subject heading with most of the disciplines of engineering listed as sub-topics. Its address is:

http://galaxy.einet.net/galaxy.html

If you prefer to use an electronic library with an organization like the one found at many universities, then you will probably be most comfortable with the Clearinghouse for Subject-Oriented Internet Resource Guides. Let's call it Clearinghouse for short. The Clearinghouse comes from the

University of Michigan, and its emphasis is on information generated by other universities. You find Clearinghouse at:

http://www.lib.umich.edu/chhome.html

The addresses of other library-like sources that may be useful in one of your searches are:

http://www.nova.edu/Inter-Links/
http://www.uwm.edu/Mirror/inet.services.html
http://gnn.com/
http://duke.usask.ca/%7Elowey/encyclopedia/index.html
http://www.mid.net/NET/
http://library.microsoft.com
http://www.bizweb.com/

We suggest that you try using the Web as an information source. It is available to you at little or no cost, and the hours are much better than those posted by the Engineering library. Finally, you will not be frustrated to find the item you wanted is either missing or checked out to someone on sabbatical leave in West Africa. True, you may be frustrated by the information overload, but you will not be annoyed by the absence of a suitable reference.

13.8 SUMMARY

Writing is a difficult skill to master, and most engineers experience significant problems when they prepare reports early in their careers. Unfortunately, the writing experiences in college do not correspond well with the writing requirements in industry. In school, we write for a knowledgeable instructor. In industry, we write for a wide range of people with different reading abilities. Moreover, the audience often varies from assignment to assignment. In both college and industry, we write to meet deadlines imposed by others. While writing may not always seem to be fun, at least early in your career, there are many techniques that you can employ to make writing much easier and more enjoyable.

The first technique is to organize the report; we have suggested an outline for a typical report. Gather information for the report from a wide variety of sources, and generate notes that you can use to refresh your memory as you write. Sort the notes and transpose the information to expand the outline for your report.

As you organize the outline, but before you begin to write, determine as much as possible about your audience. They may be technically knowledgeable regarding the subject or they may be functionally illiterate. The language that you use in the report will depend on the reader's ability to understand. Also be clear about the objective of your document. Is it a report, an extended memo, a proposal or an instructional manual? Styles differ depending on the objective, and you must be prepared to change accordingly.

There is a process to facilitate the preparation of any document. It begins with starting early and working systematically to produce a very professional document. Divide the report into sections and write the least difficult sections (appendices and technical detail portions) first. Defer the more difficult sections, such as the executive summary and the introduction, until the other sections have been completed.

The actual writing is divided into four different tasks—composing, revising, editing and proofreading. Keep these tasks separate:

- Compose before revising
- Revise before editing
- Edit before proofreading
- Proofread with great care.

Multiple drafts are necessary with this approach, but the results are worth the effort.

Word processing software does not substitute for clear thinking, but it is extremely helpful in preparing professional documents. Word processing saves enormous amounts of time in a systematic editing process. It enables a mix of art, graphics, and text neatly integrated into a single document. Word processing software has a thesaurus and word search features that are helpful in editing. It essentially turns a computer and printer into a print shop so that you have wide latitude in the style and appearance of your professional documents.

Searching for information is an essential part of writing a good technical report, although you will usually complete your search **before** beginning to write. We described techniques for using the traditional libraries to seek specific information, and then introduced the electronic libraries that reside on the World Wide Web. We briefly described the Internet, and the World Wide Web. Browsers and search engines that are essential as you learn to move about on the Web were described. Finally, addresses for search engines and a number of different electronic libraries were provided.

REFERENCES

1. Eisenberg, A., Effective Technical Communication, 2nd ed., McGraw Hill, New York, NY, 1992.
2. Goldberg, D. E., Life Skills and Leadership for Engineers, McGraw Hill, New York, NY, 1995.
3. Elbow, P. Writing with Power, Oxford University Press, New York, NY, 1981.
4. Levine, J. R., C. Baroudi, and M. L. Young, The Internet for Dummies, 4th Ed., IDG Books Worldwide, Chicago, IL, 1997.
5. Hoffman, P. Netscape and the World Wide Web for Dummies, 2nd Ed., IDG Books Worldwide, Chicago, IL, 1996.
6. Hoffman, P. Netscape Communicator 4.5 for Dummies, IDG Books Worldwide, Chicago, IL, 1998.
7. Hanrahan, T., "The Best Way to Search Online," The Wall Street Journal, December 6, 1999, p R 25.

EXERCISES

13.1 Prepare a brief outline of the organization for the final report describing the development of a human-powered water pump.

13.2 Prepare an extended outline of the final report for the development of a human-powered water pump.

13.3 Write a section describing one of the subsystems in the human-powered water pump.

13.4 Write a section covering the calibration of the receiving tank for measuring the amount of water pumped during the prototype test period.

13.5 Write the Introduction for the final report on the development of a human-powered water pump.

13.6 Write the Executive summary for the final report on the development of a human-powered water pump.

13.7 Revise the Introduction that another team member wrote.

13.8 Edit the Introduction after it has been composed and revised by others.

13.9 Proofread the Introduction after it has composed, revised, and edited by others.

13.10 Go to the Engineering Library and develop a one-page description of the materials used in the die casting process.

13.11 Using the Internet and the Web, develop a one-page history of the Ford Motor Company.

13.12 Using the Internet and the Web, develop a one-page history of the largest manufacturing company in the State in which you reside.

13.13 Using the Internet and the web address http://collegehousebooks.com, write a brief history of this publishing company.

13.14 Prepare a listing of the web addresses of at least ten different search engines. Share this list with your team members.

13.15 Write a brief description of the procedure for performing a patent search using the Internet and the Web.

13.16 Write a brief description of the procedure for performing a patent search using the Engineering Library.

13.17 Using a computer reference such as Microsoft Bookshelf®, find a suitable quotation for this class.

CHAPTER 14

DESIGN BRIEFINGS

14.1 INTRODUCTION

We communicate by writing, speaking and with a number of different visual methods. All three modes of communication play a role as we try to convey our thoughts, ideas and concerns to others. All three modes are important. Let's focus our attention on speaking in this chapter. We use speech almost continuously in our daily life—why do we need to study about design briefings? There are several reasons. We usually speak with our friends and family in an informal style. We know them and feel comfortable with them. They know us. They are concerned with our well being, genuinely like us and are usually interested in what we have to say. A professional presentation is different. It is a formal event that is usually scheduled well in advance. The audience may include a few friends, but usually they are strangers. The time we have in which to convey our thoughts is very limited, and the audience may not be interested in our message. A person or two in the audience may be managers who control our advancement in the company. There are many reasons for tension headaches as one prepares for a professional presentation.

The design briefing is extremely important to both the product development process and to your career. Information must be effectively transmitted to your peers, management, and any external parties involved in the project. Clear messages, that accurately define problems, which the development team will effectively address, are imperative. On the other hand, ambiguous messages are often misunderstood, hinder the definition of the problem, and lead to delays in implementing timely solutions. Clearly, such messages should be avoided. The design briefing provides an opportunity to review the status of a specific product development process. It also permits peers to share their ideas with you, and it affords management an open forum for assessing the quality of your work and the progress made by your development team. Since the professional presentation is critically important, let's become knowledgeable about some great techniques for accurately delivering our messages to a mixed group of strangers, peers and acquaintances.

14.2 SPEECHES, PRESENTATIONS AND DISCUSSIONS

To begin, let's distinguish between three types of formal methods of oral communication—the speech, the presentation and group discussion.

14.2.1 Speeches

The speech is the most formal of the three. In the year of the presidential election, we always have many examples of speeches. We are besieged with political addresses that clearly illustrate the key features of speeches. They are given to large audiences in huge rooms, stadiums or coliseums. The setting is usually not appropriate for visual aids although the speaker often uses a teleprompter. The audience is diverse with a wide range of concerns. They are of widely different ages with different interests and persuasions. The speech is carefully scripted, and the speaker usually reads or closely follows the script. Ad-libbing is carefully avoided. Communication is one way—from the speaker to the audience. Questions are usually not appropriate and generally not appreciated. Time is strictly enforced unless you are the President of the United States giving the State of the Union address. Fortunately, engineers are rarely called upon to make speeches; hence, we will not dwell on this topic.

14.2.2 Presentations

Professional presentations differ from speeches. Presentations are made to smaller audiences in smaller rooms that are usually equipped with electric power, light controls and projection devices. We depend on visual aids, demonstrations, simulations and other props to help convey our message during a typical presentation. The audience is knowledgeable (about our topic) and usually has many common characteristics. The presentation is very carefully prepared because of its importance. It has order and structure, but it is not scripted. The flow of information is largely from the speaker to the audience; however, questions are permitted and often encouraged. The speaker is considered the expert, but discussion, comments and questions give the audience an opportunity to share their knowledge of the topic. Time is carefully controlled and is often insufficient from the speaker's viewpoint. The professional presentation is a vitally important mode of oral communication that you must quickly learn to master.

14.2.3 Group Discussions

Group discussions are also very important to the design engineer. The audience is smaller with much more in common. The group may all be members of a development team. The topic being discussed is narrowly focused. The speaker (central person) serves as a moderator, and he or she is an expert on the topic being considered. However, members of the discussion group (audience) may be as knowledgeable as the speaker (moderator). The moderator works in two modes. He or she may act as a presenter giving brief background information to frame an issue that prompts any member of the group to discuss the topic. The moderator may also direct questions to a group member known to be the most knowledgeable on the issue being addressed. The speaker (moderator) controls the flow of the information, but the flow is clearly multi-directional—from speaker to the group and from one group member to another. In-group discussions, time is difficult to control and the content and the range of coverage is strongly dependent on the skill of the moderator. Group discussion is very important in industry because the leader of a development team will often use this method of communication to identify problems and to initiate efforts directed toward their solution. We will not address discussion group methods in this textbook; however, we recommend that you watch *Washington Week in Review* on the PBS television network to gain some insights. While the participants are journalists, we can adopt many of their clever techniques to guide engineering discussions.

14.2.4 Design Briefing

A design briefing is a type of professional presentation. It is the method you will use in speaking before the class and will be an educational objective emphasized in this course. Indeed, you probably will be required to make presentations describing your product development on at least two occasions—the preliminary and the final design reviews. In these two design reviews, you will employ professional visual aids to assist in delivering the information required to describe the status of your project with conciseness and clarity.

14.3 PREPARING FOR THE BRIEFING

A design briefing is far too important to take casually. You should prepare very carefully to insure that you will accurately report the status of your teams' development and to identify problems or unresolved issues. There are three aspects that you should consider in your preparations:
- Identify the audience
- Plan the organization of the presentation and your coverage of each topic
- Prepare interesting and attractive visual aids.

14.3.1 Identify the Audience

In a typical college classroom, it is much too easy to identify the audience. You have your classmates, the instructor and perhaps a visitor from industry. In an industrial setting, the audience will be more difficult to identify because it will be much more diverse. The size and diversity of the audience depends on the magnitude of the development project. A small briefing will have 10 to 20 people in attendance with most participants from within the company. A large briefing may include 50 to 100 representatives who are both internal and external to the company. The characteristics of the audience are important because you must adapt the content of the presentation to the audience. Classify your audience with regard to their status, interest and knowledge before you begin to plan the style and content for your presentation.

The status of the audience refers to their position in the various organizations they represent. Are they peers, managers or executives? If they are a mix, which is likely, who is the most important? If you are preparing a design briefing for high-level management, it must be concise, cogent and void of minor detail. Executives are busy, stressed and always short on time. High-level managers are impatient and rarely interested in the technical details that engineers love to discuss in their presentations. Recognize these characteristics and adjust the content and the timing of your presentation accordingly. The executives are interested in costs, schedule, market factors, performance and any critical issue that will delay the development or increase projected costs of either the development or the product. Usually you will have only 5 to 10 minutes to convey this information at an executive briefing.

First and even second level managers are more human. They usually are more interested in you and the designs that you are creating. If you prepare the presentation for this lower level of management, you can safely plan for more time (10 to 20 minutes). These managers are likely to be engineers, and they will share your knowledge of the subject. In fact, they probably will be more experienced, expert and knowledgeable than you. They will want to know about the schedule and costs because they share responsibilities with the higher-level management for meeting these goals. However, they are also interested in the important details, and you can enter into discussion of subsystem performance with them. The first and second level managers usually control the resources for the development team. If you need help, make sure that they receive the message in enough detail to provide the assistance required. Do not hide the problems that your team is encountering; managers do not like to be surprised. If you have a problem, make sure they understand it and are able to participate in its solution. If you hide a problem from management and it causes a delay or escalation in costs, the manager will take the heat and you will be in a very unpleasant doghouse.

14.3.2 Peer Reviews

Design briefings for peers are often less tense, and you will have more time (20 to 30 minutes). You will address schedule and cost issues because everyone needs to know if you are meeting the milestones on the Gantt charts. However, the main thrust of your presentation will be on the design details. If you are working on one subsystem that interfaces with other subsystems, you will cover your subsystem in sufficient detail for all to understand its geometry, interfaces, performance, etc. It is particularly important that the interface issue be fully addressed in the presentation. If four subsystems are to be integrated in a given product, many details about all four subsystems must be addressed to ensure that the integration goes smoothly. Suppose, for example, that we are developing a power tool with a motor that draws a current of 10 amps, and plan to employ a switch to turn the power on/off that is rated at 6 amps. The switch will function satisfactorily in the short-term tests with the prototype, but it will malfunction in the field with extended usage sometime after the product has been released to the market. Clearly, the two persons responsible for the interface between the motor and the switch did not communicate properly. The integration of the two subsystems failed.

The most important purpose of design briefings with peers is to make certain that all of the details have been addressed. We strive to integrate the subsystems without problems arising when the prototype is assembled and tested.

Peer reviews in the absence of management are very beneficial in the development process. The knowledge of most of the team members is about the same although some members are more experienced than others. The topic is usually a detailed review of one subsystem or another. The audience is fresh and capable of providing a critical assessment of the technology employed. Your peers may check the accuracy of your analysis, comment on the choice of materials, discuss methods to use for manufacturing and assembly, and relate their experiences with similar designs on previous products. Peer reviews afford the opportunity for the synergism that makes fine products even better. They also provide a forum for passing on important lessons from the more experienced designers to the beginning designers in a friendly, stress-free environment.

The subject of the presentation is self-evident in a design review. We are developing a product, and the content of the design briefing will deal with one or more issues regarding the development. There is a choice of topics and considerable latitude regarding the content. Although we have given advice in previous paragraphs about matching content with the interests of your audience, you must understand the importance of knowing who will be in your audience. Executive reviews serve a different purpose than peer reviews, and the content and the time allotted for the presentation is adjusted accordingly.

For every type of review, there are two absolute rules that you must follow. First, you must know the material absolutely cold. It only takes an audience about two seconds to understand that you are faking it. Also, some presenters fail to rehearse adequately, and the audience often misinterprets this lack of preparation for insufficient knowledge. In either event, when they realize that you are not the expert you allege to be, your presentation fails. You have lost the opportunity to effectively communicate your message. Second, be enthusiastic. It is all right to be calm and cool, but do not be dead on arrival. You must command the attention of your audience, or they will quickly turn off. You control the attention of the audience only if you are enthusiastic and knowledgeable in presenting your material (story).

14.4 PRESENTATION STRUCTURE

There is a well-accepted structure for professional presentations. If you are familiar with PowerPoint®, a graphics presentation program (GPP) marketed by Microsoft, you are already aware of the templates included for various types of presentations. An excellent structure for a design briefing is included in the outline for a professional presentation shown below:

- Title
- Overview
- Status
- Introduction
- Technical Topics
  - First Topic
    - Background
    - Status
  - Second Topic
    - Background
    - Status
  - Final Topic
    - Background
    - Status
- Summary
- Action Items

Let's examine each of these topics individually and discuss the content that will be included on the visual aids that accompany your presentation. Recognize that the visual aids (35 mm slides or overhead transparencies) control the flow and the information, which you will present. For this reason, we will address the topics listed above in the context of information to be placed on the presentation slides.

14.4.1 Title Slide

The title slide[1], illustrated in Fig. 14.1, obviously carries the title of the presentation. However, it also states your affiliation and the names of the team members that contributed to the work. Remember, since you may be reporting on the work of the entire team, it is necessary for you to acknowledge their contributions on the title slide and in your opening remarks. A brief title—ten words or less—is a good rule to follow in drafting the title for your presentation. It is also a good idea to use descriptive words in the title. For example, you could use **Preliminary Design Briefing: HUMAN POWERED WATER PUMP**. Seven words and you have told us the topic (a design briefing or review), the type of briefing (preliminary), and the product being developed (a human powered water pump). A significant amount of information is successfully conveyed in seven words.

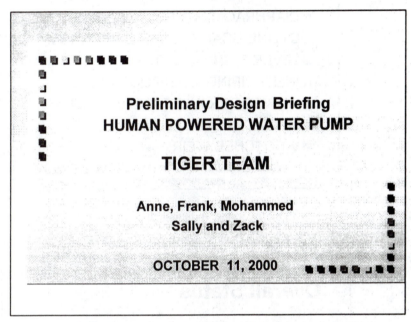

Fig. 14.1 Illustration of an informative and descriptive title slide.

Below the title, the team name together with the name of each team member involved in the development is listed. It is essential that you share the ownership of the material covered in your presentation. Often a single member of the team covers the work of several people in a design briefing. It is not ethical to make the presentation without acknowledging the contributions of others. As students, we typically use first names in identifying team members, but in industry the complete names of the team members are used. If necessary, division and department affiliations are also stated.

It is also a good idea to date the title slide and to cite the occasion for the presentation. If you make many presentations, this information is useful when referring to your files, allowing you to easily reuse some slides after selective editing and/or updating.

14.4.2 Overview Slide

The second slide is an overview. The overview for a presentation is like a table of contents for a book. You list the main topics that you intend to cover in the presentation. In other words, you tell the audience what you plan to tell them in the next 10 to 20 minutes. The number of topics should be limited. In a focused

---

[1] We will refer to slides in this discussion; however, the information described is also true for overheads.

presentation, you can cover about a half dozen topics. If you press and try to cover ten or more topics, you will encounter difficulty maintaining the attention of the audience throughout the presentation. Constrain the tendency to tell all—focus on the important issues. In a design review, the topics usually are organized to correspond with the subsystems involved in the product under development. For a component redesign, we concentrate on the issues guiding the component modification being considered. An example, of an overview slide for a preliminary design review for a human powered water pump, is illustrated in Fig. 14.2.

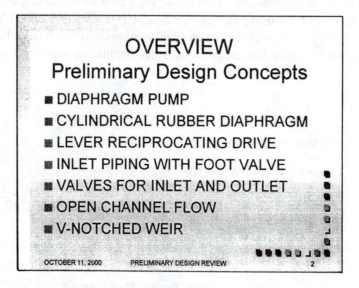

Fig. 14.2 The slide with the overview indicates the content in your presentation.

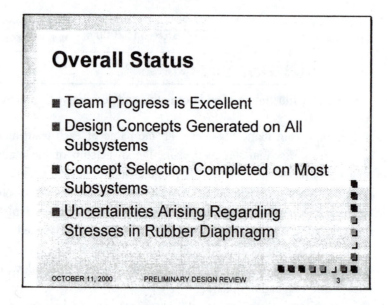

Fig. 14.3 Status slide indicates overall progress and accomplishments while signaling potential problems.

14.4.3 Status Slide

The third slide is used to present the product development status. You should report on the progress made by the team to date. It is a good idea to incorporate a Gantt chart showing the development schedule either on the status slide or a separate slide. The progress of the team on each task is defined on the chart and is immediately conveyed to the audience. If there are problems or uncertainties, this is the time to air the difficulties that the team is encountering. An example slide, shown in Fig. 14.3, illustrates uncertainties that the team has with regard to the high stresses imposed on the neoprene rubber diaphragm.

14.4.4 Introductory Slide

The fourth slide covers the introductory material such as background, history or previous issues. It is important that the audience understand the product, the main objectives of the development, the role the team has in that development and the most pressing of the current issues. This introductory slide permits you to set the stage for the body of the presentation that follows. The audience responds better to your presentation when they know in advance the background and the topics that you will cover. They are better prepared mentally to receive your message.

14.4.5 Technical Topics

The body of the presentation usually deals with five or six technical topics. In most instances, the topics selected correspond to the more critical subsystems involved in the design of the product. In the redesign of a component, the topics deal with the issues arising when considering new design concepts. Avoid trying to cover a large number of topics; it is better to report on a limited number of topics thoroughly than to rush through a dozen topics with incomplete coverage. We suggest two slides for each topic. The first slide describes the progress made by the team in generating design concepts and indicates the criteria employed in selecting the best concept. The second slide covers the status of developing the selected concept. The type of information reported on this slide includes accomplishments, outstanding issues, lessons learned, etc. An example of the status of the early stages of the design of a diaphragm pump is presented in Fig. 14.4.

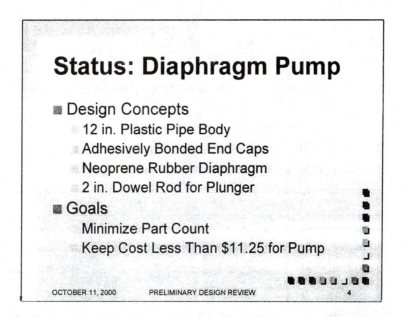

Fig. 14.4 Information presented in describing the early design of diaphragm pump.

14.4.6 Concluding Slides

After the technical topics have been covered, you must conclude the presentation with a decisive pair of slides. Suggestions for two concluding slides are presented in Figs. 14.5a and 14.5b. The next to last slide in the presentation, shown in Fig. 14.5a, is titled "Key Issues." This is your opportunity to identify very important issues (questions) that your team has discovered. Do not hesitate. This is the time to introduce the uncertainties and to come forward and seek help. Design reviews are not competitive. We seek help regardless of our status or role as a member of the audience. Managers will arrange support for your team if it is required. Peers will make suggestions and introduce fresh approaches to unresolved problems that the team may find useful. The design review is a formal process in the product development in which everyone participates. Take advantage of this review to identify uncertainties and to seek help whenever your team needs assistance.

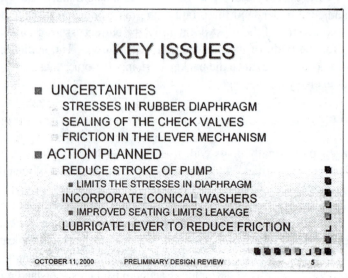

Fig. 14.5a The "key issues" focuses the team and management on your most significant problems.

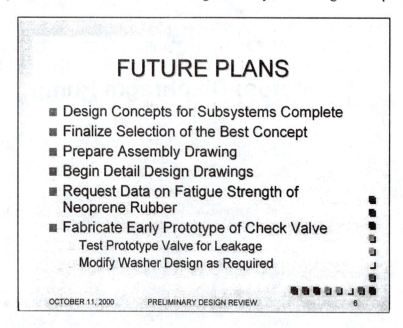

Fig. 14.5b The "future plans" slide gives you the opportunity to describe your approach for solving problems that the team is encountering.

The final slide presented in Fig. 14.5 is to present your plans for the future. Management wants to know how you intend to solve the issues that you have disclosed and the amount of time and money required for the solutions. Your peers will be interested in the technical approaches that you intend to follow. You should also define any requirement that you have of anyone in the audience. If you need help from your peers or more resources from the manager, be certain this assistance is clearly indicated on this final slide. You should distinctly identify those individuals who have agreed to provide the requested support. If action items are listed, clearly identify the responsible individual and estimate the time to completion. The planning incorporated in the final slide is very important. It is the prescription for your team's get-well program.

14.5 TYPES OF VISUAL AIDS

Because visual aids control the content and the flow of the information conveyed to the audience, they are essential for a technical presentation. They provide visual information that reinforces the verbal information. With visual aids you communicate using two of a person's five senses. Take time to carefully select the type of visual aid that will most effectively convey your message.

In selecting visual aids, we are usually limited—first, in our ability to produce the visual aids within the time and financial constrains and second with regard to the availability of equipment and the room in which the presentation is to be made. Is the room equipped with a 35 mm slide projector, an overhead projector or a computer-controlled projector? Another important consideration is the control of the light intensity in the room. Most rooms have light switches used to control the overhead lights. However, some rooms do not have blinds and bright sunshine will cause problems with the visibility of 35 mm slides when they are projected.

14.5.1 Computer Projected Slides

Computer projected slides are an excellent choice if the room can be darkened. Computer projection of slides has several advantages:

- The presentation is easy to prepare using PowerPoint.
- The slides are in color that is an improvement over black and white illustrations.
- Advanced visual effects (slide transition and slide building feature) enhance the impact of the presentation.
- It is also easy for you to copy your presentation onto a floppy disk.

A computer is attached to a digital projection system to load your program, and you employ a mouse to click through your slide line by line. Finally, low-cost handouts with four or six slides per page can be prepared, which help the audience follow your presentation. The only disadvantage of computer-projected slides is the dark room required for projection. With a darkened room, eye contact with the audience or even reading their body language is not possible. You lose your visual contact with the audience that is so important in reading their response to your presentation style.

14.5.2 Overhead Transparencies

Projecting overhead transparencies is probably the most common method of presenting visual material. Overhead projectors are relatively inexpensive (a few hundred dollars compared to a few thousand dollars for the computer controlled projectors), and for this reason they are available in most rooms. Overhead projectors are also the best choice if the room cannot be darkened. The projectors are sufficiently bright to provide quality images without extinguishing the lights. Bright rooms have an advantage because they help you keep the audience alert. Overhead transparencies are also easy to make on a copy machine. If you are willing to sacrifice the benefits of color, utilize inexpensive black and white transparencies. If you want to use color in the presentation because it is more effective, the costs increase significantly when using a color laser printer or a color copier to prepare the overheads. However, color ink-jet printers can be used to produce good quality color transparencies at a reasonable cost.

14.5.3 35-mm Slides

The 35 mm slide presentation is a third option for visual aids; however, the equipment required is not always available. Projectors for 35-mm slides are common, but in an engineering college, they usually are much more difficult to find than the overhead projector. A slide presentation is the preferred option if you have many photographs to project because the quality of colored 35-mm slides is excellent. The slides are inexpensive and easy to prepare if you have a camera and the time necessary for the film processing. It is possible to prepare colored slides from your PowerPoint files, but conversion from the digital format to the 35-mm format requires special equipment. Again a significant disadvantage is that the room must be darkened to view the image from a 35-mm projector. Also, the relatively long distance required from the typical projector to the screen precludes the use of very small conference rooms for presentations.

Although ineffective, sometimes speakers will employ two or three different types of projectors. Such a tactic should be avoided as the switch from one projector to the other is disruptive. If possible, use only one type of projector. If you must use more than one type of projector to properly present your message, try to minimize the number of times you switch from one device to the other.

14.5.4 Other Types of Visual Aids

There are several other visual aids sometimes used in professional presentations. Videotapes are common; unfortunately, the smaller size of TV monitors often makes viewing difficult for a large audience. However, if you want to show motion or group dynamics, video clips are clearly the best approach. Video cameras are readily available, and after practice you can become a reasonably good video producer.

Motion picture films, particularly of older material with historical interest, are very effective in developing a long-term prospective. However, finding an 8-mm or 16-mm motion picture projector may be difficult—allow time in your schedule for making the necessary arrangements.

Hardware, or other materials to be used in the product development, is sometimes passed around the audience during a presentation. The hands-on opportunity for the audience is a nice touch, but you pay for it with the loss of attention by some members of the audience during the inspection period. We recommend that you defer passing materials to the audience during your presentation. Instead, after the presentation, invite the audience to inspect exhibits that are placed on a strategically located table. This approach maintains the attention of the audience throughout the presentation while giving those interested in the hardware time for a much more thorough examination.

14.6 DELIVERY

Excellent presentations require well-prepared slides and a smooth, well-paced delivery. There are many aspects to the delivery part of the presentation process including: dress, body language, audience control, voice control and timing.

14.6.1 Attire

Let's start with dress. Broadly speaking, there are four levels of dress. The highest level, black tie for men and formal gowns for women, is a rare event and is never appropriate for design briefings. The next level, business attire for men and women, is sometimes appropriate for design briefings. A conservative Eastern company may have a dress code requiring business attire; whereas, a less formal Western company would encourage more casual attire. Casual attire should not be confused with sloppy attire. Casual attire is neat and tasteful, but without suits, ties and white shirts for men and suits for women. Although becoming more common in the workplace, sloppy attire (old jeans and a sweatshirt) is strictly taboo. We suggest that you select either business or casual yet tasteful attire when you dress for the presentation. We recognize that most students these days prefer a relaxed, collegiate style of dress, but resist the impulse to dress that way. If you look like a bum or a homeless street lady, who will believe your message?

14.6.2 Body Language

Posture is another important element in the presentation. In the words of former President Reagan, "stand tall." Your body language signals your attitude to the audience—you are the presenter and in control. Make sure you are calm, cool and collected. Nervous gestures with your hands, rocking on the balls of your feet, scratching your head, pulling on your ear, etc. should be avoided.

14.6.3 Audience Control

Before you begin your presentation, take control of the audience. One approach is to pause a moment before projecting the title slide and immediately make eye contact with the entire audience. How do you look at everyone in the house? You scan the audience from left to right and then back looking slightly over their heads. Occasionally drop your eyes and make eye contact for a second or two with one individual and then another. Pause long enough for the audience to become silent (10 to 20 seconds). If someone is rude and continues to talk, walk toward them and politely ask for their attention. When you have everyone's attention, project the title slide and begin your delivery.

If you have rehearsed, you will not need notes. The slides carry enough information to trigger your memory. If they do not, you have not rehearsed long enough to remember the issues. Scripting the presentation is not recommended. People who script will eventually start to read their comments, which is a deadly practice. Rehearse until you are confident. Make notes to help you rehearse, but do not use them during the presentation.

14.6.4 Voice Control

When you begin to speak, be certain everyone can hear you. If you are not sure of how far your voice carries, ask those in the back of the room if they can hear. If you use a microphone with an amplifier and speakers, be careful; the tendency is to speak too loudly. Try to control the loudness of your voice within the first minute of the presentation when you introduce the topic with the title slide. The audience will read that slide with anticipation, and they will bear with you as you adjust the volume of your voice.

Do not mumble. Speak slowly, clearly and carefully enunciate each word. It is not necessary to speak as slowly as Vice President Gore, but it is better to present your material too slowly than too rapidly. If you can improve your enunciation, you will be able to increase the rate of your delivery without losing the audience.

Don't run out of breath while speaking. Learn to complete a sentence, pause for breath (without a gasp) and then continue with the next sentence. There is nothing wrong with an occasional pause in the presentation, provided the pause is short.

14.6.5 Timing

If you forget some detail, do not worry. Skip it, and move to the next point that you are trying to make. If the detail is critical, count on one of your team members to raise the issue in the question and answer period. We all forget some of the material that we intended to present and introduce some additional items on an ad-hoc basis. Usually, the audience will never be aware of our omissions or additions.

Studies have clearly shown that people can listen to conversation faster that the presenter can speak. The trouble with listening to someone who speaks very rapidly is not their rate of speech, but their enunciation. Many people who speak rapidly tend to slur their words, and we have trouble understanding poorly pronounced words. The trouble we have in listening to Al Gore is to stay with his message; while you are waiting for his next word, your mind goes off on a mental tangent. You anticipate that your mind will return to the speaker in time for the next word, but unfortunately the mind is sometimes tardy. We often miss keywords or even complete sentences before our attention returns to the speaker.

Listening is a skill, and many of us have not developed it properly. As speakers, we have the responsibility to keep the listener, even the poor ones, on the topic. We use several techniques to maintain the attention of the audience. Employing the screen is probably the most effective tool for this purpose. As you project the slide, walk to the screen and point to the line on the slide that corresponds to the topic that you are addressing. If a few members in the audience are coming back from their mental tangents, you reset their attention on the current topic.

When referring to the image projected on the screen—and good speakers use this technique—position yourself correctly as shown in Fig. 14.6. Stand to the left side of the screen being careful not to block anyone's view. Face the audience and maintain eye contact. When you need to look at the slide, turn your head to the side and read over your shoulder. The 90° turn of your head and the glance at the slide should be quick because you want to continue to face the audience. Under no circumstance should you stand with your back to the audience and begin reading from the slide as if it were a script.

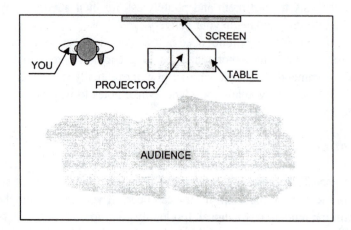

Fig. 14.6 Typical room arrangement for a presentation. Position yourself to the left of the screen and face the audience.

As you develop the discussion, point to key phrases to keep the audience focused on the topic. Engage both of their senses. Deliver your message through their eyes and ears simultaneously. When you point to a line on the slide, point to the left side of the line. People read from left to right—so start them from the initial point on the left side.

As you speak, modulate your tone and volume. Avoid both a monotone and a singsong delivery. A continuous tone of voice tends to put the audience asleep. A singsong delivery is annoying to many listeners. If you have a high-pitched voice, try to lower the pitch because a high-pitch tone is bothersome to many people.

If a question is asked, answer it promptly unless you intend to cover material later in the presentation that addresses the question. The answer should be brief, and you should try to avoid an extended dialog with some member of the audience. If a member of the audience is persistent, simply indicate that you will speak with him or her off-line immediately after the presentation. The timing of your presentation can be destroyed by too many questions. Some questions are helpful because they permit audience participation, but excessive questions cause the speaker to lose control of the topic, the flow and the timing of the presentation. Too many questions cause the format of the communication to change from a presentation to a group discussion. Group discussions have a purpose; however, they are different than a design briefing.

14.7 SUMMARY

The design briefing is extremely important both to successful product development and to your career. Information must be effectively transmitted to your peers, management and to external parties involved in the project. Clear messages, that accurately define the problems, which the entire development team can effectively address, are imperative. The design briefing provides the opportunity to inform management and our peers of our progress and problems.

There are three types of formal communication: the speech, the professional presentation, and the group discussion. The characteristics of these three types of communication are described. For the engineer, the professional presentation is the most important and the formal speech is the least important. The main emphasis of the chapter is on the professional presentation with a focus on the design briefing.

Preparation for a presentation is essential. Accomplished speakers often spend the better part of a day preparing for a 20-minute briefing. There are three important elements in the preparation. First, know the characteristics of your audience, and be certain that the content that you include corresponds to their interests. Second, organize the presentation in a manner that is acceptable to the audience. Ensure that the organization is efficient so that you use the allotted time effectively. Prepare high-quality visual aids (slides) that will help you convey your message and rehearse until you command the subject material.

A structure for the presentation is recommended that is commonly employed in design briefings. This outline includes four front-end slides—title, overview, status and introduction. The body of the presentation is devoted to technical details with from eight to twelve slides. Two closure slides are for a summary and the action items.

The type of visual aid selected is important because the visual transmission of information will markedly affect the outcome of the presentation. Computer projected slides have some significant advantages, but the availability of the equipment and a suitable (darkened) room often precludes their use. The overhead transparencies are the most common medium. However, if you have a presentation that includes many photographs and a large room in which to deliver the presentation, 35 mm slides are the most suitable medium.

The delivery is the make or break part of the presentation. We have provided you with two pages of details about how to do this—and not that. However, the best way to learn to deliver a design briefing is by practicing your presentation. On your first two or three attempts, have a friend video tape the event. Then review your delivery style, technique and appearance during a trail presentation. You will identify many problems of which you were not aware. Another suggestion is to use one of your free electives to take a speech course. The experience gained in a typical communications course will not help much in crafting the content to include in a design briefing, but it will help you to develop very important delivery skills.

REFERENCES

1. Eisenberg, A., Effective Technical Communications, 2nd ed., McGraw Hill, New York, NY, 1992.
2. Goldberg, D. E., Life Skills and Leadership for Engineers, McGraw Hill, New York, NY, 1995.
3. Wilder, L. Talk Your Way to Success, Simon and Schuster, New York, NY, 1986.

EXERCISES

14.1 Write a brief description of the characteristics describing your classmates in this course. Focus on the characteristics that will influence the language and content in your design presentations.

14.2 Prepare an outline for a preliminary design review. Include in the outline the titles of all of the slides that you intend to use.

14.3 Prepare the title slide for your preliminary design review.

14.4 Prepare an overview slide for your preliminary design review.

14.5   Prepare a status slide for your preliminary design review.

14.6   Prepare the "Key Issues" slide for your preliminary design review.

14.7   Beg, borrow, but do not steal a video camera. Video tape a practice presentation by a teammate. Critique his or her presentation with good taste.

14.8   Locate all of the projectors that you are permitted to borrow for the presentations to be made in this class.

14.9   Practice the delivery of your presentation employing the most suitable projector that you can borrow.

14.10   Prepare a group of slides for a design briefing that utilizes the slide transition and slide building features of PowerPoint.

PART VI

ENGINEERING AND SOCIETY

CHAPTER 15

ENGINEERING AND SOCIETY

15.1 INTRODUCTION

It should be made clear from the beginning of this chapter that engineering and society are closely coupled together. Engineering is an integral part of society, and consequently, we must treat engineering as a social enterprise as we continue to develop sociotechnical systems.

When engineers develop a new product or improve an existing system, we change society. The magnitude of the change differs from one product to another, and the effect on society may be trivial or enormous, but no matter what, there is an effect. When a few companies introduced a random orbit sander a few years ago, the time, effort, and cost of producing a flat, smooth, scratch-free surfaces was reduced. These sanders helped several hundred thousand workers in the finishing business to produce high-quality finishes on furniture, cabinets, counters and floors. The results also benefited many millions of customers who purchase furniture each year. This new product had a modest, but beneficial impact on society. There was, however, a small negative effect in that minor emissions to the environment occurred during manufacturing of the product and some virgin resources were consumed. Overall though, society has clearly gained by the introduction of this product.

When transistors were first developed from new ultra pure semiconductors, the impact was revolutionary. While the first transistor was developed in the mid 1950s, the integrated circuit was introduced in 1959, and the microprocessor was released in 1971, the technological revolution in microelectronics is still underway. Micro-electronic based products are markedly affecting the way in which society lives today. We call it the "information age". New products, which are introduced almost daily, are driven by technology. Products based on new advances in microelectronics will continue to change the way we work and play in a major way for at least another generation.

We hear much about new products and new life styles. Television, radio and even newspaper advertising keep us well informed of the availability and prices of both old and new products. However, we rarely hear much about the actual technology involved even when the product is new and revolutionary. The media and the public usually are not interested in learning about the technology employed in developing a new product; this is unfortunate because it would be easier to introduce new or modified systems if the public were more technologically literate.

Engineers are, in part, to blame for this apparent wall between the public and the understanding of technology. We often speak in such technical terms that we leave those rare individuals interested in our work in a confused state. As engineers, we must recognize our responsibilities to effectively communicate to society. We can begin by learning how to explain why and how our products work in terms more easily understood by the general public. We must also be more enterprising in understanding society and their changing views relative to technology. Lastly, we need to think on a more global scale. Engineers develop sociotechnical systems, which are usually large and complex. These systems rely on hundreds of products and many different types of services to function efficiently. Even though, as an engineer, we may be dedicated to developing a single product that is only a small part in a much larger system, we must understand the entire system. Those teams that have a thorough understanding of all aspects of the complete sociotechnical system develop the superior designs.

Let's consider an example of a sociotechnical system—the commercial air transport system. Do we have technical products involved in this system? **YES**, literally hundreds of advanced state-of-the-art products. Do we provide technical services for the airlines? **YES**, again. Does an engineer working for General Electric in the development of a new jet engine need to understand what Boeing is doing in the design of a new aircraft or modifications of an existing aircraft? Does that same engineer need to know about the sound levels that the public will accept as the plane lands and takes off? Is the engineer working for Boeing required to understand the causes of each and every airline accident? Again, **YES** to all.

From these simple questions, it should be evident that our sociotechnical systems are extremely complex. It will require significant effort for you to understand the detail of the interaction of the many products incorporated in these systems with the public. Nevertheless, engineers that understand and appreciate the improvements, which both the customer and society seek, will develop the most successful products.

Finally, as an engineer, you must understand that the systems we develop are adaptive. They are dynamic and change with time in response to several different forces. First, society changes. What society and the customer wanted in the 1950s is a far cry from what either will accept today. The customers, a segment of society (more on this fact later), typically expect that the new model of the product will be much improved over the older model and cost less. Others in society, not necessarily the customers, will expect (perhaps demand) that you manufacture the product without degrading any aspect of the environment.

An example is in order to distinguish between the customer and society. Let's consider the automobile as the product. The customer of an automobile demands style, comfort, mobility, convenience, affordability, reliability, performance and value. Society requires (through federal regulation) safety, fuel efficiency and reduced emissions and pollution. The fact that the requirements from the customer and society often conflict should be recognized. Conflicting requirements challenge engineers to create improved designs that advance the state of the art in all areas of technology.

The governments (Federal, State, County and City) are also involved in the relationship between engineers and society. Various government agencies continually issue regulations that affect both the sociotechnical systems and the product. Consider the commercial air transport system again. Obviously, safety is a very serious issue. The Federal Aviation Administration (FAA) is responsible for issuing regulations to insure safe operations of commercial aircraft. However, 1996 was not a good year for the FAA or our commercial airlines because several very serious accidents occurred with significant loss of life. As a consequence, of the fire in the cargo bay, which caused the crash of a Boeing 737 aircraft[1] in the Florida Everglades and the loss of all aboard, the FAA is now requiring the airlines to retrofit all planes with a fire suppressant system. Regulations from government agencies drive a portion of the changes to any sociotechnical system.

The understanding of the relationship between engineering and society is relatively new. If we examine the five-volume tome, *A History of Technology*, [1] published by Oxford University Press over the period from 1954-58, we find a chronological, deterministic catalog of technological development. However, the editors almost completely ignore the relation between engineering and society. In his review of this tome, Hughes [2] states that the editors belatedly conclude their five-volume treatise with an afterthought essay on technology and the social consequences. Today, the relationship is more clearly recognized although we have not yet managed to include adequate treatment of this very complex relationship in the typical engineering curriculum. We will attempt to show a few examples of the relationships between engineering and society in this chapter. For a more complete treatment, we encourage you to read reference [3].

---

[1] An investigation showed that the operator Value Jet had stored supposedly depleted oxygen bottles in the cargo bay, which were a significant factor in the fire that resulted in the crash.

15.2 ENGINEERING IN EARLY WESTERN HISTORY

Technology has impacted the way that we live since the beginning of recorded history. Long before the first engineering college was founded, we had intelligent, hard-working men[2] working as engineers developing new products and processes. We can look at history and cite many developments that changed society, and for the most part, benefited society.

Prior to about 6000 BC, there was little evidence of technology. Men and women were hunters and gatherers. They scrounged for food and lived in caves if they were lucky and shacks or huts if they were not. The significant development in the period from 6000 to 3000 BC was in agriculture with the domestication of animals and the cultivation of grains. With time, it was possible to insure a relatively stable, if limited, food supply and people began to congregate together in villages and small cities.

The first technology involved the building of roads, bridges and boats for local transportation. People clustered in villages and small cities wanted to transport food, water and other materials over short distances. The geographic regions where technology was evident were extremely local. Early Western history involved only Mesopotamia, Egypt, Greece and then Rome. In a few select cities, there were aqueducts to transport water and limited sewerage. In most regions, only the ruling class lived well. The great monuments of Egypt and Greece were built with slave labor. In fact, slaves (people who were on the losing side in a war or those born into slavery) provided most of the power required for building, mining and agriculture for many thousand more years. The rise of religion, after the coming of Christ, is often credited with the reduction in the prevalence of slavery. However, the steam engine developed by Thomas Savery, Thomas Newcomen and James Watt in the early 18th century signaled the beginning of the end of the need for vast amounts of human power. As we enter the 21st century, the development of the computer and software programs portends another beginning of the end for the need for huge numbers of people performing repetitive tasks or clerical work.

15.2.1 The Egyptians

The Egyptians built many huge monuments and were masters in moving large blocks of stone from distant quarries to a construction site and then elevating these blocks to their respective positions in the structure. Some estimates indicate that 20,000 to 50,000 men were required to drag the 20 to 50 ton blocks of stone used in the construction of these monuments. In spite of the majestic grandeur and durability of the monuments, the new innovations introduced by the Egyptian engineers were not outstanding. They constructed temples in a simple fashion using only uprights (columns) and cross pieces (short, deep beams). While the arch was known during this period, it was constructed with mud bricks and never adapted to stone construction. Classical treatments of the history of technology tell us very little about the impact of these developments on society. However, it is apparent that the construction of a pyramid or a temple required a significant segment of the working population for many decades. This effort probably did little to improve the welfare of even the **free** segment of the population. It is also evident that the effort required of the slaves to move the countless large, heavy stones must have been brutal.

---

[2] Historical accounts of engineering achievements indicate that the profession was male dominated. The author is sensitive to the gender issue in engineering education, but it must be recognized that the presence of women in the profession is very recent when considered relative to a historical span of about 8000 years.

15.2.2 The Greeks

Although they had very capable engineers, the Greeks are better known for their contributions to science. Thales, Euclid and Aristotle are recognized for their contributions to the development of the basics of geometry and physics. This new knowledge of geometric relations was applied almost immediately in architectural engineering. However, scientific understanding during this period was so limited that nearly 2,500 years elapsed before this knowledge was expanded sufficiently to be useful to the engineering community.

Greek engineers built cities, roads, water (hydraulic) systems and some machinery. Archimedes and Hero were innovators who used the pulley, screw, lever, and hydraulic pressure to develop cranes, catapults, pumps, and several hydraulic devices. The Hellenistic period brought better buildings and local roads, which improved living conditions for the upper class. However, the construction of the city infrastructure was still performed largely by slaves because the use of other forms of power was nonexistent. The Hellenistic world was divided into extremely independent city-states. Politics of the period did not foster cooperation between these governmental entities; as a consequence, few long roads were built. Transportation was largely by sea in boats equipped with oars powered by a galley of slaves. Some of the boats were equipped with a square sail, but these were so poorly designed that they were effective only on down-wind tacks.

15.2.3 The Romans

The Roman engineers followed the Greeks, and while they were less inventive, they were masters of detail. As the Roman Empire spread from the Middle East to Scotland, the Roman engineers built long roads connecting the distant cities; roads so well constructed that they were in service for many centuries. The large cities constructed in the conquered lands had paved streets and heated homes with water supplies and sewer systems. The Roman engineers had little or no theoretical knowledge, but they had developed practical methods of construction that served their purposes well. During this period, the subjects of statics and mechanics of materials were not understood. Beam theory would not be developed for another 1,700 years. The Roman engineers based their construction on experience, and they used very large safety factors. These empirical methods, while crude by today's standards, sufficed to produce many bridges, roads, aqueducts and cities. A great bridge over the Tagus river in Spain supported a road nearly 200 m long with six spans 60 m above the river. Considering it was completed at about 100 AD, this bridge was a remarkable engineering feat. The durability of the Roman construction is well documented by their structures some of which are still in use today—2,000 years after they were placed in service.

The Roman engineers deviated from the Egyptian and the Greeks in their construction by using much smaller building stones and bricks. They were able to produce massive structures with the smaller blocks by utilizing mortar and cement. The use of smaller stones and bricks and cement clearly impacted society. Buildings could be constructed with much less effort since thousands of slaves were not necessary to move a single block of material. Life for the slaves clearly improved although the construction work was still very demanding. Also, the upper classes benefited from the more rapid construction of buildings and houses for their cities.

While the Roman engineers were outstanding in the construction of civil infrastructure, they did very little to advance the use of power in driving the few machines in use during that period. Humans were still used in most cases to power pumps, mills and cranes. In rare cases, water wheels were used at power mills to grind grain, but these were of the undershoot design and much less effective than the overshoot design that was developed later. While references on the history of technology are quiet on the issue, the life of a worker powering a pump or a mill must have been very difficult.

Eventually, the vast Roman Empire crumbled, and we entered a period that the historians classify as medieval civilization from 325 to 1300 AD. The fact that barbarians stripped and destroyed the cities does not imply that progress stopped, but it sure did slow progress to a snails pace. When Rome collapsed, the Byzantine Empire followed, and significant engineering achievements were made in developing dome structures and curved dams. The author has omitted discussion of these developments to keep this historical treatment brief.

15.2.4 Early Agriculture and the Use of Animals

In spite of the slow development of engineering and the lack of the construction of monuments during the Middle Ages, we experienced an early agricultural revolution. White [4] has described the changes that occurred as agriculture moved from the dry sandy soil of Italy into the heavy alluvial and wet soil of northern Europe. The plows that were effective in the light sandy soils would not turn the sod in the heavy, moisture-laden soil of the north. A new, heavy-wheel-plow was invented that would cut through the grass sod and turn over the soil. However, this plow required eight yoked oxen to provide the large force necessary to pull it through the heavy sodded soil. Very few farmers of the day owned eight oxen. They had to combine their land, share their oxen and cooperate in plowing and planting. The combination of the farmers and their land holdings eventually led to a manorial society.

It may be difficult to imagine that the invention of a new plow made such a major change on the way people lived. Today we are able to produce all of the food needed in the U.S., while exporting excesses in significant quantities, with less than 2% of our population. However, before the advent of engines and motors (stream, gasoline and electrical), men and women struggled to grow enough food to keep from starving. In the Middle Ages, 90% of the population worked in agriculture, and they had much less to eat than we do today.

Oxen were used as a source of power for plowing and hauling on the farms. However, oxen are very slow and they consume large quantities of food. At that time, horses were of limited use because of inadequate harnesses and saddles and the frequent problem of splitting hoofs. (Early harnesses fitted about the horse's neck tended to strangle the animal.) A newly designed horse collar which transferred the load to the shoulders of the horse was introduced—together with iron horse shoes—alleviated these problems. Horses gradually replaced the oxen because they were 50% faster, worked longer and ate less.

In early warfare the horse was not very effective in frontal attacks because stirrups for the rider did not exist. Consequently, horse mounted warriors were of limited value because they could not adequately thrust their lance and stay mounted on the horse. With the advent of the stirrup, the horsemen could stand-up, lean forward and thrust the lance through the defenders' shields without losing their seat. The simple addition of stirrups to saddles had a profound influence on society. Feudalism evolved with an aristocracy based on warriors conquering and defending landholdings. A new battle strategy with armored knights mounted on large strong horses was dominant for several centuries [2].

In early history, small technological improvements produced major changes in the form of government, the type of society and the way people of all classes lived. It is remarkable that for about 5,000 years, humans were the prime source of power for most activity. Progress in agriculture and engineering proceeded together to provide more food, better housing, roads and bridges and eventually horse- and oxen-powered plows and wagons.

15.3 ENGINEERING AND THE INDUSTRIAL REVOLUTION

Civilization has always been power limited; even today, our space travel is severely constrained by the inadequacies of rocket power. Until the last three centuries, animals, windmills and waterwheels provided the only power available for agriculture, construction, milling, mining, and transportation. Progress was often slow because power was not available when and where it was needed. Life, in the power starved 17th century, was rather bleak for the general population. They worked from dawn to dark as farmers or craftsmen. Only those endowed with landed estates and/or money lived what we would consider a comfortable life.

15.3.1 The Development of Power

The first of several breakthroughs in developing new sources of power occurred in the 18th century. In Great Britain, Savery, Newcomen and Watt developed stationary steam engines. While these early steam engines were woefully inefficient (only 0.5%), they were produced in relatively large numbers (500 existed in 1800) and were employed to pump water and power mills (textile, rolling and flour).

Since the steam engines were replacing horses as power sources, James Watt conducted experiments to measure the power that a "brewery horse" could provide. His measurements indicated this horse generated 32,400 (foot-pounds/minute) of power. The horse's output was rounded to 33,000 (foot-pounds/minute) and this value is still used today as the conversion factor for one horsepower (HP).

Most of the 18th century was devoted to innovations directed toward improving several different models of steam engines. Applications were largely limited to stationary power sources because engines, of that day, were large, heavy and inefficient. However, with improved metallurgy and machining methods, it was possible in the 19th century to reduce the size and the weight of the engines while maintaining their output. These improvements permitted the steam engine to effectively power riverboats, ocean-crossing ships, rail locomotives and steam-powered automobiles.

The development of the steam engine and its adaptation to several modes of transportation formed the foundations of the industrial revolution. The life style of almost everyone was markedly changed with the emergence of steam-powered factories and much more effective transportation. Fewer people were employed in agriculture, but many more were working under appalling conditions in factories and mines. Twelve-hour days, seven days a week, with only Christmas as a holiday during an entire year was the norm. Manufactured products, such as clothing and housewares, were available to a larger share of the population at reduced prices, but the margin between earnings and the money necessary for a factory worker, miner, or farmer to stay alive remained very narrow. Steam power had relieved many thousands of men and horses from brutally hard labor, but they were released from one dull job to another. On a more positive note, the progress in agriculture and engineering enabled the population to triple during this period (1660 to 1820).

15.4 ENGINEERING IN THE 19TH AND 20TH CENTURIES

For the eight thousand years of recorded history prior to about 1800, the advance of technology was extremely slow. We lived in an agricultural society with most of the population growing food, which was consumed locally. Factories were being developed and simple products needed on the farm, in the home, or by the military were produced. Transportation of goods was usually limited to sailing ships on the seas or horse drawn barges on rivers or canals.

During the 19th and 20th centuries, technology literally exploded. The many scientific discoveries of the preceding millennium sufficiently expanded knowledge to allow countless engineering applications. To illustrate this explosion, let's define modern technology to consist of the following four elements:[3]

1. Abundant available power.
2. Transportation by air, land and water with associated infrastructure to provide rapid, convenient and inexpensive access.
3. Communication between everyone, anywhere, anytime, and at a reasonable cost.
4. Information (knowledge) storage, retrieval, and processing in seconds at any location, anytime, and at reasonable costs.

---

[3] We assume in this listing that housing, food and housewares are available to all from the technological-based industrial complex already existing in developed countries. A significantly different listing would be necessary for third-world countries.

While the advances in technology have been astonishing, particularly in the last 50 years, there are still significant deficiencies. Rocket power for launching space vehicles currently constrains space travel. Traffic jams in large cities extend the working day for many millions of people every workday. We still suffer nearly 45,000 deaths by accident on the highways every year. Communication in many parts of the world is still not available to the public. (Five years ago the author vacationed near a small fishing village in Mexico and learned that the radiophone at the tourist hotel was the only phone in the entire area.) We are just beginning to explore the different ways we can use the massive amount of information that has recently become available to people fortunate enough to own a personal computer and a modem.

Take some time during the summer or spring recess and read about the technological developments of the last two centuries. Several key developments with approximate dates are listed below. Think about these inventions, and how they have affected your life.

- Batteries 1800.
- Steam locomotive 1825.
- Electric generator 1831.
- Steamship 1835.
- Electric motor 1835.
- Telegraph 1837
- Baltimore to Wheeling railway 1850.
- Telephone 1876.
- Internal combustion engine 1876.
- Incandescent lamps 1880.
- Central electric power generation 1882.
- Steam turbines 1889.
- Alternators (ac generators) 1893.
- Radio 1896.
- Automobile 1900.
- Airplane 1903.
- Vacuum tubes 1906.
- Television 1935.
- Jet engines 1940.
- Atomic bomb 1945.
- Nuclear bomb 1950.
- Digital computer 1950.
- Transistor 1953.
- Integrated circuit (IC) 1959.
- Nuclear power 1960.
- Communication satellite 1970.
- Microprocessor 1971.
- Cellular phones 1982.
- Internet 1990.
- Electronic commerce 1996

Obviously, this list is not complete, as many more important innovations could have been included. The idea is to show you that the period from 1800 to 1990 was packed full of new technology. This technology markedly impacted the way that we live, work and play.

We can also debate the exact year of the invention. However, the exact date is rarely important since an invention usually evolves into a commercially successful product after several iterations with improvements or modifications by several designers. You should not worry about the exact dates. Instead, you should recognize the continuous progress and observe the shift in emphasis from one period to another. In power, we moved from animals, to steam, to internal combustion (gasoline) and to nuclear energy.

Power, from gasoline fueled internal combustion engines, became mobile and electricity was widely distributed. In transportation, we improved from the horse and buggy and sailing ships to automobiles, airliners and nuclear powered submarines. In communications, we progressed from the pony express to the telegraph, telephone, fax machines, cellular telephones, pagers and the Internet. In information, we are currently evolving from a hard copy library system to a digital multi-media information bank available at any time, to everyone, at nearly zero cost.

Have these advances in technology changed society? We believe that they have had a profound beneficial effect. We produce an overabundance of food with less than 2% of our population working in agriculture. We manufacture most of our consumable products with about 20% of the work force. Our workweek is shorter, and we have many more affordable products. Most families own at least two cars and home ownership is at an all time high. We do hear concerns regarding the improvements in our standard of living over the past 20 years, particularly for those people at the lower end of the income spectrum. Technology has probably not helped this group as much as others. Global trade has permitted many of the semi-skilled tasks to be moved to third world countries where the cost of labor is about 10 to 15% as much as in the developed countries. Consequently, many of the previously well-paid factory jobs in manufacturing in the U.S. have been moved offshore.

15.5 NEW UNDERSTANDINGS

In the discussion of the historical development of technology, we have been optimistic about the very positive effects of technology on society. In the past the public accepted, without serious questions, all forms of technology and the risks that were involved. However, the situation has changed to a remarkable extent in the past 40 years. Our leadership (engineers included) have lost the public trust. An excellent example of this loss pertains to the technology employed in the generation of electricity using nuclear power. Since the Three Mile Island incident in 1979, the general public in the U.S. wants no part of nuclear power. There is a valid public perception that the technological risks associated with nuclear power were significantly understated. John O'Leary, a deputy director for licensing at the Atomic Energy Commission (AEC), stated that "the frequency of serious and potentially catastrophic nuclear incidents supports the conclusion that sooner or later a major disaster will occur at a major generating facility" [6]. This statement was made several years prior to the accident at Three Mile Island. In spite of this warning of the likelihood of a serious accident, federal, state or local governments have not planned for orderly emergency evacuation of nearby residents or for containment measures to limit the contamination of many of our reactors.

The Chernobyl accident in 1986 proved that O'Leary was correct. The No. 4 reactor at the Chernobyl facility exploded sending a radioactive plume in the air so high that an alert nuclear plant operator in Sweden detected it. Many died in this radioactive explosion and many more will die in the future from radiation poisoning. High radiation levels affected countries as far away from Chernobyl as Italy.

The nuclear industry and the government agencies regulating it, world wide, have not been candid with the public. Nor has the public been very interested in intelligent debate—resorting instead to large, ugly demonstrations to express their displeasure. We need an accurate and honest assessment of the risks of any nuclear power project by the industry, the regulators and experts representing the public. The risks must be weighed against the benefits to society. Sometime in the future fossil fuels will be come unavailable, scarce or costly. When the fuel crises occurs, we may find that nuclear power is the only viable alternative for power generation from carbon burning plants.

We are not here to argue the case for or against nuclear power. We use it as an example of the fact that understating or poorly assessing risk brings a severe penalty—loss of public trust followed by a ban on technical development in the subject area.

As engineers, we must significantly broaden our perception regarding technological risk. Today, many engineers dismiss failure due to human, operator or pilot error because they are not machine errors. This attitude is absolutely wrong. The human operator and the machine constitute a system and affect public safety accordingly. The machine and its operator must be included in system design since they both constitute an interacting set of weaknesses and capabilities [6].

Another new understanding pertains to geography. Prior to the Second World War (WW-II), international trade was very limited. We produced what we consumed, mined our minerals, and pumped our own oil. Geography was very important because it constrained trade by law and by distance. After the war, laws were changed and trade agreements (GATT, NAFTA, etc.) were arranged. International trade was encouraged so much so that the U.S. now consistently imports more goods and materials than it exports. Our annual deficit in the trade balance, exceeding two hundred billion dollars, is the equivalent of 5,000,000 high-income jobs paying $40,000 a year.

Previously, the cost of shipping limited trade among countries separated by large distances. However, following WW-II many technological improvements were implemented in the shipping business: we developed the super tanker—greatly improving the efficiency of shipping huge quantities of oil. We also developed very large containerized ocean vessels reducing the cost and time of loading and unloading cargo. Large efficient diesel engines were installed on the ships permitting increased speed and significant reductions in shipping time.

With the time and costs of shipping greatly reduced and the legal barriers to trade removed, we evolved into a global marketplace. Engineers compete worldwide to develop world-class products or services. Products designed in the U.S. may be produced anywhere in the world, or products designed in Europe or Asia may be produced in the U.S. Today, design and production activities are located to minimize costs and to maximize benefits to the customers and/or to the corporate entities.

The third new understanding is in the role of the governments (federal, state, county and city). Prior to WW-II, governments were small and taxes were relatively low. The primary role of the federal government was to insure national security. The state governments worried mostly about transportation infrastructure, and the local governments concerned themselves with education. Today, the situation is markedly different. Governmental agencies, at all levels and in response to their interpretation of the law (either old or new), issue regulations which pertain to issues of concern to society including the protection of the environment, safety, health and energy conservation.

We can debate the wisdom of many of the regulations, but that is not the point. The regulations currently exist and new ones will continue to be issued with an alarmingly high frequency. As an engineer, you must be prepared to serve society-at-large as reflected by these governmental regulations. If you question the wisdom of some of the regulations, the best approach is to become involved in the political process so that you may have the opportunity to influence them. Engineers tend to resist constraints because they limit the freedom of their designs and add to the cost of the product. We tend to think of regulations as government imposed constraints, but they are really barriers imposed by society-at-large. Society controls the regulatory process because it has the voting power to change the decisions of the politicians drafting the laws and of the bureaucrats formulating the regulations.

15.6 BUSINESS, CONSUMERS AND SOCIETY

Engineers serve three constituencies—business, the customer and society-at-large. The business leaders (management) look to engineers to develop products, services, and processes that meet global competitive challenges. Consumers seek more convenient, reliable, enhanced and value-laden products at reduced prices. Society-at-large, through elected politicians and public interest groups, demand action leading to solutions of problems involving safety, health, energy conservation and preserving or improving the environment. It is an overlapping set of constituencies, as shown in Fig. 15.1, because society encompasses business, governments, and consumers.

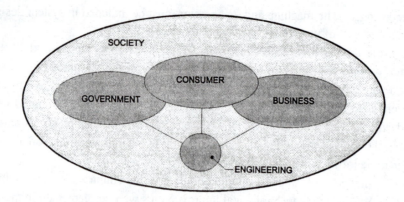

Fig. 15.1 Schematic illustration of the overlapping constituencies served by engineers.

To discuss the issues raised by serving three constituencies, we will use the U.S. automotive industry as an example. Marina Whitman has described many aspects regarding the three-way demands. Since we will draw from her excellent contribution, you may want to study reference [8] in depth.

Business demands are simple although they are often difficult to meet. Any business must be profitable to continue to exist. Losses can occur over the short term, but if the red ink is prolonged the company is forced by its creditors into bankruptcy proceedings. Making a well-defined profit every quarter is the usual goal of management. For many years following WWII, it was relatively easy for U.S. businesses to make money. Bombing and shelling during the war had destroyed essentially all the manufacturing capability in every developed country in the world except in the U.S. It was easy to be a leader when the competition was without factories or infrastructure. For the automobile industry, the golden years began to erode in the late 1960s, and survival became a life or death battle in the 1980s. Problems for the auto industry came from global competition (mostly Japan). The Japanese firms provided much more reliable automobiles at lower costs than those produced in the U.S.

To show the extent of the market penetration by imported automobiles, recall that the imports accounted for only 4% of the market in 1962, but they controlled 30% of the market by 1990. This devastating loss of market share was not limited to the auto industry; several other industries including optics, consumer electronics, ship building, steel, etc. suffered even more serious losses.

Engineers and management in the U.S. firms had lost the competitive edge after decades of modest competition from the Europeans. The Japanese automobile companies (Toyota, Honda, Nissan, etc.) provided more affordable, reliable, appealing, fuel efficient and value-laden cars than either the American or European companies. In recent years, the loss market share has been reversed by improved participative management and technical methods (concurrent design and systems engineering[4]), but regaining the market share of the 1960s will be a very long and difficult process.

The second constituency, the consumers, has been well recognized in the past decade. The importance of keeping the customer pleased is foremost in the minds of both management and engineering. Systematic methods for establishing the consumers' needs have been developed and placed into practice [9]. Many companies in the U. S. currently follow a product development method that begins with the customer interviews and proceeds through all of the phases of product development. This procedure is briefly described in Chapter 12. We do not want to underestimate the importance of the second constituency, but engineers have usually kept the customer in mind during the design of new products. The newer methods [9] tend to insure that the customers' requirements are more accurately defined by the product specification before beginning a development.

---

[4]Systems engineering means many different things to different groups. We define it as a cost and time efficient method to specify, design, develop and integrate all of the individual subsystems so that the total system for the product meets the requirements of customers, business and society-at-large.

The third constituency, society-at-large, generates the most challenging requirements. Society-at-large is represented by government agencies and a variety of public action groups. Officials of the government agencies (bureaucrats) produce regulations that define requirements for automobiles that may affect safety, fuel conservation or the environment. The public action groups create pressure and influence politicians in setting public policy that eventually generates new requirements.

Engineers often have some difficulties with regulations and/or proposals from the public action groups. Lawyers write the laws, and public servants write the regulations. Because lawyers and public servants are often technically illiterate, they sometimes seek technical support prior to drafting binding regulations. However, sometimes, regulations are placed into effect by government bureaucrats prior to thorough study by knowledgeable representatives of the public and engineering communities. The difficulty with this procedure is that the demands of the customer, including style, comfort, affordability, reliability, performance, convenience, etc., may be in conflict with societal imposed regulations on safety, fuel efficiency and emissions.

A very recent example illustrates what many believe to be an ill-advised regulation [10]. The National Highway Transportation Safety Administration (NHTSA) mandated requirement for dual airbags in every vehicle licensed in the U.S. Most people consider air bags a valuable safety feature because several thousand lives have been saved in serious frontal accidents. However, since 1990 many children and small adults (mostly women) have been killed by rapidly inflating airbags. As of May 6, 2000, 158 people have died by rapidly deploying air bags since the use of this safety feature was mandated on the passenger side [11]. The number of deaths of women, children and senior citizens continues to increase with time.

In a very recent ruling that modifies the existing standard, NHTSA reduced the impact speed of the crash test from 30 to 25 MPH [11]. NHTSA will also require vehicles be tested using a family of dummies—an average size man, a small woman and 1, 3 and 6 year-old children. This new standard will be phased in over a three-year period beginning in 2004. Hopefully these changes will enable the protection of large adults while eliminating the dangers of air bags to small women and children.

The killing of women and children by a safety device is a **very serious concern**. Moreover, the regulation prohibits anyone other than the car's owner to disable even the passenger side air bag[5]. If you are a concerned parent, you can disable it yourself, but few owners have the required skills. Also, if you ruin the mechanism and/or the bag, the cost of replacement is about $1000. Because of their high costs, air bags have replaced stereo systems as the theft of choice if your car is robbed.

The issue is not whether we should have air bags or not. They have proven to be a very effective safety device for most adults either driving or riding as passengers. The issue is with the rigidity of the federal mandates and the failure of the regulators to:

1. Update the regulations to accommodate different behavior of drivers.
2. Provide more design flexibility in the requirements.
3. Provide a means of permitting the driver to activate or deactivate the passenger side bag.

The air bag regulation was written in 1977 when only 14% of the population even bothered to use their seat belts. By 1994, when the mandate was fully effective, 49 of 50 states were enforcing seat belt laws and seat belt usage had risen to 68%. While seat belts do not eliminate the need for air bags, they do

---

[5] A recent ruling from the NHTSA allows auto dealers to install an on-off switch to control the passenger side air bag; however, NHTSA requires an extremely complex process for drivers to obtain a waiver, which is necessary before the dealer will perform the necessary modifications to the deployment system. First, you have to obtain a NHTSA request form. Then you must swear—under threat of criminal penalty—that you are too short to be safe with an air bag deployment, or cannot always put the children in the back seat, or have one of five NHTSA approved medical conditions. If you option for the medical conditions you must have a letter from a medical doctor confirming the condition. To date relatively few people (11,838 requests were filed as of October 30, 1997) have fought the bureaucratic battle with NHTSA and obtained waivers.

increase the time allowed for the deployment of the air bag. The NHTSA did not change the regulation to accommodate the increased usage of seat belts until early in 1997. Responding to the controversy about the risk of air bags, NHTSA issued a new regulation permitting automakers to install airbags 25% to 30% less powerful than the current bags. The new regulation gives the automakers two years to implement the change. In the meantime on many cars manufactured before 1999, air bags continued to be deployed at a speed of 200 MPH to protect an unbelted adult male. This is great for the big guys; however, for kids or smaller women, a 200 MPH kick in the face is extremely dangerous. When seat belts are used, there is no need for the 200 MPH deployment of the bag because the belt constrains the passenger providing more time for deployment of the bag.

Industry representatives recognized these problems and warned the NHTSA of the dangers to children. Rather than change the regulation to provide more flexibility in the design of the air bag systems, NHTSA drafted a warning label and industry added instructions in the owner's manuals explicitly cautioning against placing a front facing child safety seat in the car's front seat. The manuals indicate that it is safe to place a child in the passenger seat if the child is buckled-in with a seat belt, **BUT** the seat must be in the full rear position. The manual also states that children should never lean over with their faces near the air bag cover while the car is in motion. The warning labels and instructions in the owner's manual are a result of poor design that was forced by an inflexible regulation. We now have in place a safety device (air bag) that is dangerous to small children. Warning labels and instructions in manuals reduce the consequences of litigation (there is another law indicating that the user must be warned of possible dangers), but they are totally inadequate to protect children, too young to read, riding in the passenger seat.

It would be relatively easy to modify the design of the control system to include an on/off switch with a warning light on the control panel of the automobile. With this feature the driver could deactivate the passenger side air bag whenever he or she wished to do so. Another design approach entails a modification of the pneumatic system to reduce the explosive force with which air bags deploy. This modification would probably have taken place several years ago when seat belt usage finally became widespread. However, the federal regulation did not permit modification then, and only recently has it allowed modification after the death of 82 (and counting) helpless women and children.

Another problem with air bags that is not understood by the public is related to sodium azide, which is used as the propellant. Sodium azide is very toxic. When the air bag is deployed, you may be contaminated, and usual practice is a trip to the hospital for decontamination.

The public is concerned, if not angry. The NHTSA has received many letters from owners requesting the waivers required to have trained mechanics at the dealers to disconnect the dangerous air bags. Many mechanics refuse to disconnect the airbags for fear of future litigation. A recent survey of 700 repair shops indicated that two-thirds refused to install the on-off switches. There appears to be a significant amount of confusion about the remedy.

The big three automakers do not have a ready solution. GM proposes cut-off switches in existing vehicles. Ford is in favor of permitting only certain categories of people to disconnect the system. Chrysler is reconsidering its opposition to cut-off switches. The dealers and independent mechanics do not want to touch the problem because they fear they might be sued sometime in the future. With the new regulation, it appears that all of the automakers will be reducing the power of the deployment of the airbags, which should mitigate the problem in time.

The current disarray is alarming. The interaction of the government agencies representing the public-at-large, business and the consumer is woefully inadequate. We need a solution, and we need it quickly. Yet, the regulatory system is too awkward to provide rapid solutions. The awkwardness comes from inflexibility on the part of the government and inadequate cooperation between the automakers. We can understand competition on the part of business to increase market share, but safety features should be excluded from the normal competitive secrecy. The government agencies must become much more flexible and recognize that they cannot decree away all danger.

The difficulties between the government and business are not new. An interesting summary of the regulatory relationship between government and business was recently published by Jasanoff [12]. We quote her directly:

Studies of public health, safety and environmental regulation published in the 1980s reveal striking differences between American and European practices for managing technological risks. These studies show that U.S. regulators on the whole were quicker to respond to new risks, more aggressive in pursuing old ones, and more concerned with producing technical justifications for their actions than their European counterparts. Regulatory styles, too, diverged sharply somewhere over the Atlantic Ocean. The U.S. processes for making risk decisions impressed all observers as costly, confrontational, litigious, formal, and unusually open to participation. European decision making, despite important differences within and among countries, seemed by comparison almost uniformly cooperative and consensual; informal, cost conscious, and for the most part closed to the public.

The assessment by Sheila Jasanoff gives us clear direction. We must reduce the adversarial relationship between business and government. As engineers working for both government and business, we should be an important element in the future to produce a regulatory system with more flexibility, cooperation, cost consciousness, and consensus.

15.7 CONCLUSIONS

The standard of living in the U.S. will be determined by the interplay of three powerful influences:

- New and rapid technological advances.
- Business (management) response to global consumer demands.
- Social demands as evidenced by legislative and regulatory requirements.

Engineers have always provided the leadership in technological advances, and we expect to continue to provide business and the public-at-large with cutting-edge, world-class technology.

Business requires capital, technology and good management to remain competitive. In addition to providing the technological base for a company, engineers frequently serve in management. If this career path appeals to you, plan on extending your education in a business school. The combination of a Bachelor of Science degree in engineering and a Master of Science in business provides a very solid foundation for a career path in a technically oriented business.

In the past, engineers usually have not been actively engaged in societal issues. We lack patience in dealing with the public and become irritated by their lack of technical literacy. We fume at the inefficiencies of government agencies and their lack of flexibility. We become outraged at the arrogance of bureaucrats and at their lack of concern for time and costs. We retreat into our caves and work on new developments and new products. In the meantime, the public-at-large grows wary of many important sociotechnical systems and the government resorts to litigious processes, which impedes real progress.

It is unfortunate that engineers have not been a significant force in dealing with societal issues. As society, business and the customer create conflicting demands, engineers are essential in the crafting of well-balanced solutions. The conflicting demands provide new opportunities and complex challenges for engineers. To take advantage of these opportunities, the engineers of tomorrow will require enhanced communication skill, greater disciplinary flexibility, a better understanding of the mechanisms of regulatory agencies, and a much wider perspective relative to societal demands.

REFERENCES

1   Singer, C. E. J. Holmyard, A. Hall, and T. Williams, Eds. A History of Technology, Oxford University Press, New York, five volumes, 1954-1958.
2   Hughes, T. P., "From Deterministic Dynamos to Seamless-Web Systems," Engineering as a Social Enterprise, ed. Sladovich, H. E., National Academy Press, Washington, D. C. 1991, pp. 7-25.

3 Sladovich, H. E., ed. <u>Engineering as a Social Enterprise,</u> National Academy Press, Washington, D. C. 1991, pp. 7-25.

4 White, L., Jr., <u>Medieval Technology and Social Change</u>, Oxford: Clarendon Press, 1962

5 Kirby, R. S., S. Withington, A. B. Darling, and F. G. Kilgour, <u>Engineering in History</u>, Dover Publications, New York, 1990.

6 Ford, D. F., <u>Three Mile Island: Thirty Minutes to Meltdown</u>, Viking, New York 1982

7 Adams, R. McC. "Cultural and Sociotechnical Values," <u>Engineering as a Social Enterprise</u>, ed. Sladovich, H. E., National Academy Press, Washington, D. C. 1991, pp. 26-38.

8 Whitman, M. v. N., "Business, Consumers, and Society-at-Large: New Demands and Expectations," ed. Sladovich, H. E., National Academy Press, Washington, D. C. 1991, pp. 41-57.

9 Clausing, D. <u>Total Quality Development: World-Class Concurrent Engineering</u>, ASME Press, New York, 1994.

10 Payne, H. "Misguided Mandate," Scripps Howard News Service, Knoxville News-Sentinel, January 12, 1997, p. F-1.

11 "New Air Bag Rules Meant to Help Kids, but Critics Say They May Harm Adults", Knoxville News Sentinel, May 6, 2000.

12 Jasanoff, S. "American Exceptionalism and the Political Acknowledgment of Risk," Daedalus, Vol. 119, No. 4, 1990, pp. 61-81.

EXERCISES

15.1 Recall a product that you or your family has purchased in the past month or so. Write a brief paper describing both the positive and negative impacts of that product on society.

15.2 The Federal Aviation Administration enacted a regulation affecting the mailing of packages weighing more than 16 ounces. The regulation requires you to present the package to a postal clerk for mailing and prohibits the mailing of the package from a post box. Write a paper covering the following issues:

- Describe the regulation in more detail.
- Why is the FAA writing regulations affecting the U.S. Post Office?
- What effect (cost) does this regulation have on the individual and on small business.
- Do you believe that the regulation will be effective for its intended purpose? Please give arguments supporting your viewpoint.
- What actions will the post offices (40,000 of them) have to take to make the regulation effective?
- What actions do the post offices actually take with regard to the regulation?
- Will these actions be costly? Estimate the costs to both the public and the individual. Assume that a postal clerk is paid $10.00/hour and that an individual considers his or her free time worth $10.00/hour.
- Give your assessment of the cost to benefit ratio for this regulation?

15.3 Suppose that you were a slave in the time of Rameses II and were assigned to the task of constructing his statue. For those not up to date on Egyptian statues, this one weighs 1,000 tons and was 56 feet tall. Write a paper describing your daily tasks.

15.4 Horses were not used extensively to relieve man from brutal work or to provide a significant advantage in military actions until nearly 1000 AD. Write a brief paper describing both the social and technical reasons for the very long time required to effectively employ the horse in either military or commercial enterprise.

15.5 Suppose that you were living on a manor in England in about 1300 AD. Describe your lifestyle and indicate how technology affects your work and play. Select in which of the two classes of society that you existed.

15.6 Describe where power is available in the U.S. today. Define the type of power in your response. Also, include an example where you personally suffered because of a lack of power.

15.7 Write a paper comparing the lifestyles, as you imagine them, for a man living in 2000 and 1000. Did technology make a difference in the quality of life?

15.8 Repeat Exercise 15.7 replacing the man with a woman.

15.9 Why is the public-at-large opposed to power generation with nuclear energy? Is the public-at-large correct in their collective assessment?

15.10 Why do we have a global marketplace today? Does global trade improve our standard of living? What does our current trade deficit have to do with wages for factory workers? Does technology help or hinder our balance of trade deficit?

15.11 If you were the director of the National Highway Transportation Safety Administration (NHTSA), what action would you take regarding the public concern about air bags?

- Explain the reasons for your actions.
- Explain how you would convince the big three automakers to agree with you.
- Prepare an outline you would follow in the press conference announcing this action.
- Describe how you would handle a public interest group disagreeing with your ideas.
- Describe how you would handle the public interest groups that agree with you.

15.12 Prepare a paper describing the information provided by NHTSA on air bags. A significant amount of information is given on its Web Site at http://www.nhtsa.dot.gov/airbags.

15.13 In preparing your paper for Exercise 15.12, you noted that the data for lives saved and lost was for the period, which closed on November 1, 1997. Explain why the NHTSA has not updated its Web Site on this important topic.

15.14 What social science courses should you take during your undergraduate program to broaden your prospective and aid you in dealing with sociotechnical issues?

15.15 Is there a technical elective in your department that deals with sociotechnical issues?

CHAPTER 16

SAFETY, RISK AND PERFORMANCE

16.1 LEVELS OF RISK

When engineers design a new product or modify an existing one, we often introduce an element of risk in society-at-large. Sometimes the risk is minimal, and the resulting damage from a malfunction or failure is small. However, in some instances, the risk may be large and the damage may be catastrophic.

Some simple examples are in order to illustrate the concept of risk with attendant benefits to society. Suppose we design a new model of an electric, pistol-grip drill that is powered from the standard 120 volt, 60 cycle single-phase power supply from the local utility company. The operator is always at risk for an electric shock. It is our responsibility as engineers to minimize this risk while maintaining the advantage in performance of an electric-powered drill.

16.1.1 Acceptable Risk

How do we protect the operator from electric shock? The operator is holding an electric motor with 120 volts across the armature coils and the field coils adjacent to his or her hands. The answer in this instance is to build the case for the drill from a tough, durable plastic that is structurally strong, but also an excellent electrical insulator. The case keeps the operator from touching any part of the electrical circuit powering the motor. In fact, Black & Decker® a major manufacturer of small power tools, uses double insulation to provide two independent insulation barriers to prevent the operator from making contact with any part of the electrical circuit. Operators may still manage to receive an occasional shock, but it will not be easy. Accidental contact of the operator's hands with a live electrical circuit is a rare event. In fact, the risk of electrical shock is so small that we routinely pick up a power tool and employ it to perform some task without even a passing thought about the possibility of getting shocked. Operating a power tool does involve risk, but it is acceptable because the operator is not consciously concerned about his or her safety while using the tool.

16.1.2 Voluntary Risk

Let's move up the risk ladder and consider flying in a commercial airliner from point A to point B, some 1,200 miles distant. Is it safe for us to make this trip? Most of us appreciate that it is reasonably safe to fly commercial airliners, but realize that there is a slight probability of a crash. The statistics show the probability of a fatality in an airline accident to be about 1 in a billion passenger miles flown. So if you make the round trip from point A to B, your probability of getting killed due to an airline accident is:

$$P_k = (2)(1200)/10^9 = 2.4 \times 10^{-6} = 0.00024\%$$

This is a very low probability for a fatal accident; consequently, most people do not worry much about the possibility of dying in a crash when they board an airliner. However, if you are a sales engineer flying 100,000 miles a year, every year for 20 years, your total mileage accumulates to 2×10^6 miles. Your probability of getting killed in a crash increases to:

$$P_k = 2 \times 10^6/10^9 = 2 \times 10^{-3} = 0.2\%$$

The probability has increased significantly if you accumulate the miles flown by a traveling professional over a 20-year period. Knowing there is one chance in 500 that you will die in an airline crash, would you still want to be a sales engineer flying weekly to meet with customers? Could you tolerate this level of risk? Would you be apprehensive?

Everyone has some tolerance for risk. We weigh the speed, convenience, cost and risk of flying against that of traveling by train or by car. In almost all instances, travelers select the plane for long distances and the car for short distances. We select the train only in those rare instances when it combines relatively good service (high speed with frequent trains) to convenient (downtown) locations. Most of us do not contemplate the risks involved in travel because, except for driving a car, fatal accidents are rare events.

Unfortunately, fatal accidents in automobiles are not rare events. Each year about 45,000 fatalities occur on U.S. highways. In addition hundreds of thousands of serious injuries occur in these accidents. It is clearly more dangerous to drive than to fly when you compare the probability of fatalities and/or injuries on a per mile basis. How do we handle the higher risk associated with driving? Rationalization for one—I am an excellent driver, and it will not happen to me. Not necessarily a true statement, but the rationalization of one's superior driving skills alleviates the worry and concern about a fatality or injury producing accident. The real risk still remains.

When we drive, we are an active factor in determining the risk. Our driving habits (high speed, reckless steering, tailgating, etc.) and driving skills affect—to a large degree—the level of risk involved. When flying, we are only a participant albeit a passive one. We have voluntarily decided to fly and sometimes even have a choice of airlines. We can pick the airline with the best safety record. (Southwest Airlines is the best with no accidents in their entire 26-year history). However, sometimes, even knowing the facts, we are placed at risk, and we have little or no choice in the matter.

16.1.3 Involuntary Risk

Suppose you live 40 miles northeast of a nuclear power plant. Is 40 miles sufficiently far away to avoid significant fallout from a serious explosion involving the nuclear reactor at the power plant? Remember the winds are usually from the Southwest, so that the fall-out plume will be pointed in your direction in the event of an accidental release of radioactive gasses or particles.

The risk associated with radioactive fall-out, air pollution, toxic chemicals, polluted ground water, etc., is a serious concern to almost everyone; people do not like to be exposed involuntarily. They expect governmental regulators to control the environment and to reduce the level of the risks of accidental exposure. Unfortunately, governmental agencies are not effective in protecting the population from these exposures. Accidents have occurred in the control of reactors, and citations for safety violations by inspectors of the Nuclear Regulatory Commission are commonplace. From a public perception, the risk of a serious malfunction of a nuclear reactor became unacceptably high in the U.S. and many other European countries. After the accident at the Three Mile Island facility in 1979, the commercial nuclear industry died in this country. Some plants that were under construction were completed and others were converted to fossil fuels, but no new nuclear plants have been started since. With the more recent and very serious accident at Chernobyl in 1986, any new plans for power generation using nuclear energy in the U.S. have been placed on hold for the foreseeable future.

There is a very large difference between the acceptable level of risk depending on whether the risk-producing activity is voluntary or involuntary. The degree to which we personally control the risk is also extremely important. Activities like skiing, scuba diving, horseback riding, hang gliding, mountain climbing, dirt biking, etc. carry extremely high levels of risk. Yet, intelligent people swarm to the ski resorts and pay big bucks to potentially break their bones. Why? They have voluntarily decided that the thrill, or other pleasures, derived from the activity is worth the risk. Also, they control the level of risk. They can chose the "bunny" slope and minimize the risk or the "black diamond" slope to maximize the thrill.

As engineers, we must recognize that some level of risk is involved in almost all the products we produce. It is imperative that we minimize this risk while maintaining an acceptable level of performance. Also, it is essential that this risk be acceptable to our customers and to society-at-large. An appropriate balance between risk and performance must be achieved. It is important that we cooperate with business and governmental regulatory agencies to provide a realistic assessment of the risk level. Finally, the public should be made aware of the risks involved and should not be surprised by news releases describing the gruesome details of victims involved in these accidents.

16.2 MINIMIZING THE RISK

An engineer minimizes risk by preventing failure of each and every component of the system. This is not an easy task as there are several different ways in which components can fail. Parts fail by breaking and aging, by corrosion or fatigue, by overload and burning, etc. In some instances, we can anticipate these failures and replace the parts before they malfunction. We call this controlled replacement of finite life parts—scheduled maintenance. In most cases, we try to design each component so that failure will not occur during the anticipated life of the product.

16.2.1 Safety in Design

A complete treatment of methods to avoid failure would require more than a single chapter or even an entire textbook. However, an analytical procedure for designing tension members will be introduced to provide you with an illustration of a conservative design, which insures safety. These tension members will be sized (made sufficiently large) so that they exhibit a safety factor. In other words, they will accommodate a service load higher than anticipated in service over their entire life cycle.

Let's begin by introducing a tension member as shown in Fig. 16.1.

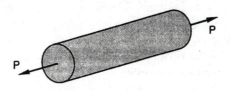

Fig. 16.1 A tension member is a long thin rod subjected to axial load P.

When the rod is subjected to an axial load P, an axial stress σ develops which is uniformly distributed over the cross section of the rod. This uniformly distributed stress acting on the rod is illustrated in Fig. 16.2.

We determine the magnitude of this stress from the formula:

$$\sigma = P/A \qquad (16.1)$$

where P is the load in lb or Newton
A is the cross sectional area of the rod in in^2 or m^2.

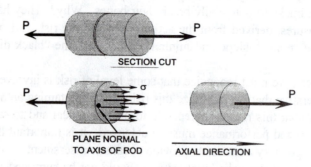

Fig. 16.2 The stress σ is distributed uniformly over the cross-sectional area of the tension rod.

Two simple examples for computing stresses are given below:

EXAMPLE 16.1

If the axial load P = 10,000 lb, determine the axial stress σ in a circular rod with a ½ in. diameter.

Solution

The cross sectional area is given by:

$$A = \pi r^2 = \pi (0.25)^2 = 0.1963 \text{ in}^2$$

From Eq. (16.1), the axial stress σ is:

$$\sigma = P/A = 10,000/0.1963 = 50,930 \text{ psi}$$

where psi is the abbreviation for lb/in^2.

At this stage of the analysis, it is not possible to interpret the significance of our answer of 50,930 psi. We need more information pertaining to the strength of the material from which the tension member is fabricated. When the strength is known, we make a comparison of the magnitude of the stress with this strength and remark on its significance.

In this first example, we have used the U.S. Customary Units expressing the axial load in pounds, the cross sectional area in square inches, and the stress in psi or pounds per square inch. In the next example, we will use the International System of Units (SI) where the load is expressed in Newton, the cross sectional area in meters, and the stress in Pascal.

EXAMPLE 16.2

If the axial load P is 50,000 Newton (N), and the radius of the circular rod is 5 mm, find the axial stress σ.

Solution

First, determine the cross sectional area A as:

$$A = \pi r^2 = \pi(5 \times 10^{-3})^2 = 78.54 \times 10^{-6} \text{ m}^2$$

Then, from Eq. (16.1), we can determine the axial stress as:

$$\sigma = P/A = 50 \times 10^3 \text{ N}/ 78.54 \times 10^{-6} \text{ m}^2 = 636.6 \times 10^6 \text{ Pa} = 636.6 \text{ MPa}$$

where the abbreviation Pa stands for Pascal—the unit used for stress in the SI system.

A Pascal is equal to a Newton per square meter (N/m^2). When calculating stresses in the SI system, we usually determine very large numbers for the stress when it is expressed in Pascal. For this reason, we usually use MPa (mega Pascal) where one MPa $= 10^6$ Pa $= N/(mm^2)$.

Okay! We can compute the stress on a tension member if we can ascertain its size and the load that it must resist. The stress we determine is a number with associated units—either psi or MPa. As a number, it is not very useful until we compare it to the strength of the material from which the tension member is fabricated.

Let's suppose that the tension rod used in this example is machined from a very strong alloy steel with a yield strength of 100,000 psi. Do you think that the tension rod is safe? Will it retain its shape under load? Will it fail by yielding (stretching under load and not recovering completely when the load is removed)? We answer these questions by a simple comparison of the applied stress σ and the strength of the material S. In this example, the axial tensile stress $\sigma = 50,930$ psi and the strength of the material from which the rod was machined is S = 100,000 psi. The comparison shows that the yield strength of the rod is greater than the applied stress; therefore, the rod will not fail by yielding or breaking. But how safe is the rod? We know it will not fail; however, some measure of the safety of the rod should be established.

16.2.2 Safety Factor

To respond to the safety issue for the rod, determine the ratio of the strength divided by the stress, and define it as the safety factor (SF):

$$SF = S/\sigma \qquad (16.2)$$

For the stress imposed on the rod described in Example 16.1, the safety factor is given by Eq. (16.2) as:

$$SF = 100,000/50,930 = 1.96$$

The safety factor (SF = 1.96) indicates that the tension rod is almost twice as strong as it must be to resist yielding under the applied load. We have a comfortable safety factor and can be confident that the rod will perform safely in service. Our only concerns are the accuracy with which the load has been predicted and the quality control for the material used in manufacturing the tension rods. If the factors listed below are satisfied:

- We have extensive experience with the application.
- We are certain of the magnitude of the load.
- We have confidence in our manufacturing division in tracking their materials.
- We have verified the strength of the materials received from suppliers.

Then we usually can be satisfied that a safety factor of about two is adequate. However, if we are not certain about the load and/or the materials, a safety factor of two is not sufficient. We increase the safety factor to accommodate our ignorance of the applied load or our lack of control over the material employed.

Safety factors ranging from two to four are commonly employed in design. Safety factors of less than two require considerable care and expense. The use of relatively low safety factors is justified only for very high performance applications or for very high volume components. In these special situations, the engineering analysis, prototype testing, quality control inspections, maintenance inspections and documentation need to be extensive.

16.2.3 Margin of Safety

The margin of safety (MS) is sometimes used to describe the degree of safety incorporated into the design of a component. The margin of safety should not be confused with the safety factor as they are different quantities. The margin of safety is defined as:

$$MS = (S - \sigma)/\sigma = SF - 1 \qquad (16.3)$$

In Example 16.1, the margin of safety MS = 1.96 − 1 = 0.96 or 96%. The tension rod has a margin of safety of 96%, which implies that the strength of the material exceeds the applied tensile stress by 96%. If the load on a tension member can be determined, it is easy to size (adjust the cross sectional area) the rod to provide any margin of safety that you deem necessary to satisfy management and to meet professional obligations to society-at-large.

16.3 FAILURE RATE

Sometimes components fail by burning out, aging or wearing out—not necessarily by breaking or yielding (excessive permanent deformation). In this type of a failure analysis, the safety factor is not the relevant parameter. We must cope with wear, aging, or burn out by using other methods. To begin, let's recognize that all components do not wear out at the same time. Some people drive their car 50,000 miles before the brakes wear out and others may drive 60,000 miles or more. Some people can use their personal computers for several years before a component fails; yet others have problems after only a few months in service. To analyze these differences in service before failure, we introduce in Fig. 16.3 what is known as a mortality curve for both mechanical and electronic components.

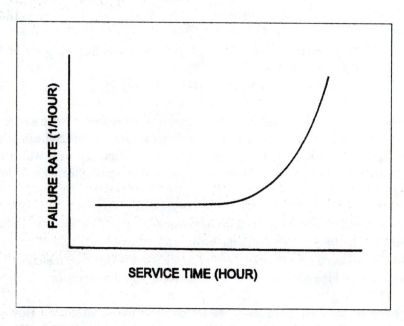

Fig. 16.3a Mortality curve for mechanical components showing failure rates with time in service.

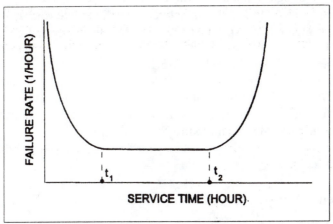

Fig. 16.3b Mortality curve showing the failure rate for electronic components with time in service.

To introduce the concept of failure rate FR, we maintain records of the performance with time-in-service of a large number of components. We begin our log of performance with N_0 components that are placed in service at time $t = 0$. After some arbitrary time t, some number N_f will have failed and the remainder N_s will have survived. Clearly at any arbitrary time:

$$N_f + N_s = N_0 \qquad\qquad\qquad (16.4)$$

With time, N_f increases and N_s decreases, but the sum remains constant and equal to N_0. The failure rate FR is defined as:

$$FR = N_f / (N_0 \times t) \qquad\qquad \text{when } t > 0 \qquad\qquad (16.5)$$

When we plot the results of Eq. (16.5) with respect to time, the component mortality curves presented in Fig. 16.3 are obtained. For mechanical components (see Fig. 3a), where testing is performed on the assembly line, the failure rate is small when the component is new. However, the failure rate increases non-linearly with time in service because parts begin to wear, corrode, or they are abused.

The failure rate for electronic components can be characterized by the well-known bathtub mortality curve presented in Fig. 16.3b. This curve exhibits three different regions or interest—each due to a different cause. The high failure rates in the first region, $t < t_1$, are due to components manufactured with minute flaws that were not discovered during inspection and testing at the factory. In the center region of the curve, for $t_1 < t < t_2$, the failure rate FR_0 is very low and nearly constant with time. Clearly, this is the best region to operate if we are to maximize the reliability of the product. Near the end of the service life, when $t > t_2$, the failure rate increases sharply with time as the components begin to fail due to the effects of aging.

For high-reliability systems, it is important to eliminate the inherently defective components associated with early times of operation. For mechanical systems, inspecting and testing the system's performance to eliminate flawed components produced in manufacturing accomplishes this objective. Life is not so easy for the electronic component manufacturers because the feature sizes on the components are so small that inspection requires very high magnification, and even then not all of the flaws can be detected. A better procedure to eliminate the few inherently flawed components is to conduct what is known as burn-in testing prior to incorporating the components into an electronic product. With burn-in testing, all of the components that are produced are operated for a time t_1 (as defined in Fig. 16.3b). The flawed components burn out (fail) and are eliminated. The surviving components will then function with the much lower failure rate FR_0.

Clearly, we want to design components with low failure rates. If the failure rates are low, then the time between failures will be high. In fact, manufacturers sometimes cite what is known as the mean time between failure (MTBF) in their product specifications. The MTBF and the failure rate are related by:

$$MTBF = 1/FR \qquad (16.6)$$

Okay! We understand the concept of failure rates for components and how they vary over its life. How do we use this information about the failure rate to establish the reliability of our components to perform safely over the anticipated life of our product?

16.4 RELIABILITY

16.4.1 Component Reliability

To answer the question of component and system reliability, we must introduce the concept of probability or chance of occurrence. In considering probability, let's go back to the test conducted to determine the failure rate and make use of the data collected for N_f, N_s, and N_0. We can then determine the probability of survival P_s as:

$$P_s(t) = N_s(t)/N_0 \qquad (16.7)$$

Then the probability of failure P_f is:

$$P_f(t) = N_f(t)/N_0 \qquad (16.8)$$

Both the probability of survival and the probability of failure are functions of time. With increased time in service, the probability of survival decreases and the probability of failure increases.

From Eqs. (16.7) and (16.8), it is evident that:

$$P_s(t) + P_f(t) = 1 \qquad (16.9)$$

Substituting Eq. (16.8) into Eq. (16.9) yields:

$$P_s(t) = 1 - [N_f(t)/N_0] \qquad (16.10)$$

If we differentiate Eq. (16.10) with respect to time t and rearrange the results, we obtain:

$$[N_0/N_s(t)][dP_s(t)/dt] = [-1/N_s(t)][dN_f(t)/dt] = - FR(t)dt \qquad (16.11)$$

Note, we consider the term $[1/N_s(t)][dN_f(t)/dt]$ to be the instantaneous failure rate FR(t) associated with a sample size N_s at time t.

When we integrate Eq. (16.11) and let FR(t) = FR_0, a constant, we obtain a relation for the probability of survival as a function of the failure rate.

$$P_s(t) = e^{-(FR_0 t)} \qquad (16.12)$$

where e = 2.71828 is the exponential number.

Let's consider another example using Eq. (16.12) to determine the reliability of a mechanical component.

EXAMPLE 16.3

Suppose we have determined from our records the number of failures of a certain component over a long period of time. Analysis of the data indicates that our failure rate is essentially constant with 1 failure per 10,000 hours of service. Does that sound good to you? Will the reliability be adequate? Let's determine the probability of survival (the reliability) of our mechanical component with time.

Solution

Substituting $FR_0 = 1/10,000h$ into Eq. (16.12), gives the results for the reliability shown in Table 16.1.

Table 16.1
Reliability $P_s(t)$ of a Mechanical Component
With Increasing Service Life ($FR_0 = 1/10,000$ h)

Time (1000 h)	Time* (years)	$FR_0 \times t$ (unitless)	Reliability $P_s(t)$
1	0.5	0.1	0.905
2	1	0.2	0.819
5	2.5	0.5	0.606
10	5	1.0	0.367
20	10	2.0	0.135
50	25	5.0	6.738×10^{-3}
100	50	10.0	4.540×10^{-5}
200	100	20.0	2.061×10^{-9}

*The conversion from hours to years of service life is based on 2,000 hours/year.

The reliability varies significantly depending on service life. For a short life, say a year, the reliability is reasonable with 81.9% chance of surviving. However, for long life, say 10 years, the reliability drops to only 13.5%. For very long life 50 to 100 years, failure is almost certain.

While the failure rate of 1 in 10,000 hours looks good in an initial assessment, the reliability, which results from this failure rate, is disappointing. You cannot be certain that you will have 10,000 hours of service prior to a failure. In fact, you have only a probability of 36.7 % to survive for 10,000 hours. If you require 90 % reliability for a service life of 10,000 hours, the failure rate FR_0 must decrease to about 1 failure in 100,000 hours.

16.4.2 System Reliability

The failure of a component may or may not cause the failure of the system. A system may be comprised of several components and its reliability will depend on the probability of failure of the individual components and their arrangement. There are two possible arrangements—series or parallel. Consider your automobile to illustrate both the series and parallel arrangement of components. For lighting the highway during the night, an auto is equipped with two headlights. This is a parallel arrangement because we have two identical components (the headlights), which essentially perform the same function. If one headlight burns-out, you can still drive. Your visibility is impaired to some degree, the system has been compromised but it has not failed.

On the other hand, suppose you want to start your engine. Consider the components involved in this action:

- Ignition switch
- Battery
- Wiring
- Solenoid relay
- Starter motor
- Starter motor clutch and gear
- Engine ignition components (spark plugs and points)

If any of these components should fail, the engine will not start when you turn the key. You need a complete series of successful components for the system to function. To start your engine, every component must properly function since the failure of a single component will result in a failure of the entire system.

16.4.2.1 Reliability of Series Connected Systems

Let's consider the reliability of a system involving three components that are in a series arrangement, as shown in Fig. 16.4.

Fig. 16.4 A system comprised of a series arrangement of three components.

Since all three of the components in this series arrangement must operate successfully for the system to perform correctly, the probability of survival of the system (P_s^s) is a product function given by:

$$P_s^s = P_{s1}\, P_{s2}\, P_{s3} \tag{16.13}$$

where the superscript s refers to the entire system of components.

If we combine Eqs. (16.12) and (16.13), it is possible to express the system reliability in terms of the failure rate of the individual components as:

$$P_s^s = e^{-(FR_1 + FR_2 + FR_3)t} \tag{16.14}$$

As the number of components in a system is increased with a series arrangement, the system reliability is markedly decreased. Also the probability of survival continues to decrease with time in service.

EXAMPLE 16.4

Consider the case where we have n components in a series arrangement where n increases from 1 to 1000. Let's also assume that the reliability of each of the n components is the same ($P_{s1} = P_{s2} = \ldots\ldots = P_{sn}$) at some time during the operating life of the system. Consider three different values for the component reliability P_s equal to 0.999, 0.99, and 0.90.

Solution

The results obtained from Eq. (16.13) are shown in Table 16.2.

Table 16.2
Reliability of a System with n Series Arranged Components
As a Function of Component Reliability P_s

Number of Components, n	$P_s = 0.999$	$P_s = 0.990$	$P_s = 0.900$
1	0.999	0.990	0.900
2	0.998	0.980	0.810
5	0.995	0.950	0.590
10	0.990	0.904	0.349
20	0.980	0.818	0.122
50	0.950	0.605	*
100	0.905	0.366	*
200	0.819	0.134	*
500	0.606	*	*
1000	0.368	*	*

* System reliability of less than 1%.

Examination of the results presented in Table 16.2 clearly shows the detrimental effect of placing many components in a series arrangement. For components with a high reliability (0.999%), the system reliability drops to about 90% with 100 series arranged components. However, with components with a lower reliability (90%), the system degrades to a reliability of less than 1% when the number of components approaches 50. The lesson here is clear—if you connect a large number of components together in a series arrangement to develop a system, then the individual component reliability must remain extremely high over the entire service life of the system.

16.4.2.2 Reliability of Systems with Parallel Connected Components

Recall the description of a highway lighting system on an automobile that includes a pair of headlights. This system is an example of a parallel arrangement because both headlights must fail before the system fails. We can drive with one light although the State Troopers might warn us of the dangers of doing so. The human body has several parallel systems; we have two hands, two eyes, and two ears, which increase our reliability to function in the event of a mishap.

A system with a parallel arrangement of three components is represented with the model shown in Fig. 16.5.

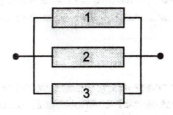

Fig. 16.5 A system with a parallel arrangement of three components.

These parallel systems are redundant because all of the components must fail before the system fails. Accordingly, we can write the relation for the probability of system failure as:

$$P_f^s = P_{f1}\, P_{f2}\, P_{f3} \qquad\qquad (16.15)$$

Substituting Eq. (16.9) into Eq. (16.15) gives:

$$1 - P_s^s = (1 - P_{s1})(1 - P_{s2})(1 - P_{s3}) \qquad\qquad (16.16)$$

Expanding Eq. (16.16) and reducing the resulting expression gives:

$$P_s^s = P_{s1} + P_{s2} + P_{s3} - P_{s1}\, P_{s2} - P_{s2}\, P_{s3} - P_{s1}\, P_{s3} + P_{s1}\, P_{s2}\, P_{s3} \qquad (16.17)$$

Let's examine Eq. (16.17) to determine if redundancy (the number of parallel components) improves reliability.

EXAMPLE 16.5

Suppose we consider P_s for all the components in a parallel system to be the same, but we treat P_s as a variable increasing from 0.1 to 1.0. Show the effect of redundancy by considering a single component, two components and three components in parallel.

Solution

The results obtained from Eq. (16.17) are shown in Table 16.3.

Table 16.3
Effect of the Degree of Redundancy on Reliability
$$P_{s1} = P_{s2} = P_{s3}$$

Component Reliability	Single Component	Two Components	Three Components
0.1	0.1	0.19	0.271
0.2	0.2	0.36	0.488
0.3	0.3	0.51	0.657
0.4	0.4	0.64	0.784
0.5	0.5	0.75	0.875
0.6	0.6	0.84	0.936
0.7	0.7	0.90	0.973
0.8	0.8	0.96	0.992
0.9	0.9	0.99	0.999
1.0	1.0	1.00	1.000

An examination of the results presented in Table 16.3 shows that the degree of redundancy improves the reliability of a system with a parallel arrangement of components. If you have a relatively poor component reliability, say 40%, and employ two of these components in a parallel arrangement, the system reliability improves to 64%. Add a third component to the parallel arrangement and the reliability increases to 78.4%. Continue adding components, and you continue to improve the reliability, although it is a case of diminishing returns.

The improvement in reliability by employing parallel (redundant) components in building a system is a distinct advantage. However, we rarely, if ever, have a free lunch. The corresponding disadvantages are the increased costs and the added power, weight and size of the system. These are significant disadvantages, and redundant design is used only when component reliability is too low for satisfactory system performance or when a failure produces very serious consequences.

For a more complete discussion of component and system reliability, see reference [1].

16.5 EVALUATING THE RISK

When the space shuttle Challenger exploded during launch on January 28, 1986, a Presidential Commission [2] was established to:

1. Review the circumstances surrounding the accident to determine the probable cause or causes of the accident.
2. Develop recommendations for corrective or other actions, based on the Commission's findings and determinations.

Richard Feynman, a Nobel Prize winning physicist from the California Institute of Technology, was appointed to the Commission. Dr. Feynman, near the end of his career and his life, took the appointment very seriously and devoted his entire time and energies to the investigation. One of his many contributions to the review was to determine the probability of failure of the space shuttle system.

During the course of this investigation, Dr. Feynman asked three NASA engineers and their manager to assess the probability of failure P_f for a mission due to the failure of one component—the shuttle's main rocket engine. Secret ballots by the engineers indicated two estimates of $P_f = 1/200$ and one of $P_f = 1/300$. Under pressure, the manager finally estimated the risk at $P_f = 1/100,000$. There was such a large difference between the manager's and the engineers assessment that Dr. Feynman concluded that "NASA exaggerates the reliability of its products to the point of fantasy" [3].

The safety officer for the firing range at Kennedy Space Center, who had been under considerable pressure to remove the distruct charges from the shuttle, did not believe the reliability figures cited by NASA. He had collected data for all of the 2900 previous launches using solid rocket boosters. Of this total, 121 had failed. This data provide a very crude estimate of $P_f = (121)/(2900) = 0.042$ or about one chance for failure in every 24 launches. The safety officer considered this high risk of failure to be an upper bound because improvements made since the early launches improved the reliability of the solid booster motors. Also, the pre-launch inspections on the shuttle were very thorough. His estimate of risk accounting for taking these improvements was $P_f = 1/100$. Dr. Feynman [3], after extensive interviews with engineers, reliability experts, managers from NASA and many subcontractors, concluded that the shuttle "flies in a relatively unsafe condition with a chance of failure of the order of 1%".

Risk assessment is a difficult task. Each component in a complete system must be evaluated to ascertain its failure rate. Then the components are placed in either a series or parallel arrangement, or some combination of the two, to model the system prior to determining its reliability. The system reliability may be lower or higher than the component reliability depending on the arrangement of the components.

Testing to determine component reliability is possible in some instances where the components are relatively inexpensive. However, establishing reliability estimates by testing requires a very large number of tests that often destroys the component. If you want to show a probability of failure less than $P_f = 1/1000$, you must test considerably more than 1000 components to failure. Obviously, it would not be possible to test more than 1000 booster rockets when the cost of a single test is of the order of several million dollars.

Often the probability of failure of a component must be estimated based on previous experience with similar applications. In some instances, we can compute probability of failure; however, these calculations require considerable knowledge of the spectrum of loading and the ability of the material to resist fracture.

When analytical methods are inadequate and engineering judgment is required to assess the probability of failure, the estimate should be made by a senior engineer with considerable experience and expertise. Even then, the estimate should be pessimistic rather than optimistic. A frank and honest estimate of P_f, based on all of the data and knowledge available, is much better than unrealistic appraisals that give an unwarranted feeling of safety.

16.6 HAZARDS

When we design a product, there are usually risks involved in either the production of the product or in its use in the marketplace. The public-at-large is exposed. What can you do to minimize the risk? In previous sections, we briefly discussed the concept of safety factor, component and system reliability, and evaluating the risk. There is one more topic to consider—developing an ability to recognize the hazards involved. You must clearly recognize the hazards before taking the necessary precautions in your designs to minimize the risks associated with them.

16.6.1 Listing of Hazards

Mowrer [4] has a complete list of hazards and an extended discussion of each. You are encouraged to read Mowrer's chapter in reference [4] and to use the extensive checklists incorporated in his coverage. A very brief excerpt from this material is provided to assist you in recognizing the many hazards that should be considered when designing a product.

The list of hazards includes:

1. Dangerous chemicals and chemical reactions
2. Exposure to electrical circuits
3. Exposure to high forces or accelerations
4. Explosives and explosive mixtures
5. Fires and excessive temperature
6. Pressure
7. Mechanical hazards
8. Radiation
9. Noise

16.6.2 Dangerous Chemicals

Chemicals can be nasty and you must appreciate the extreme dangers of exposures to certain toxic substances. Do you remember the leak that developed in a storage tank in Bhopal, India, in 1984? The storage tank contained the toxic chemical methyl isocyanate used in the manufacture of pesticides. It was a big leak and about 80,000 pounds of the chemical was released to the environment. Three thousand nearby inhabitants were killed, 10,000 permanently disabled and another 100,000 injured. The problem was not in the design of the tank but in the training of the individuals responsible for the plant maintenance and operation. Nevertheless, a catastrophic accident occurred because several workers and managers, in positions of responsibility, did not adequately understand the dangers of this very toxic chemical.

16.6.3 Exposure to Voltage and Current

What about exposure to electrical circuits? Almost everyone has been shocked by the standard 120 volt, 60 cycle electrical power supplied by the local utility company. Why worry? You should worry because electrical shocks are **dangerous**. **Yes!** Even the 120-volt supply can cause big problems. Your body acts like a resistor and limits the current flowing from the voltage source through your hands, arms, legs, etc. The problem is that the resistance of your body is variable. It depends on the moisture on your hands, or the type of soles on your shoes or even the moisture on the ground. If your hands are dry, your shoes have rubber soles and you are standing on a dry floor when you touch one wire of the circuit with only one hand, then you will probably not feel much because you have arranged a very high resistance path to ground. The current flow through your body will be very small. However, if your hands are wet and you touched both of the wires (the white and the black) from the electrical supply, one with the left hand and the other with the right hand, you have placed 120 volts across your heart. You can be electrocuted quite well with 120 volts under these conditions where the moisture on your hands greatly reduces the effective resistance of the body.

Do not take chances with electricity. Insulate the operator from the circuits preferably with two independent layers of insulation. High voltages are even more serious than low voltages because the currents flowing through one's body increase dramatically. When the current flow through the human body increases to about 10-100 mA (milli-amperes), there are very serious consequences to the respiratory muscles. Higher currents of 75-300 mA produce problems in the operation of your heart. We can increase the electrical current I flowing through the body in two ways: first, by increasing the applied voltage V, and second, by decreasing the resistance R offered by the body. Ohm's law gives the relation among the voltage, current and the resistance:

$$V = IR. \hspace{4cm} (16.18)$$

Using adequate electrical insulation in the design of products with electrical power, markedly increases the resistance R and decreases the current I to negligible amounts.

Some people think that low voltages (5 or 10 volts) are safe because you barely feel a tingle when you touch a low voltage circuit. However, some low voltage circuits particularly on high performance computers carry substantial (100+ amps) currents. If you short a circuit with high current flows, you will strike an arc and generate significant amounts of heat. Also, the flash of the arc can damage your eyes and the heat may produce serious burns. In your designs, insulate and shield even low voltage circuits.

16.6.4 The Effects of High Forces and Accelerations

High forces and high accelerations (or decelerations) go hand in hand. Newton's second law requires the connection (**F** = **ma**). If we suddenly apply the brakes on an automobile, the car decelerates and the passenger (without seat belt or air bags) is thrown into the windshield. You must always be concerned with acceleration or deceleration because of their effects on the human body. Military pilots are trained to withstand high accelerations (high Gs). A good pilot can pull 7 or maybe 8 Gs before losing consciousness. However, a civilian will become irritated at less than 2 Gs. If you want to feel an acceleration thrill, go to an amusement park and ride the roller coaster. There is no need to incorporate high accelerations into the design of most new products. A reasonably hot sports car that accelerates to 60 MPH in six seconds requires an acceleration of only 0.46 Gs. You should be careful to keep both acceleration and decelerations low when you design products that move and change velocity.

16.6.5 Explosives and Explosive Mixtures

With the bombing of the Federal Building in Oklahoma City, we have all become aware of the disastrous effects of a large explosion. The gas pressures generated by the blast destroy very substantial buildings, break glass in a very large region and kill or injure many people. We also are aware of the common explosives like dynamite and ANFO (ammonia nitrate and fuel oil). In most designs, we do not encounter a need to accommodate these explosives. What we must be aware of are other less apparent agents that act like explosives under special circumstances. Fuels like natural gas, propane, butane, etc. can leak and combine with air to produce an explosion when ignited. Boating accidents are common when gasoline leaks in an engine compartment. The resulting mixture of gasoline fumes and air explode when subjected to a small spark, destroying the boat and killing or injuring the passengers. Still another unusual source of fuel for an explosion is dust. When handling large quantities of a combustible solid, dust (fine particles) is generated. If these particles are suspended in air, the resulting mixture will certainly explode when ignited. A grain elevator exploded in Westwego, Louisiana, in 1977 killing 35 people when an explosive mixture of combustible particles (dust) from the grain and air was ignited. More recently on June 8, 1998, the De Bruce grain elevator in Haysville, Kansas exploded killing several workers.

16.6.6 Fires

Over a million fires occur in the U.S. every year. Some are vehicle fires (400,000) and others are structures (650,000), which begin to burn. Some people die in the fires (4,700), and many more are injured (28,700). Clearly, fires are a serious problem. People are killed, injured or traumatized and property is lost (eight or nine billion dollars worth per year). Some of these fires are, simply put, stupid. For example, he or she who smokes in bed may some night fall asleep and set the house on fire. A surprisingly large number of fires (about 100,000 per year) are deliberately set—apparently to collect the fire insurance, to take revenge or as a sick kind of diversion.

Engineers have a responsibility to decrease the number of fires that result from the products that we design. Did these fires start with an appliance or a motor that overheated for some reason? Determine the reason for the overheating and redesign the product so that excessive heat will not be generated. Why did the automobile burst into flames? Did a fitting on the gasoline line leak? Redesign to eliminate the fittings or specify a fitting that will not fail under the prevailing conditions. There are many solutions to the problem of fires in the U.S. As a nation, we are far too casual about fires. We tolerate carelessness in personal practices, poor design in products intended for the home, and fraud (arson) to collect fire insurance reimbursement for lost property.

16.6.7 Pressure and Pressure Vessels

We use pressurized fluids for many good reasons, and in most cases, our pressure vessels (the containers that hold the fluids) perform very well. It was not always the case. In Boston during the winter of 1919, a huge tank about 90 feet in diameter and 50 feet high fractured. It contained two million gallons of molasses, which flooded the local area. Twelve people died and another 40 were injured in the accident.

It is relatively easy to design a pressure vessel today. In fact, the American Society for Mechanical Engineers (ASME) has developed a code that engineers follow in their design to produce pressure vessels that can be certified as safe for service. However, we still occasionally encounter tanks that fail in service. The difficulty is usually with the steel plates that are welded together to manufacture the tank. Both the plates and the welds have to be tough at low temperatures and the welds must be free of large flaws. If you have the responsibility of designing a pressure vessel, follow the ASME code, make certain that the welding procedures followed in manufacturing produce crack-free welds, and employ a steel for the plate and welding rods, which is sufficiently strong and tough. If you intend to become a mechanical engineer, you will have the opportunity to learn how to design pressure vessels, select suitable materials for their construction and specify welding procedures in courses presented later in the curriculum.

16.6.8 Mechanical Hazards

Mechanical hazards are features, which exist on products that may cause injury to someone nearby. An example might be a fan used to cool the room. Is the fan blade adequately guarded, or can you stick your finger into the rotating blades? The cabinet that you have designed to hold a special tool has a sharp corner at hip level. Can someone walk into the cabinet and break skin on that corner? You have designed a center post crane to lift material and move it over a 25-foot diameter area. Will the center post be stable under all conditions of loading or will it collapse? You have designed a new pizza machine that rolls the dough into sheets exactly 0.120 inch thick. Have you provided protection for the entrance to the rolls that prevent the operator from inserting his or her fingers in the machine? You have designed a wonderful guard that prevents a person from exposing their hands and arms in operating a punch press. However, the guard is attached to the machine with two screws. Have you used locknuts and/or safety wires to insure that the screws will not loosen during the operation of the press?

There are many mechanical hazards that we encounter in designing equipment and products. Always examine each component, and look for sharp points, cutting edges, pinch points, rotating parts, etc. Do not count on peoples' good judgment. If they can stick a finger into the machinery, even if it is foolish to do so, you can be certain that sooner or later someone will do so.

16.6.9 Radiation Hazards

Radiation hazards are due to electromagnetic waves to which we are exposed. The damage produced depends on many different factors such as:

- The strength of the source.
- The degree to which the emitting radiation is focused.
- The distance from the source.
- The shielding between the source and the object being radiated.
- The time of exposure to the radiation.

We have divided the radiation spectrum into several different regions—very short wave length, visible light, infrared, and microwave radiation. The short wave length radiation is the most dangerous (x-rays, gamma rays, neutrons, etc.) with serious health risks for overexposures. The hazards due to UV, visible light and IR are usually due to excessive exposure where serious burns to the skin can occur. For very intense radiation, even short exposures are detrimental to the eyes. Microwave radiation is absorbed into the body and may result in the heating of one's internal organs. We are not certain how the organs dissipate this heat, nor of the effect of the localized increase in temperature because of this heating. It is prudent to avoid exposure to microwaves. In the U.S., OSHA (Offices of Safety and Health Administration) has issued a regulation limiting the power density of microwave energy to 10 mW/cm^2 for an exposure of 6 minutes or more.

16.6.10 Noise

Noise levels that occur in the environment may be damaging to our hearing, may interfere with our work or play, and may degrade the quality of our life-style. Most of the noise we hear is man made although occasionally we encounter a storm and Mother Nature provides the sounds of thunder and wind. If you listen occasionally to a band playing rock and roll music, there is a temporary shift in the threshold of the hearing level. However, if you play in the rock and roll band almost every night for an extended period, the shift in the threshold hearing level becomes permanent.

Noise is a pressure disturbance that propagates through some medium such as air. The velocity of propagation through air is 344 m/s at room temperature. The pressure disturbance is oscillatory usually with many different frequencies present. The frequency content of the pressure waves depends on the source of the sound. A note from a violin will have much higher frequencies that a note from a tuba. We measure the intensity of the noise with a sound level meter that consists of a microphone, amplifier, and a display meter that provides a reading in decibels (dB).

As a general guideline, the threshold for audibility is less than 25 dB[1] before a person is considered handicapped. In addition, there is a threshold for feeling noise-generated pressure at about 120 dB and another threshold for pain between 135 and 140 dB. When we design a product the noise level is a serious

---

[1] What constitutes normal hearing differs from one authority to another. We are citing a reference by the American Academy of Ophthalmology and Otolaryngology.

consideration. Clearly, we must avoid the feeling and pain levels of noise intensity, but what levels are satisfactory. The U.S. Environmental Protection Agency (EPA) has established standards, which provide guidance to engineers in the design of products. For example, the results presented in Fig. 16.6 show the relation among the sound pressure level, the communicating distance and the degree of speech intelligibility.

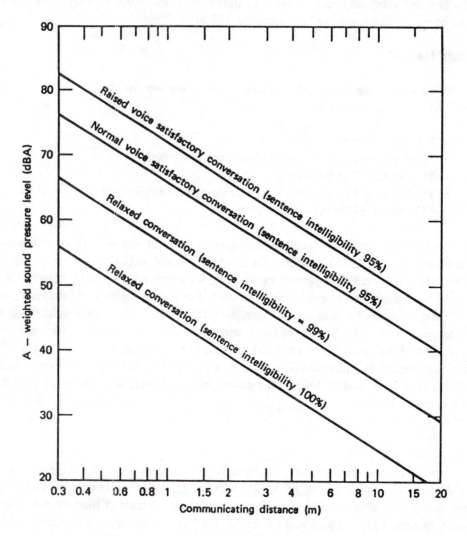

Fig. 16.6 Outdoor distances for intelligible conversation with a background of steady noise [6].

16.7 SUMMARY

When we introduce a product in the marketplace, there is usually some risk involved to the workers making the product, our customers, and society-at-large. Hopefully, the risks will be small and acceptable to all concerned because of the significant benefits produced by using the product and/or the sociotechnical system. In determining if the risk is acceptable, the public needs an accurate and honest assessment of the consequences of a failure and the probability of the occurrence of an accident. It is the responsibility of the engineering profession to assist business and governmental agencies in these assessments.

While we must accept risk, we must also make every attempt to minimize it within reasonable constraints on cost and performance. There are many excellent engineering methods for ensuring safe design. We have briefly introduced the concepts of stress, strength, safety factor and margin of safety to illustrate one approach to minimizing risk. We have also recommended a range of commonly accepted safety factors and have provided the issues that should be considered in selecting the safety factor to employ when sizing components.

Sometimes we know that failures will occur and it is necessary to determine the probability of the failure event. To illustrate methods for determining probability of failure, we first introduced the concept of failure rate and the mean time between failures. We indicated that the failure rate could be determined simply by keeping records of the failures of components as a function of time after they were placed in service. These concepts are important, but they should not be confused with the probability of failure or survival.

We introduced component reliability, and showed how it could be computed from the failure rate. In developing this relation, Eq. (16.12), we assumed the failure rate was a constant over the service life. It is important to observe that the probability of survival decreases as the service life is increased and the probability of failure increases with time in service. For reliable service for a long period of time, the failure rate must be extremely low.

System reliability is different than component reliability. A system is usually composed of many components. If the components are arranged in series and they must all function for the system to operate correctly, the reliability of the system is given by Eq. 16.13. It is evident from the data presented in Table 16.2 that the number of components arranged in a series markedly lowers the reliability of a system. Sometimes a system must function all of the time. We simply cannot consider a system failure because of its very negative consequences. In these instances, we design with redundant components arranged in a parallel system. Redundancy improves system reliability, as shown in Table 16.3, but at increased cost and the requirement for more power, weight and size.

We have presented the example of the explosion of the space shuttle Challenger to illustrate the importance of evaluating the risk. It is often a very difficult problem, and an analytical solution for the risk is frequently not possible. Nevertheless, it is a professional responsibility to prepare an intelligent, accurate, honest and frank estimate of the risk and to insure that all of the principals involved are aware of the consequences of a failure.

Finally, we have introduced a very brief listing of hazards that cause injury or death. Unfortunately the list is relatively long. It is important that you recognize the hazards and be vigilant in your designs to avoid them. Avoidance often does not require extensive calculation from elaborate formulae, but rather a detailed assessment of each component in the system and a good measure of common sense.

REFERENCES

1. Dally, J. W., Packaging of Electronic Systems: A Mechanical Engineering Approach, McGraw Hill, New York, 1990, pp. 288-296.
2. Lewis, R. S. Challenger: The Final Voyage, Columbia University Press, New York, 1988.
3. Feynman, R. "Personal Observations on the Reliability of the Shuttle," Appendix to the Presidential Commission's Report, Ayer Co., Salem, 1986.
4. Mowrer, F. W., Introduction to Engineering Design: ENES 100, McGraw Hill, New York, 1996, pp. 149-172.
5. Cunniff, P. F., Environmental Noise Pollution, Wiley, New York, 1977, pp. 101-115.
6. "Information on Levels of Environmental Noise Requisite to Protect health and Public Welfare with an Adequate Margin of Safety," EPA Report, March 1974.

EXERCISES

16.1   Have you been involved in one or more automobile accidents since you began to drive? Were you or anyone else injured or worse yet killed? Estimate the total mileage you have driven over this period and calculate your accident rate. What were the reasons for the accident or accidents? Comment on your driving behavior and its influence on the accident rate.

16.2   Are the benefits of driving worth the risks of injury or death? Determine your probability of being killed in a fatal accident while driving this year. Hint: Statistics on fatal accidents are always listed in the World Almanac. If you do not have a recent edition, use an older edition for your data. Make any assumptions necessary for your analysis, but justify each of them.

16.3   Your co-worker designs a tie rod (a tension member) from a ¾ inch steel bar with a strength of 60,000 psi. If the team leader has indicated that he or she wants to maintain a safety factor of 2.8, determine the maximum load that can be applied to the tie rod.

16.4   If a component is designed with a safety factor of 2.6, what is its margin of safety?

16.5   If we are so smart and we have all of these gee whiz computer programs to determine stresses, why do we need to specify a margin of safety or a safety factor when we design a structural member?

16.6   NASA's space shuttle has two booster rockets fueled with solid propellant to provide the thrust required for launch. Is this a redundant system? State your reason for this conclusion.

16.7   If the probability of failure of one of the solid propellant, booster rockets on the space shuttle is 1/500, determine the reliability of the solid booster system.

16.8   You are designing a very large computer system to contain the database for a huge reservation system. It is estimated that at any instant 600 operators will be accessing the database, and another 3400 operators will soon be ready with their requests for computer availability. The mainframe computers that you plan to employ each have a MTBF of 6,000 hours. Present a design of the computer system that will insure that the 600 operators have a 99.9 % probability of being served. In your design show all the calculations, and carefully list your assumptions. Justify the costs involved if each mainframe computer employed in the system is valued at $400,000.

16.9   Jane is an engineer in the transmission department of the Fink Motor Corp. Her job is to record data on the mileage prior to failure of the new lightweight transmission that has been placed in 150,000 of the new 1999 model of the Clunker. She records the following data:

Mileage (1000 miles)	No. Failures, N_f
0 - 5	68
5 - 10	37
10 - 20	20
20 - 30	22
30 - 40	21
40 - 50	35
50 - 60	97
60 - 70	564
70 - 80	1,485
80 - 90	5,677
90 - 100	13,592

Determine the failure rate as a function of service life expressed in terms of mileage, and prepare a graph showing your results.

16.10   Using the data from the table in Exercise 16.9, determine the probability of survival and the probability of failure of the transmissions as a function of service life measured in terms of mileage. Prepare a graph of your results.

16.11   If the transmissions in the Clunker were under warranty for 100,000 miles, what would be the consequences for the Fink Motor Corporation?

16.12   Prepare a paper discussing your reaction to the risk assessment of the space shuttle system by NASA. Give arguments for and against NASA's position and practices regarding safety of the shuttle and its crew both before and after the Challenger accident.

16.13   Describe an incident when you received an electrical shock. Something obviously went wrong to cause this incident. Please indicate the problem. Was there a design deficiency? What could be done to prevent the incident from reoccurring?

16.14   You are designing a cabinet-mounted, self-cleaning oven that requires very high power levels to heat its interior surfaces until they are free of all the burnt-on splattered grease and grime. As the lead engineer on this design team, what precautions can you take, to insure that the oven will not be the source of a fire during its anticipated 20-year life?

CHAPTER 17

ETHICS, CHARACTER AND ENGINEERING

17.1 INTRODUCTION

This chapter on ethics was written shortly after the 1996 presidential election, and it has been revised every year since then. During this period the newspapers and TV news and talk shows covered many ethical issues emanating from both Congress and the White House. Ethical lapses occurred on nearly a weekly basis, keeping the news services busy reporting on questionable behavior of our elected officials. These officials are the people we have entrusted and empowered to write and approve the laws governing the country. The news media has pointed out numerous instances where our leaders have skirted the truth if not the law. Questionable fundraising activities by both parties prompted the Governmental Affairs Committee, chaired by Senator Fred Thompson, to request $6.5 million to cover the extensive cost of the investigations. (The Committee was finally authorized to spend $5 million on the investigation and sought additional funds to extend the investigation). How do you feel when learning that $50,000 to $100,000 is the price for an invitation to the White House and an overnight stay in the Lincoln Bedroom?

During the period from the initial writing of this chapter to its revision in January of 2000, the ethical situation has deteriorated. Ethical issues continue to be reported by the news media with alarming frequency. For example[1], U. S. District Judge Royce Lambert cited the Secretary of the Interior, Mr. Bruce Babbitt, and the former Secretary of the Treasury, Robert Rubin, for contempt of court for withholding evidence regarding the Indian trust funds. (The government cannot account for $2.4 billion of these funds). Judge Lambert concluded that the employees of the Interior and Treasury Departments had "engaged in a shocking pattern of deception of the court. I have never seen more egregious conduct by the federal government." It is clear that the Clinton administration is usually in a state of siege with a small army of attorneys working continuously to mitigate several alleged indiscretions of the leaders of our government.

Last year, the president admitted having an affair with an intern about half his age. He initially claimed on national television that he did not have sex with **that** woman. However, when it was clear from the results of tests with that woman's dress that he had had sex, he admitted it. Previously, the president, in still another legal proceeding, denied the existence of an intimate relationship under oath. When the facts of the president's behavior became apparent, the House of Representatives impeached Mr. Clinton for lying and obstructing justice, and a federal judge held him in contempt of court for lying.

Orin Hatch, Chairman of the Senate Judicial Committee, recently summarized[2] the tactics of the current administration in dealing with these ethical failures. He stated, "We also have to acknowledge the bold, steady, instinctive use of all modern means of communication to dissemble, mislead and fool the people as well as to cover up official corruption. Such actions are really something new and something

---

[1] See the article entitled, "Shredding 162 Boxes," Wall Street journal, December 9, 1999.
[2] See article entitled, "Pertinent Questions: Will GOP Duck Clinton Character?", The Wall Street Journal, January 19, 2000.

terribly dangerous. This routine practice of political deception to hide an inner falsity—this institutionalization of the cynical deceit that you've not done anything wrong if you can talk your way out of it is the real cultural legacy of this administration. And is at the heart of what disturbs us about the corrosion in our political system."

The lax ethical practices by many of our elected officials have led to a very serious skepticism about the merits of many of the government institutions and agencies. The fact that fewer than 50% of the eligible voters took time to cast their ballots in the 1996 elections is testimony to the deep skepticism toward our government and our elected officials.

As engineers, we want to preserve, if not enhance, the confidence of the public. To gain public confidence, it is essential that we all behave in an ethical manner every day of every year. Whether we act as individuals or as professionals, consistent ethical behavior is mandatory. Our primary employer, business, must also act in an ethical manner. Since engineering developments are sponsored, financed and controlled by businesses, corporate and engineering behaviors are inseparable. Our actions and words always must be above reproach.

17.2 CONFUSION ABOUT ETHICAL BEHAVIOR

Many people are confused about ethical behavior. The results of several polls indicate that Americans are less ethical today than in previous generations. Sixty-four percent of a sample of 5000 individuals admits to lying if it does not cause real damage. An even larger percentage of those sampled (74%) will steal providing the person or business that is being ripped off does not miss it [1]. Many students (75% high school and 50% college) admit to cheating on an important exam [2]. Most students have not developed a moral code to use as a guide for their behavior. Poor behavior when the author was in high school, 1944-47, included making too much noise, chewing gum in class, talking, littering, running in the halls, etc. Today, these relatively minor transgressions have been replaced with more serious problems involving mass murder, drug and alcohol abuse, pregnancy, suicide, rape, robbery and assault. Extremely negative changes. Why?

There are many reasons for the degradation of ethical behavior over the past 50 years. Let's discuss two key reasons. First, institutions (family, church and schools) that once taught ethics are much weaker today than 50 years ago. Many families have been torn apart by an extremely high divorce rate and single parent homes are common. The concept of "love, honor and cherish, in sickness and in health" is often ignored.

The influence of our churches has deteriorated due to serious declines in membership. Unfortunately, church finances often limit their activities to current, not new, membership. Some social problems are with persons who are not well known to the church. The public has also become much more secular with no church affiliation. Our public school administrators are so concerned with possible litigation that they have largely removed the teaching of moral values from their curriculum. Given the current constrains on discipline in the public school system, it is often difficult for the administrators and teachers to maintain orderly classrooms. The consequence is that students graduating from high school have not been given an adequate opportunity to develop a personal code of ethics. The result of the failure of our institutions to teach ethics is that many of our young people are not aware of what is right and what is wrong. All actions are not gray. Many are plainly right, and many others are obviously wrong.

The second reason for the deterioration in ethics on the part of our younger population is the current state of ethical standards in society. We are exposed to cynical behavior and selfish attitudes on TV, magazines, and newspapers every day. The "bad guy" often walks away from severe crimes with minor penalties. Minor infractions are frequently ignored. Killers and child sex offenders are paroled after remarkably short periods of incarceration. Popular TV programs frequently show examples of lax ethics and self-indulgences to generate a popular but sick kind of humor. The result is to create confusion in our youth as they try to develop some system for judging right from wrong. Our youth mirrors the behavior of society leading to cynicism, selfish attitudes, and dishonesty and irresponsible conduct.

17.3 RIGHT, WRONG OR MAYBE

Christina Sommers [2] develops an interesting viewpoint regarding what is considered right, wrong and controversial by society-at-large. She argues that there are some ethical issues that are not clearly defined, and that society has not reached a consensus regarding the correctness of these issues. We can develop cogent arguments for and against a specified viewpoint for controversial issues such as abortion, affirmative action, capital punishment, etc. Sommers refers to these controversial issues as "dilemma" ethics, which should be distinguished from "basic" ethics. With "dilemma" ethics, society-at-large is not certain about what is right, wrong or gray. There are several attitudes prevalent in society—sometimes the issue is so important to segments of our population that demonstrations are organized to elevate one viewpoint or another.

We will make no attempt in this textbook to resolve any "dilemma" issues. These issues have been unresolved for decades, and they will require much more time and understanding before our society is willing to reach the consensus necessary for resolution. It is much more productive to treat "basic" ethics where right and wrong or white and black are much more clearly recognized by the majority of society.

17.3.1 Right Versus Wrong

It is easy to define right and wrong. In fact, the understanding of right and wrong dates back to Aristotle [3] and fundamental doctrine has not changed much since then. Let's start with the four classic virtues [3]:

- Prudence
- Justice
- Fortitude
- Temperance

Prudence refers to careful forethought, good judgment and discretion. In other words, think before you act and exercise care and wisdom. If you act, will your action cause problems, now or later, for you or anyone else? Are you careful about your remarks concerning others? One of my rules is not to speak ill of others under any circumstances. Is the action that you are about to undertake in your best interest? Will that action be appreciated by your family, friends, manager and peers?

Justice involves fairness, honor, keeping your word, honesty, truthfulness, etc. It is clearly wrong to lie, cheat, steal and break promises. It is also wrong to tolerate those about you who do so[3]. My behavior with regard to justice is governed by the golden rule. Do unto others as you would have them do unto you. It is a very simple rule and it works.

Fortitude is about courage and persistence. We usually relate fortitude with regard to warfare and/or battle. Warriors stand their ground and exhibit a very special brand of courage when their lives are at extreme risk. However, there are other brands of courage that must be exhibited in more typical circumstances. Will you stay with an idea, even if it is not popular, if you know that it is the "right" approach? Showing determination and resolution is sometimes difficult when the risks of failure are high or when peers encourage you to abandon your idea.

Temperance, of course, refers to moderation in drinking alcoholic beverages. However, temperance has much wider implications with regard to ethics. Temperance can also mean control of human passions such as anger, lust, hostility, exasperation, and lechery. With regard to food and beverages, it implies self-control and moderation in consumption. Clearly, much is to be gained by avoiding overindulgence in food, drink, sex, and in restraining your emotional extremes.

---

[3] Many people are willing to tolerate about them those who cheat, steal and break promises. They do not believe in imposing ethical constraints on friends or associates. This no tolerance concept is difficult to establish in many honor codes and the author may be in the minority on this issue.

There are other virtues such as loyalty and obedience that are less commonly discussed these days. With the restructuring and massive downsizing that has taken place in the business world in the past two decades, the concept of company loyalty to the employee and vice versa has been destroyed. Hopefully, loyalty to the family is an invariant, at least in those situations where a family, in the true sense exists. The concept of obedience was seriously damaged—if not destroyed—in this country by the Vietnam conflict. The government ordering young people to fight in a war of questionable value was too much for many to endure. Many young men refused to obey the orders to report for duty and fled the country to avoid persecution.

Because of these events (many others can also be added), the virtues of loyalty and obedience have shifted from the "basic" to the "dilemma" category, and arguments can be advanced on both sides of these questions. What do you think of being loyal to an employer who will eliminate your position during a period of **increasing** profits? How do you feel about completing your tax return for the Internal Revenue Service? Talk about straining one's patience!!!

17.3.2 Theological Virtues

In addition to the four basic virtues, prudence, justice, fortitude and temperance, we have three religious (theological) virtues. These are:

- Faith
- Hope
- Charity

The theological virtues are well known, so we will not elaborate on their importance. We will again emphasize that their merits are definite and irrefutable. It is good to be faithful, hopeful and charitable. If you want to be even more complete, add other ethical constants such as humility and respect to the list.

While the emphasis in most college courses on ethics is with social dilemmas where arguments and counter arguments are to be developed, almost all of our individual behavior can be judged by very clear and well-understood virtues that are part of "basic" ethics. There is no need to be confused, nor is there any reason to impair your "basic" code of ethics with issues raised in the study of moral relativism.

17.4 LAWS AND ETHICS

The governments (federal, state and local) play a role in ethical behavior, because they generate laws and set public policy. Laws (mostly at the state level) are written to aid people in their pursuit of a morally correct and safe life. The purpose of these laws is to prohibit a well-defined set of vices. For example, it is illegal to sell drugs, rob banks, commit burglary, kill or abuse another human, engage in prostitution, or pursue deviant sexual practices (rape, among other offenses).

These laws, and many more like them, serve to create a moral ecology in which we exist [4]. The laws require visible or outward conformity to a publicly accepted moral code of behavior, but do nothing to inhibit illegal behavior unless it is observed, and then reported and prosecuted. An unscrupulous person can sell drugs if he or she is not observed in the act. Even if they are observed, they can avoid penalty if they are not arrested and fully prosecuted for one reason or another. Even if they are observed, detained, prosecuted, and convicted they may receive a suspended sentence. The law does not insure that infractions **will not** occur; the law simply indicates that certain behavior **may not be tolerated**.

In some instances, laws are not a suitable approach for creating a moral ecology. An example of an unsuitable law in this country was prohibition. When it became evident that the law was causing more problems than it solved, it was repealed. The government wanted to make it illegal to consume alcoholic beverages, but society-at-large wanted to drink. Bootlegging, racketeers, and speakeasies followed bringing crime and violence to the neighborhoods. It soon became apparent that the law was not a suitable approach to solve a temperance problem. In such cases, the government uses public policy to discourage people from pursuing a vice. Policies are adapted that limit (but do not preclude) the consumption of alcoholic beverages. Hours of sale are restricted, licenses are required, taxes are imposed, the number of outlets are constrained, etc.

Public policies are conveyed by means of rules and regulations—not laws. The policies should be crafted to strengthen families, communities and churches by discouraging, but not prohibiting, irresponsible conduct.

17.5 ETHICAL BUSINESS PRACTICES

Corporations are entities that under law are treated as persons. A corporation, acting like a person, has the right to conduct business and the obligation to pay taxes. Just as individuals are judged by their ethical standards, corporations are judged by their ethical conduct [5]. The chief executive officer (CEO) for a corporation sets the moral and ethical tone for the organization. Any large business is organized with several layers of management. If the CEO operates the business with the highest ethical standards, the middle and lower level managers will follow the example set at the highest level in the corporation. However, if lax ethical standards are permitted, a pattern is set and sloppy ethics will become the corporate style and standard.

In an excellent paper, Baker [5] has developed a list of issues that occur daily in a typical corporation. The manner in which a corporation deals with these issues establishes its character. Baker's list is shown below:

1. How people, employees, applicants, customers, shareholders, and the families who live near our facilities are treated.
2. Are we free of systemic, or individual, practices of discrimination?
3. How do we spend our shareholders' money on our expense accounts as an institution and as individuals?
4. What is the level of quality that goes into our products? Do we meet our customers' expectations for quality? Do we meet our own standards for quality?
5. What is our concern for safety, not only for our employees, but also for our customers and our neighbors in communities in which we operate?
6. How "ethically" do we compete?
7. How well do we adhere to the laws of the locales, regions, and nations in which we do business?
8. What is acceptable gift giving and gift taking?
9. How honest are our communications to our employees and our public advertising?
10. What are our corporate and personal positions on public policy issues and how do we promote those positions?

This is a long list, but we certainly could add more items to Baker's catalog of concerns—sustainability and the environment for instance. Adding more items is not as important as recognizing that corporate decisions are necessary on a daily basis by the managers and employees. These day-to-day decisions determine the corporate character. The ethical judgments that serve as the basis for most of the decisions will depend on the individuals' personal values (virtues) and experience. The quality and consistency of the ethics applied in these decisions will markedly affect the success of the corporation, its management and its employees.

17.6 HONOR CODES

We will not lie, steal or cheat, nor tolerate among us anyone who does [6]. This is a fourteen word honor code that cadets learn almost as soon as they step foot onto a military academy or college campus that has strong ties to the military (Virginia Military Institute and the Citadel). Our military academies (Navy, Army and Air Force) are educational institutions that serve to train career officers for the services. Honor and integrity is a very large and important element in their educational program.

Honor codes are much more widespread than just the military institutions. There is a growing trend to introduce honor codes on campuses where they are absent and to strengthen them on campuses where they already exist [7]. An honor code typically forbids lying, cheating and stealing. When a peer notes an infraction of the code, the code requires that student to bring the case forward. A student committee hears the case. Although faculty members may testify if called upon, they have no control over the proceedings. The penalty imposed on those found guilty of breaking the honor code is determined by the student committees, and university administrators implement their decisions.

The honor codes are very well respected by the student body and fully appreciated by the faculty. Students generally support the honor codes because they also are concerned with cheating by their peers. They want to play the game fairly and on a level playing field. When the honor system is established, the students handle the responsibility for academic integrity with careful consideration. Punishments are handed out only after complete investigations involving representation of both sides of the case. In some instances, the individual being judged by the student committee retains lawyers.

The faculty is also very much in favor of the honor system. They never have appreciated proctoring exams. With the increase in cheating over the past 25 years, proctoring has evolved into policing, which is very distasteful. The advantage of the honor code to the faculty is that the unproctored examination transfers the responsibility for policing to the students. With an effective honor system, students often are permitted to take an exam at a time and place of their choosing.

Of course honor codes are not perfect. Some students cheat regardless of the code. The military academies, where the code is the strictest and the most rigorously enforced, have suffered the most notorious lapses in standards. Scandals at the U.S. Naval Academy have been the most prevalent and widespread in recent years with as many as 133 midshipmen recently involved in cheating on an electrical engineering exam. To make matters worse, the administrators at the academy appear to have interfered with the system, and in doing so, lost the trust of the midshipmen.

The lesson from the academies with regard to the effectiveness of the honor codes is that rigid codes and rules reduce cheating. However, if the students see a way of beating this very rigid system, they will occasionally take advantage of the opportunity. Honor codes are most effective when they produce an ambience of trust between the students, faculty, and the administration. Cheating, lying, and stealing will generally be reduced. For honest students, the scoring on exams will be fairer. For faculty the distasteful task of policing is eliminated. The feeling of trust between the faculty and the students is enabling. With time and prolonged success, the honor code and the trust that it engenders becomes an essential part of the educational process.

17.7 CHARACTER

The author had the privilege of teaching for a year (1995-96) at the U.S. Air Force Academy. It was a wonderful experience for me for many reasons—both personal and professional. One of the most important reasons was the opportunity to participate in a well-planned approach to incorporate character building into the educational process. Character was a key educational outcome in every class that we taught at the Academy.

It is not possible to transplant character into an individual, and it is not possible to accomplish much in efforts to explicitly "teach" character. Character comes implicitly as we as individuals develop a personal set of ethical responsibilities. In a book like this one, we can list the attributes that form the foundation for a person's character. Let's consider some of these attributes:

- **Be committed to excellence**. This attribute simply means to do the best that you can do in both your personal and professional endeavors. To make this commitment, you have to assess your capabilities. How good are your skills? How determined are you in achieving your goals? How much time can you commit without endangering your health? If you know yourself, you can measure your achievements on a realistic scale. Engage to the full measure of that scale.

- **Respect the dignity of everyone**. Respect is the foundation for all achievements. We work, live and play in a diverse world. We must appreciate everyone that we encounter in life. Race, gender, ethnicity and religion are not criteria for judgment. Recognize the potential of those with whom you work and study. Support and encourage them and carefully avoid demeaning criticism. Teamwork, so essential in today's workplace, requires that you accept the differences inevitable in our diverse population and that you fully embrace fellow team members.

- **Integrity**. A single word describes a vitally important attribute of character. Integrity means that you decide to do the right thing solely because it is the right thing to do. You often face decisions that test your integrity. Do you instinctively make the right (honorable) decision? For example, as you parked your car last night, your bumper scraped the fender of the adjacent car. No one observed your accident. Do you leave a note with your name, address, and phone number? Or do you split and drive to another nearby parking lot? How many parking lot scrapes can you find on your car? Did you find a note from the party inflicting the damage claiming responsibility? A person with integrity will consistently make the correct decision, not because it is right for their personal benefit, but because it is clearly the "right" course of action. They "walk the talk."

- **Be decisive**. When you encounter a problem in engineering or your personal life, there is an information-gathering period followed by a decision. Some folks do not want to make that decision. They prolong the information-gathering period; they procrastinate; they hum and haw; they pass the buck. When you establish the facts, evaluate them, and make a timely decision. It will be your decision if you made it in isolation. If you made it within the framework of a team, it should represent a consensus of the members of the team.

- **Take full responsibility**. Decisions are made and actions follow. Often your decisions produce a winner, but sometimes they result in a loser. If the outcome was a loser and you participated in the decision process, it is your responsibility. Step up, accept the responsibility, and lead the effort to fix the problem. Never, under any circumstances, begin to participate in a finger-pointing exercise. Finger pointing does not resolve problems; it exacerbates them.

- **Be temperate**. Self-discipline is an essential part of character. Eat, drink and be merry in moderation. Overindulgence is disgusting. No one appreciates a drunk or a glutton. Self-discipline insures control of the human passions, sensual pleasures, anger, rage and frustration. With self-control, it is possible to attain consistently high levels of achievement.

- **Exhibit fortitude**. On many occasions, we encounter significant difficulties in completing the task at hand. It is easy to give up and divert our attention to more pleasurable undertakings. We can quickly forget the missing assignment, the incomplete solution, the late review, or the unfinished drawing. Who is to know? Stamina, mental toughness, and discipline are attributes that keep us on task. Stay with the job, and not only complete the work, but do it well and with dispatch.

- **Understand the significance of spiritual values**. Many, but not all, of us have a faith. We endorse a set of theological beliefs. Your beliefs and your church may be different than mine. Nevertheless, it is essential that I respect your convictions and you should respect mine. We all need to be sensitive to the important role that religion occupies in the mental comfort of many people. We must accommodate a diversity of beliefs by supporting the right of an individual to choose his or her faith and to pursue the ceremonies offered by the church representing this faith.

17.8 ETHICS OF ENGINEERS

We have devoted most of this chapter to a discussion of issues pertaining to individual behavior, ethics, virtues, morals, and character. A personal code of ethics is the cornerstone to achieving a meaningful sense of moral values. Professional ethics are also important, but they must begin only after a person has developed a well-understood personal code of ethics. Hopefully, the preceding sections of this chapter will be helpful in any attempt that you make to establish a sense of right and wrong.

There are several codes of ethics for engineers. Most of the founding societies of engineering have developed and distributed codes, guidelines for professional conduct and faith statements. The codes of ethics for the different engineering societies are all similar. The fact that each society has their own code, is more to insure a complete distribution of the code than to pursue unique ethical issues.

Accreditation Board for Engineering and Technology*

CODE OF ETHICS OF ENGINEERS

THE FUNDAMENTAL PRINCIPLES

Engineers uphold and advance the integrity, honor and dignity of the engineering profession by:

I.   using their knowledge and skill for the enhancement of human welfare;

II. being honest and impartial, and serving with fidelity the public, their employers and clients;

III. striving to increase the competence and prestige of the engineering profession; and

IV. supporting the professional and technical societies of their disciplines.

THE FUNDAMENTAL CANONS

1. Engineers shall hold paramount the safety, health and welfare of the public in the performance of their professional duties.

2. Engineers shall perform services only in the areas of their competence.

3. Engineers shall issue public statements only in an objective and truthful manner.

4. Engineers shall act in professional matters for each employer or client as faithful agents or trustees, and shall avoid conflicts of interest.

5. Engineers shall build their professional reputation on the merit of their services and shall not compete unfairly with others.

6. Engineers shall act in such a manner as to uphold and enhance the honor, integrity and dignity of the profession.

7. Engineers shall continue their professional development throughout their careers and shall provide opportunities for the professional development of those engineers under their supervision.

111 Market Place, Suite 1050, Baltimore, MD 21202-4012

*Formerly Engineers' Council for Professional Development. (Approved by the ECPD Board of Directors, October 5, 1977)

AB-54 2/85

Fig. 17.1 ABET's code of ethics.

Let's consider the Code of Ethics of Engineers, presented in Fig. 17.1, which is sponsored by the Accreditation Board for Engineering and Technology (ABET). The code is divided into two parts. The first part deals with four fundamental principles, and the second part contains seven fundamental canons. The Engineers' Council for Professional Development first advanced this code in 1977. The fact that the code remains without modification for more than 20 years indicates that professional ethical values are as constant as personal ethical values.

The fundamental principles seek to insure that the engineer will uphold and advance the integrity, honor and dignity of the profession. These goals are common to our personal goals that were discussed in a previous section; however, the approach to achieve the professional goals is different as indicated below:

1. We are selective in the use of our knowledge and skills so as to ensure that our work is of benefit to society.
2. We are honest, impartial and serve our constituents with fidelity.
3. We work hard to improve the profession.
4. We support the professional organizations in our engineering discipline.

The seven fundamental canons in the ABET Code of Ethics of Engineers, presented in Fig. 17.1, are explained in considerable detail in guidelines that are used to expand and clarify the relatively brief statements that represent the "regulations" that shape our professional behavior. The detailed guidelines may be obtained from ABET.

17.9 ETHICS IN LARGE ENGINEERING SYSTEMS

Engineers design and build many different products each year that are included in large and complex sociotechnical systems. Usually these products and the systems are conservatively designed with adequate safety factors, carefully tested, and perform well in service for extended periods of time. They provide a much needed service without endangering either individual or public safety. However, from time to time mistakes are made in the initial design. These mistakes are usually detected in prototype testing and eliminated prior to releasing the product to the market place. In very rare circumstances, a mistake or several mistakes are overlooked, for a variety of reasons, and the system is released and placed in service with an unacceptably high probability for failure. An undetected mistake often leads to a catastrophic accident. The space shuttle system clearly falls into this category. If you are a space supporter, you may argue that the risks are worth the benefits. But if this is the case, unprepared and untrained high school teachers should not be invited to ride along to enhance the image of the space program. It is one thing to order a career astronaut into peril, but totally a different proposition to invite an uninformed civilian to participate in a very dangerous project.

In Chapter 16, we have discussed the probability of failure (or an accident) in considerable detail. Here you need to understand that there is always some risk of failure when designing high performance systems. We must accept a trade-off between risk, safety and performance. In most complex sociotechnical systems (air transportation for example), there is a small but finite risk for failure with a subsequent loss of life and property. The public must know this risk. It is the responsibility of the engineering community, industry and the government to alert the potential customers to the dangers involved. Knowing the risk, you can decide whether or not you want to use the system. Some people worry more than others and place a very high value on safety. They require a probability of failure of nearly zero. Do you know someone who will not travel by flying? (John Madden, the popular football announcer for the Fox network, travels from game to game each week on a special bus). Other folks love the thrill of a risk, and they are willing to accept a much higher probability of an accident. They think hang-gliding is great sport.

17.9.1 The Challenger Accident—a Case Study

With this background on risk, safety and performance, let's begin our discussion of the Challenger accident [9]. The Challenger was one of the original four orbiters built by the National Air and Space Administration (NASA) to serve the space shuttle system. The rocket fuel (liquid hydrogen) on the Challenger 51-L mission exploded 73 seconds after launch on Tuesday, January 28, 1986. The crew of six and a civilian passenger were killed, and the space shuttle was lost.

The Challenger accident has been selected as a case study because it illustrates several ethical issues in the engineering and management of large, complex and inherently dangerous systems: Some issues that will be raised are:

1. The design of the shuttle and the selection of the contractors involved many political considerations [9].
2. The lack of communications between key people and organizations was a significant factor in the accident.
3. The interface between the upper-level administrators (business managers) and the engineers was an important element in the decision to launch on that disastrous morning.
4. Public attention and opinion, not safety, markedly affected the decision-making process.
5. The risk potential was not known by the public and not appreciated by top administrators and politicians directly involved in the launch decision.
6. Christa McAuliffe, a high school teacher and mother of two children, was killed in the accident. Why?
7. Dr. Judith Resnik, a career astronaut and an electrical engineer, was also killed. She is very special to us because she earned her Ph. D. from the University of Maryland at College Park[4].

17.9.2 Background Information

To set the stage for the accident, we have to go back to the early 1970s. NASA had been very successful with the Apollo missions (in spite of Apollo 13), and looked forward to larger and more aggressive space endeavors. They proposed an integrated space system that would include a space station, space shuttle, space tug and manned bases on both Mars and the moon. Sounds great until you look at the price tag. The public did a quick look and wanted no part of it. A poll indicated that the public believed that Apollo had been too costly. The politicians, always driven by the polls, took note and reduced NASA's budget. NASA recognized the need for a new, cost-effective project to follow Apollo that the public (and politicians) would buy. Responding to these political pressures, NASA proposed the space shuttle that would serve the military, the scientific community and the rapidly growing commercial business of placing satellites in orbit. The space shuttle system, illustrated in Fig. 17.2, was marketed as a relatively routine space transport system. Even the name "shuttle" connected with the airline shuttle services that routinely fly every hour on the hour from one large city to another.

To make the space shuttle system a commercial success, NASA proposed a fleet of four orbiters [10], which would eventually fly on a weekly basis (NASA initially set a goal of 160 hours for the turnaround time for an orbiter.) They planned on nearly 600 flights in the period from 1980 to 1991. The early estimate of the cost of a launch was $28 million with a payload delivery cost of $100 to $270 per pound. After some operational experience, the cost estimates proved to be much too optimistic. Launches actually cost on the order of $280 million, and the cost to place a pound of payload in orbit on the shuttle was in excess of $5,200 [11]. Early experience with the space shuttle system indicated that NASA could

---

[4] The author served on the faculty at the University of Maryland at College Park for more than 20 years. Dr. Judith Resnick, a graduate of our program, was an outstanding role model for women in engineering until her untimely death.

not hold their schedules and their costs were running more than a factor of 10 higher than the original estimates. Because of its poor performance, NASA was struggling to improve its image as 1985 ended.

Prior to the launch of the Challenger on January 28, 1986, twenty-four shuttle flights had been made. These flights were all successful in that they returned to earth with all crewmembers safe. They also showed the operational capabilities of the shuttle system in placing commercial satellites into orbit, repairing satellites in space and salvaging malfunctioning satellites. However, these flights also showed that the shuttle system had many serious problems. The three main liquid rocket engines were too fragile with many critical components. Some of the tiles in the heat shield needed repair and/or replacement after every flight. The computers and the inertial navigation systems experienced occasional failures. The brakes and landing gear were stressed to the limit when the orbiter (an 80 ton dead stick glider) landed at speeds ranging from 195 to 240 MPH. Finally, the seals in the solid fuel booster rockets showed distress (sometimes extensive) in 12 of the 24 previous launches.

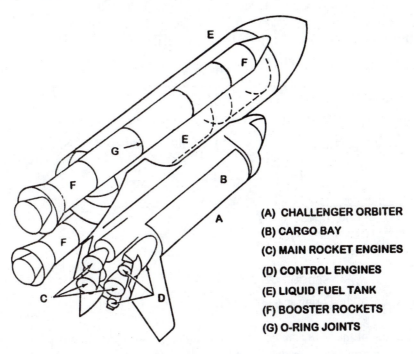

(A) CHALLENGER ORBITER
(B) CARGO BAY
(C) MAIN ROCKET ENGINES
(D) CONTROL ENGINES
(E) LIQUID FUEL TANK
(F) BOOSTER ROCKETS
(G) O-RING JOINTS

Fig. 17.2 Illustration showing the main components on the NASA space shuttle.

The record of the shuttle during the 1981-1985 period showed a consecutive series of successful launches, but with many prolonged delays to repair failing components in a very large, highly stressed system. The maintenance records showed so many problems that NASA estimated it took three man-years of work preparing for a launch for every minute of mission flight time. Clearly, the word shuttle to describe such a transportation system is a misnomer.

17.9.3 The Solid Propellant Booster Rockets

The explosion of the main fuel (liquid hydrogen) tank on the Challenger was due to the failure of the O-ring seals on the solid fuel boosters that were adjacent to the hydrogen fuel tank. To understand the seals and their purpose, it is essential that we appreciate the size and function of the two booster rockets used to provide much of the thrust necessary for the launch. They are enormous cylinders—12 feet in diameter and 149 feet tall. Each cylinder is filled with 500 tons of a solid propellant consisting of a rubber mixture with aluminum powder and the oxidizer ammonium perchlorate. When ignited, the propellant burns to produce an internal pressure in the motor case of about 450-psi (lbs/in^2) at a temperature of about 6000 °F.

The expanding gasses exit the rocket motor case through a nozzle, to produce about 2.6 million pounds of thrust from each booster. These booster rockets are essentially very big Roman candles!

Big was the problem. The solid rocket boosters were made by Morton Thiokol in Utah, but launched from the Kennedy Space Center in Florida. They were much too long to ship across the country as a single cylinder. To circumvent this problem, the motor casing was fabricated in segments—each 27 feet long. The segments were filled with propellant and shipped by train and assembled in a special facility near the launch pad. This procedure solved the shipping problem. However, it created another. The rocket casing is a pressure vessel that must contain very hot gasses at a pressure of about 450 psi. The joints, where the cylindrical segments of the motor case were fitted together, needed to be sealed so that these hot gasses would not leak and cause damage to adjacent components of the launch vehicle.

The seal was made with a pair of rubber O-rings fitted over one finger of a clevis type of joint as shown in Fig. 17.3. The O-rings, 0.280 inch in diameter, were compressed between the two surfaces effecting a seal that prevents the pressurized fluid from leaking past the joint. A putty like compound was used to keep the hot gases some distance from the O-rings. The pins lock the segments together (axial constraint only), but did not clamp the clevis fingers about the center finger.

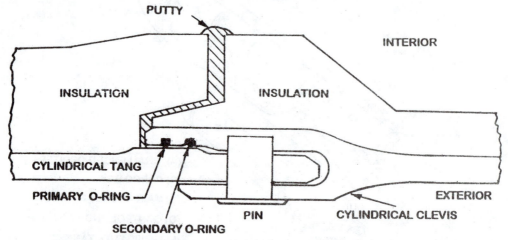

Fig. 17.3 Design of the O-ring seal used at the circumferential joints of the solid rocket motor cases.

The sealing of the segmented cylinders with the O-rings was not a new concept. A single O-ring had been employed previously on the Titan III rocket effectively sealing the circumferential joints on a smaller diameter motor case. Morton Thiokol, the contractor building the boosters, introduced the second O-ring to provide a margin of safety in the event the first O-ring failed. It all sounded good, and the initial test firings of the booster in Utah were apparently successful.

Unfortunately, one test does not validate the safety of even a simple system. To insure high reliability of a component, many tests are required. Moreover, this test firing had little or no bearing on problems associated with the reuse of the solid rocket boosters (up to 20 times). After splash down and recovery from the Atlantic, the cylinders were shipped back and forth between Florida and Utah.

17.9.4 Failure of the O-ring Seal

Motion photographs taken at the final launch of the Challenger indicated that the O-ring seals failed almost immediately after ignition of the booster. The hot gasses cut through the O-rings and the joint of the booster. The hole spread and a flaming jet of white-hot gasses escaping from the booster rocket cut through the wall of the adjacent tank containing the liquid hydrogen. The launch was effectively over. While many things happened in the last five seconds before the big explosion—all bad—the actual penetration of the adjacent tank, which contained liquid oxygen and liquid hydrogen, spelled the disastrous end of the mission.

It was a terrible day, and we as a nation were in shock while we mourned the loss of the crew and the schoolteacher/mother. It was sometime later, when the facts were brought to the public about the space shuttle program, that we learned the shuttle should not have been permitted to fly that day. The very high probability of failure of the O-ring seals was predictable. Moreover, knowledgeable engineers at Morton Thiokol tried without success to prevent the launch.

The story containing all of the facts about the failure of the O-ring seal is too long to be covered here, but you are encouraged to read references [9 – 14], where very complete and well-written accounts are given. The essential elements leading to the catastrophic failure are listed below:

1. The circumferential joint changed shape when the motor case was pressurized, and the gap, which the O-rings filled, increased markedly in size. A schematic illustration of the new gap geometry is presented in Fig. 17.4.
2. The new gap opening was so large that the back-up O-ring probably could not seal the joint. When the booster case was pressurized, the seal depended on a single O-ring—not two.
3. The hot exhaust gases had eroded the O-rings on 12 of the previous 24 launches indicating some leakage about half of the time, and on a few occasions, considerable erosion of the rings. There was also clear evidence that the putty failed to keep the hot gasses from attacking the rubber O-rings.
4. The joints moved during the early launch sequence as the orbiter engines and then the booster engines were ignited. This motion required the O-ring seals to be flexible and to reseat and reseal continuously during these movements.
5. The temperature the night before the launch dropped to 22°F and had only increased to about 28°F by the time of the launch.
6. The O-ring seals were not certified to operate below a temperature of 53°F by the contractor responsible for the solid rocket boosters.
7. Tests conducted by the contractor Morton Thiokol in 1985, six months before the accident, indicated that the O-ring seal was not effective at low temperatures. (At 50°F the O-rings would not expand and follow the movement in the joint during the period of operation of the booster.) In fact, at a normal temperature of 75°F, it required 2.4 seconds for the O-rings to reseat and effect the seal when the gap was opened.

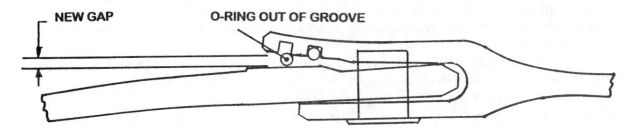

NEW GAP   **O-RING OUT OF GROOVE**

Fig. 17.4 Rotation of the joint due to cylinder pressurization opens the gap in the seal region.

The seven facts listed above show very clear evidence of two problems. First, the design of the seal in the circumferential joint was marginal and should have been fixed much earlier in the program. You do not look at a piece of burnt and eroded O-ring and walk away from the problem.

Second, the launch should have been postponed due to cold weather for several different reasons. The very low temperatures were well below the limits to which the system had been certified. The rubber in the O-rings becomes very stiff at these low temperatures and cannot respond quickly enough to accommodate joint movement. The O-rings were almost guaranteed to leak. Also, there was considerable ice on the rocket motors and the fuel tank. Pieces of this ice might damaged the rocket motors and the tiles on the heat shield of the orbiter when it fell off during the violent vibrations which always occur just after ignition but before lift-off.

17.9.5 Ignoring the Problem

Problems usually do not go away when we ignore them. They persist and sooner or later failure will occur. Although the seals had failed on 12 of the previous 24 flights, the failures were not catastrophic because the leaks were small and at locations which did not endanger the adjacent components. The boosters on the shuttle only operated for only a few minutes before they were cut loose to fall into the Atlantic for recovery at a later time. So those concerned with the launch schedule tolerated the small leaks for short periods of time. The trouble with the leak on the Challenger was that it was big, and the jet of hot gases issuing from the leak were pointed directly at the liquid hydrogen fuel tank. The jet of hot gases acted exactly like a cutting torch, and the huge fuel tank was penetrated within seconds.

The O-ring seals were inadequate and the joint needed to be redesigned and modified. In fact, during the development of the boosters in 1977 - 79, several memos were written by engineers at the Marshall Space Flight Center indicating their concern over the design of the joint. They recommended that the joint be redesigned because "the adequacy of the clevis joint was completely unacceptable." The suppliers of the O-rings indicated "that the O-ring was being required to perform beyond its intended design and that a different type of seal should be considered."

Apparently, there was a major breakdown in communication, and none of these memos and reports from the Marshall Space Flight Center was forwarded to Morton Thiokol, the designers and builders of the boosters. A problem, properly detected by engineering at the NASA Center in charge of monitoring the technical aspects of the development of the boosters, was not pursued to its logical conclusion. Instead of insisting on a redesign of the joint, Marshall Space Flight Center approved the boosters in September of 1980. The pressures of cost overruns and repeated schedule slippage sometimes make managers (and engineers) accept flawed designs. It is a very shortsighted practice and not in the best interest of safe design to do so.

Okay! Let's move on with the discussion. NASA accepted a booster rocket with inadequate seals, but the bird flew. On the first launch, with Columbia, there was no reported damage to the seals. The O-ring seals failed on the second flight. One of the O-rings was burnt with about 20% of the thickness of the ring vaporized. The ring failure did not affect the mission, but it was clear that the putty was not protecting the rings from the hot gasses. Marshall Space Flight Center reacted to the field experience by reclassifying the joint to a "Criticality 1 category." This was the official recognition on the part of NASA that a seal failure could result in "loss of mission, vehicle and crew due to metal erosion, burn through and probable case burst resulting in fire and deflagration." The failure of the O-rings by erosion continued with high frequency (50% of the missions), but NASA decided to accept the situation and did not recommend remedial action.

It appears that NASA decided to live with the seal problem at least until a new lighter booster motor could be designed with improved joints. These new boosters were to be ready about six months after the Challenger exploded. A very real lesson—in too-little-too-late.

17.9.6 Recognizing the Influence of Temperature

From the initial development phase of the boosters, the O-ring seals had been recognized as a problem. A problem that NASA classified as very serious, but one that they must have erroneously believed did not elevate the risk to unacceptable limits. In January of 1985, a year before the accident, the Challenger was launched on a cold day at a temperature of 51°F. An examination of the joints after the recovery of the boosters showed that a number of O-rings were severely damaged and evidence of extensive gas leakage was observed. The contractor, Morton Thiokol, and the monitor, Marshall Space Flight Center, reviewed the O-ring seal problems. During this review, Morton Thiokol noted that low temperatures were a contributing factor to the inability of the O-rings to seal properly. The condition was considered undesirable, but was acceptable. However, the O-ring problems persisted and NASA eventually placed a launch constraint on the space shuttle.

The launch constraint would prohibit launches until the problem had been resolved or reviewed in detail prior to each launch. Unfortunately, NASA routinely provided launch waivers for each subsequent flight to avoid delays in an unrealistic launching schedule. The problem of O-ring erosion was common, one or more joints on 50% of the launches exhibited significant erosion, and it had not caused any serious difficulty to date. The problem was expected and accepted as routine.

17.9.7 A Management Decision

The very cold weather on the night before the fatal launch was no surprise. The weatherman (or woman) was on the mark in predicting the temperature and the ice, which formed on the structures that night. The engineers at Morton Thiokol, in Utah, were very concerned about the effects of the very low temperatures on the ability of the O-rings to function properly. Teleconferences took place between the Morton Thiokol engineers and the NASA program managers. The engineers argued to postpone the launch until the temperature increased to at least 53°F.

The NASA program managers were very unhappy about the engineer's concerns. They did not want the extended delay required to wait for warmer weather. Senior vice presidents from Morton Thiokol sensed the displeasure from the customer and took over the decision process. Senior management decided to keep the customer happy and reversed the decision of the Vice President of Engineering not to launch at these very low temperatures. Senior management at Morton Thiokol signed off on the launch ignoring the fact that the temperature at launch time was expected to be 25°F lower than the lowest temperature which engineering believed the seals would be effective (53°F).

The senior management at Morton Thiokol caved in to client pressure. The engineers were unanimous in their opposition to launch. Senior management asked the engineers to prove that the O-ring seals would fail. Of course, they could not state with 100% certainty that the rings would fail. Failure analysis is in terms of probabilities. The risk had elevated as the temperature decreased, but the engineers could not prove conclusively that the seals would fail in a manner that would detrimentally affect the mission. Senior managers at Morton Thiokol and program managers at NASA concluded erroneously that the risks were low enough to proceed with the launch.

17.9.8 Approvals at the Top

NASA has a four-level approval procedure to control the launch of the shuttle. This sounds good, but for a multilevel approval process to be of any value, information must flow freely from the bottom of the organization to the top of the chain of command. The technical discussions concerning the ability of the O-rings to function at the very low temperatures took place between NASA administrators and engineers at level 4 and Morton Thiokol. Program managers at Marshall Space Flight Center, level 3 administrators, were also included in these discussions. While the discussion was extended, with clear polarization between engineering and management, not a word of these concerns was conveyed to the top two levels of management at NASA.

Approvals for the launch were given by the level 2 administrator for the National STS Program in Houston, TX, and the level 1 administrator at NASA's Headquarters in Washington D.C. These approvals were essentially automatic because the administrators in charge had been isolated. They were not privy to the management decision to fly with super cold O-ring seals. They were not informed of the engineering recommendation to postpone the launch and to wait until the temperature increased to at least 53°F.

The lesson here is clear—approvals by the very high level managers are worthwhile only if they are informed decisions. If complete information is not presented to the super bosses for their evaluation, then their approval is meaningless. These administrators are not knowledgeable, and, as such, they add no value in an informed decision making process.

17.10 SUMMARY

We have briefly described the degradation of ethical behavior during the past two or three generations and given some of the probable reasons for these changes. While there are many controversial issues that can be classified as "dilemma" ethics, there are several "basic" characteristics where the difference between right and wrong is clearly defined. The list of basic virtues include:

- Prudence
- Justice
- Fortitude
- Temperance
- Faith
- Hope
- Charity
- Humility
- Respect

We have laws written to prohibit, by punishment, a well-defined set of vices. These laws control behavior, but only the extremely repulsive actions on the part of individuals or corporations.

Individual behavior is important, but as we learned in the Challenger explosion, it is not sufficient. Corporations (business) must operate with the highest ethical standards. The chief executive officer (CEO) establishes the standards, and the lower level managers act to establish the corporate character.

Honor codes exist on some college campuses to guide ethical behavior (lying, cheating and stealing), and the students and the faculty respect them. Character development and honor codes go together in producing a meaningful educational experience. We list here some of the more important attributes in character development:

1. Excellence
2. Respect
3. Integrity
4. Decisiveness
5. Responsible
6. Temperate
7. Fortitude
8. Understanding

We treated professional ethics by introducing the Code of Ethics of Engineers that is sponsored by the Accreditation Board for Engineering and Technology (ABET). The code is short with four fundamental principles and seven basic canons.

Finally, we have discussed the fatal accident of the Challenger space shuttle, which exploded during the launch in January 1986. Many ethical issues dealing with individual and corporate behavior are apparent as we review the events leading to this fatal accident. We hope that this example will aid you in developing both your individual and professional character.

REFERENCES

1. Patterson, J. and P. Kim, The Day America Told the Truth: What People Really Believe about Everything that Really Matters, Prentice Hall, New York, 1991.
2. Sommers, C. H., "Teaching the Virtues," Public Interest, No. 111, Spring 1993, pp. 3-13.
3. Woodward, K. L. "What is Virtue," Newsweek, June 13, 1994.
4. George, R. P., Making Men Moral: Civil Liberties and Public Morality, Oxford University Press, NY, 1994.

5. Baker, D. F. "Ethical Issues and Decision Making in Business," Vital Speeches of the Day, 1993
6. United States Air Force Character Development Manual, USAF Academy, C0, December 1994.
7. Abramson, R., "A Matter of Honor," *Los Angeles Times*, April 3, 1994.
8. Lickona, T. A. Educating for Character, Bantam Book, New York, 1991.
9. Jensen, C., No Down Link: A Dramatic Narrative about the Challenger Accident, Farrar, Straus and Gitoux, New York, 1996.
10. *Time,* February 10, 1986
11. Lewis, R. S., The Voyages of Columbia: The First True Space Ship, Columbia University Press, New York, 1984.
12. *The New York Times* April 23, 1986.
13. Feynman, R. P. What Do You Care What Other People Think? Bantam, New York, 1988.
14. Report to the President by the Presidential Commission on the Space Shuttle Challenger Accident, Ayer Co., Salem, 1986.

EXERCISES

17.1   Our political leaders are often attacked by the media for lax ethical behavior. Please write a short paper describing three recent lapses of ethical behavior on the part of the leadership in either the State or Federal government. Did these officials break the law? Is it important that they did or did not break the law?

17.2   List what you consider poor behavior in a college classroom.

17.3   Have you ever cheated on an important exam in high school? What about a college exam? Do you know of someone who cheated? What was your attitude when you observed this cheating?

17.4   Is there an honor code at the University where you are pursuing your studies? Have you read it? Do you abide by the rules?

17.5   Write a paragraph or two explaining your position regarding:

- Prudence
- Justice
- Fortitude
- Temperance

17.6   Write a paragraph or two describing your feelings about:

- Faith
- Hope
- Charity

17.7   If you could write a law that would go on the books tomorrow and be strictly enforced, what behavior would it prohibit?

17.8   Why does the CEO of a corporation set the standard for ethical behavior? What are some of the issues that arise on a daily basis that develop corporate character? Is it possible for a corporation with tens of thousands of employees to develop a character like an individual?

17.9   Write a 400-word essay describing your character. Include a discussion of the weaknesses as well as the strengths of your character.

17.10   Examine the code of ethics for engineers given in Fig. 17.1, and describe your opinion of it. Can you live with the expectations if you become a professional engineer? Do you believe it is adequate, or should it be revised to reflect a more modern viewpoint of what is right and wrong?

17.11   Suppose that you were a lead engineer working for Morton Thiokol on the evening of January 27, 1986, involved in the discussion of the O-rings. What would you have done?

- Early in the teleconference.
- At the critical stage of the teleconference.
- After management had taken over the decision process.

There is no right or wrong answer to these questions. We want to place you in a professional dilemma. Someday you may be placed in a similar situation where there is a trade-off between safety and corporate business interests. Think about your response in advance and be prepared to deal with such a dilemma and its consequences.

CHAPTER 18

SUSTAINABLE ENGINEERING

18.1 ENVIRONMENTAL ISSUES

The environmental problem threatens our ability to sustain future generations with an improving or even constant standard of living. The environmental problem is difficult to define because it has many facets and many different players. In this country, societal concerns, societal consumption, corporate objectives, corporate behavior, local, state and federal agencies, political agendas, political structure, Congress, and the Executive exacerbate the problem. To put it simply—we have many different issues and many players on all sides of every issue. Let's begin by discussing an environmental assessment formulated by Paul Hawken in his well-recognized book—*The Ecology of Commerce* [1].

The fundamental issue is sustainability. Will the Earth provide the resources (food, materials, and energy) to support the population now and in the future? Others have raised this question much earlier in history. Thomas Malthus, in 1798, in *An Essay on the Principle of Population* [2], argued that the population tends to increase faster than the food supply. The population would starve unless war, famine and disease limited population growth. Today, this argument seems preposterous, at least in the U. S., because we produce enough food to more than feed our population and to export large quantities of meats and grains with less than 2% of our population engaged in agriculture. However, abundant food supplies were not always available. Henry Weaver [3] describes many civilizations and countries that failed because they could not feed their people. The Roman Empire collapsed in famine. French peasants were dying of hunger when Napoleon Bonaparte stormed Europe. As late as 1846, the Irish starved during the potato famine. Food is in abundant supply today in most places worldwide, even with significant increases in our population, because of marked advances in agricultural machinery and in plant and animal sciences. However, serious food shortages still exist in many localities. About one out of six people in the world live on the brink of starvation surviving on international handouts. Usually these food shortages are due to war, natural disasters, or political strife.

18.1.1 Finite Resources

Yet even with the abundance of the 1990s, many intelligent, concerned, and dedicated individuals raise the question of sustainability. The world population is growing exponentially. From 1800 to 1900 the population increased by about 600 million with an average growth rate of 6 million/year. From 1900 to 1950 the world population grew at a rate of 18 million/year. From 1950 to 1975 the growth rate increased to 60 million/year. Today the population growth rate is approaching 100 million/year and the world population exceeds 5.8 billion. As we produce more and more food from less and less land, the population appears to grow at a rate to ensure its consumption.

Another problem associated with an expanding population is urban growth. Current population trends indicate that by 2050, 6 billion people will be living in cities, which is three times more than today. The increase, 4 billion people, represents an average population increase of 80 million per year. To accommodate this urban population growth, will require enlarging existing megacities or developing new megacities. To grasp an idea of the size of this growth, 80 million people per year is equivalent to the formation of 10 new megacities with a population of 8 million each year for 50 years [4].

Consumption is another critical issue. The resources of the world are finite, and our understanding of the resource reserves is poor. Oil, gas and coalfields develop on a geological time scale (tens of millions of years per tick mark on a graph). The large corporations, supplying the world's gasoline and fuel oil requirements, will pump all but the very rich fields nearly dry in a decade or two. It is clear that the world will run out of fuel even without additional population growth. The only question is when the supplies of crude oil will begin to decrease[1].

The supply of nearly all high-grade ores, from which metals are smelted, is also limited. For example, during World War II, the reserves of high-grade iron ore in Minnesota were seriously depleted as iron and steel production was increased to provide war materials. Minnesota still provides over 70% of the iron ore for the blast furnaces in the U. S., but it is obtained by processing a low-grade ore known as taconite.

A final example of limited resources pertains to water supply. The Ogalala Aquifer, which underlies the Great Plains, is the largest body of fresh water in the U. S. Each year farmers pump 20 billion more gallons of water from this aquifer than is replaced by rainfall. It is predicted that the aquifer will become exhausted in thirty to forty years at the current rate of pumping. What will the farmers in the Great Plain states do when their irrigation wells run dry?

Why, knowing these constraints, do we continue to utilize resources at these very high rates of consumption?

18.2 CONSUMPTION AND WASTE

18.2.1 Municipal Solid Waste

The people of the world are consumers. This fact is implied by the annual production of $21 trillion of products and commodities. In the U. S., we are the champion consumers as we purchase nearly twice as many goods and services per person as any other people on the face of the Earth. A typical person in the U. S. consumes 136 pounds of resources per week. In addition, about a ton of wastes are generated for each person each week in the process of producing the goods and services consumed. All of this consumption not only depletes the Earth's finite resources, but also leads to significant quantities of wastes that must be disposed of without polluting the air or the ground water.

Most of us are familiar with municipal solid waste. Our wastes are picked up once or twice a week and trucked away to a landfill—someplace that we never want to see or smell. We are often encouraged by municipal governments to separate the waste into categories—newspapers, metal cans, glass, plastics, etc. We hope that these items are recycled because it is incumbent upon all of us to at least believe that we are good environmental citizens. But, what really happens to the glass, plastics or metal trash. Is it recycled or do we have two or three trucks carrying wastes (that we carefully sorted and separated) from our housing units to the local landfill? Check with your municipality to determine the effectiveness of its recycling program.

---

[1] There is some evidence that a few oil fields may be regenerating apparently supplied from very deep reservoirs. However, this regeneration capability has not been firmly established for even a single field.

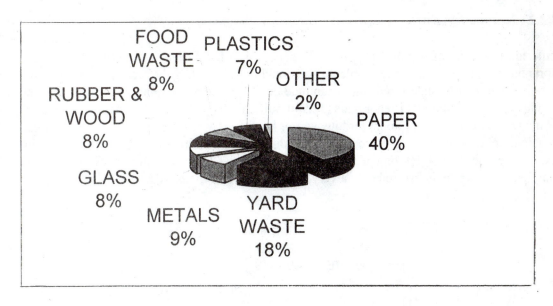

Fig. 18.1 Municipal solid waste generated in the U. S. in 1994. Annual total 180 million tons.

In 1994, we produced about 180 million tons of municipal solid waste in the U.S. The composition of this solid municipal waste is shown in Fig. 18.1. It is evident that most of these wastes consist of paper (40%) and yard trash (18%). The paper, mostly newsprint and junk mail, can be recycled if the paper mills are prepared to accept it, and if the market will tolerate a lower standard for brightness and lower strength for paper products produced with a large percentage of recycled newsprint. Many states have passed laws requiring newspaper publishers to employ newsprint that contains a specified percentage of recycled paper. In this instance, the law drives the demand for recycled newsprint that probably would not exist with normal market forces.

Most yard trash can be eliminated from landfills. First, grass clippings can be mulched with the lawn mower and dropped in place as the lawn is cut. Grass clippings should not be accepted in municipally collected wastes. Tree branches can be chipped when collected and converted to mulch. The mulch is of sufficient value for residents of the neighborhood to retrieve from conveniently located distribution stations.

Metals, glass and plastic are also recyclable, but is it economically feasible to collect sort, ship and process these types of wastes? Recycling aluminum cans is profitable because it takes 10% less energy to smelt scrap aluminum than to reduce bauxite to aluminum. Steel is also recyclable and about 25% of scrap steel has been employed in producing most of the common alloys of steel.

When it is less costly to ship many types of solid wastes to a landfill, the wastes with marginal value are usually discarded. Since tipping fees in the U. S. are still relatively low by world standards[2] at an average of $31 per ton in 1996, valuable resources are often discarded in landfills. Unfortunately, the economic arguments for tipping instead of recycling do not account for costs associated with cleaning up toxic wastes found in landfills, and in health hazards related to ground water pollution by fluids leaching from the wide variety of discarded products[3]. In the coming years, we must be more creative in producing a wide array of products from recycled materials. We will describe a few examples later in this chapter indicating successful approaches currently practiced by some companies for recycling materials into new products.

---

[2] Tipping fees at landfills in Japan range from $300 to $400 per ton.
[3] For example, a discarded television contains nearly a pound of lead, thousands of other chemicals and a vacuum tube that may implode.

18.2.2 National Wastes

While the problems of disposal of municipal solid waste have been well publicized, and recycling efforts have been organized, the problem of national waste has largely been ignored. Consequently, the population poorly understands the magnitude of the problem. National waste is mostly produced by industry, and the amount is enormous (12,900 million tons per year). National waste, as illustrated in Fig. 18.2, is classified as nonhazardous waste, hazardous waste, mining waste, oil and gas waste, municipal solid waste and other debris from industry and construction. Note that municipal solid waste makes up only 1.4% of the total national waste. Clearly the public, with its attention on municipal solid waste, is focused only a very small segment of the problem.

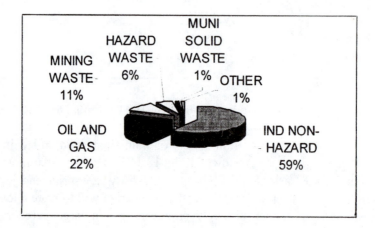

Fig. 18.2 Waste generated in 1994 in the U. S. The annual total is 12,915 million tons.

Industrial wastes are not usually shipped to municipally owned landfills; nevertheless, the refuse goes into what bureaucrats call land disposal units—some located above ground and others out of sight below ground. About 59% (7,600 tons per year) of this industrial waste is classified as nonhazardous, but even nonhazardous wastes often cause harm. Heavy metals, PCBs, pesticide containers, and organic compounds are dumped into these disposal sites. While classified as non-hazardous, these are not very friendly dumping sites.

18.2.3 Hazardous Waste

The hazardous waste sites are much worse than the industrial disposal sites. Unfortunately they are numerous—today there are 90,000 hazardous waste sites in the U. S. Moreover, 1,200 of them have been designated as priority cleanup sites under the Superfund law.

How do we dispose of hazardous waste? Generally, we try to enclose it by placement in a watertight container such as a barrel or tank. This procedure is effective for some period, but eventually the barrel begins to corrode, holes develop and the container begins to leak. At that time, the toxic wastes are released and they may enter a stream or the ground water. When possible a much more effective disposal solution is to chemically convert the toxic substance to a new chemical compound that is non-hazardous.

The amount of toxic waste produced each year by the chemical industry is huge. For example, in 1986, EPA's inventory of toxins released to the environment by the 50 largest chemical companies in the U. S. was an impressive 270 million tons. Suppose that the chemical industries incinerated[4] this waste instead of dumping it at its disposal sites. At the current rate for incineration of about $100/ton,

---

[4] Incineration as a disposal method also has serious deficiencies that will be discussed later.

the cost for burning these hazardous wastes totals $27 billion. This amount is larger by a factor of almost 10 than the entire profits accrued by the 50 large chemical companies in 1986 [1]. **The answer is not to dispose of the hazardous wastes by dumping or burning, but to reduce and eventually eliminate their production during the manufacture of chemical products.**

Reducing toxic waste is not a trivial matter. Syntex, a pharmaceutical company, with a manufacturing faculty in Boulder[5], Colorado was well recognized as the worst polluter in the city. The toxic release inventory (TRI), issued annually by the Environmental Protection Agency (EPA), identified Syntex as the worst offender in Boulder County and the second worst offender in the entire state of Colorado. The company was under intense public pressure to clean up its chemical processes. They responded with large investments in pollution prevention measures, a vigorous employee-training program, and a serious effort to communicate with the environmental organizations in Boulder. They agreed to reduce the total volume of air emissions by 50% over a four-year period. Clearly, progress can be made in reducing toxic wastes, but it is costly and significant reductions may require many years of dedicated efforts by concerned citizens and corporations.

18.2.4 Radioactive Waste

Radioactive waste is a much more significant problem than hazardous waste. Of course, both are hazardous but in different ways. For instance, dioxin is a very toxic chemical, which is stable and does not degrade. It accumulates in algae, insects and fish; eating these fish exposes one to very high levels of a known carcinogenic. Radioactive waste is less prevalent because there are fewer plants generating these wastes than there are chemical facilities producing toxic chemicals. However, while there are fewer sources of this contamination, they are not rare—most of us can identify a storage site for radioactive waste within our state. Also, the amount of radioactive waste stored at some of these facilities is immense. For instance, at the nuclear weapons research facility in Hanford, Washington, there are enough liquid wastes stored to form a lake 35-feet deep twenty-five square miles in size. Recent news reports indicate that some of the storage tanks containing these radioactive fluids are leaking and polluting the ground water. In time, this seepage will reach the Columbia River and the pollution will spread to large regions in both Washington and Oregon.

There are two significant problems in disposing of radioactive wastes. First, many of the radioactive elements have a half-life of thousands of years or more; thus, producing a long-term storage requirement that exceeds the period of recorded history. Second, a technique for permanent disposal of these wastes has not been developed. The current practice is to place the wastes in containers and maintain them in a retrievable storage facility. Spent reactor fuel rods are maintained in water pools at the reactor sites of nuclear power plants across the country. The longer-term plan is to store the containers housing the radioactive wastes in deep, geologically stable underground tunnels[6]. Unfortunately, no one wants to have this tunnel in his or her locality.

Recently[7] (March 26, 1999) the Department of Energy (DOE) opened the nation's first nuclear waste disposal facility in Carlsbad, NM. The $1.8 billion facility is intended for the permanent storage of radioactive material from 23 sites, which are located in 16 different states. The facility is composed of a series of storage rooms located off of access tunnels that are cut into an ancient salt dome about 2,500 feet underground. The facility will handle only radioactive wastes from DOE laboratories and

---

[5] Boulder has more environmentalists per unit area than any city in the U. S.

[6] Currently, the Department of Energy is exploring the feasibility of the Yucca Mountain Site in Nevada for long-term storage of highly radioactive materials. The site is remote—100 miles from the nearest sizable population center (Las Vegas). The climate is dry with less than six inches of rainfall per year and the water table is 800 to 1000 feet below the proposed repository. However, questions regarding the geological stability of the area remain.

[7] The opening of the facility followed nearly a quarter of a century of studies, protests, and lawsuits.

their contractors. Radioactive wastes from our utility plants and medical facilities are still located in temporary storage facilities.

18.2.5 Incineration

Incineration is often employed in many areas of the country where landfills are not readily available. The concept is simple—burn the trash to significantly reduce its volume. In some instances the incinerators are an integral part of a power unit with the heat from burning trash used to produce the steam that powers the turbines generating electricity, which is consumed by the local municipality. Unfortunately, the trash-to-electric power systems are rare, and even those systems generate a small fraction of the energy required to produce the solid waste consumed in the incinerator.

While incinerators reduce the volume and weight of refuse that eventually is shipped to a landfill by a municipality, it does so by an airborne redistribution process. A modern incinerator burning 1,000 tons of a typical municipal-solid-waste converts this waste to a number of different elements and chemical compounds as indicated in Table 18.1.

Some of the heavy metals are trapped in the fly ash; consequently, this fine sand like material must be encased with a plastic liner when it is initially disposed of in a landfill. Otherwise, the heavy metals will leach out and pollute the ground water. Even more serious are the gases and small particulate matter released to the atmosphere. The area over which these compounds are distributed depends on the height of the smokestack and the direction and velocity of the prevailing wind. With the large quantity of sulfur dioxide, hydrogen chloride and sulfuric acid released to the atmosphere, those unfortunate people residing downwind from the incinerator should anticipate acidic rainfall.

Table 18.1
Emissions from a Modern Incinerator
1,000 tons of Municipal Solid Waste [1]

Substance	Amount (tons)
Lead	2.22
Mercury	7.55
Cadmium	0.13
Nitrous Oxide	999
Sulfur Dioxide	379
Hydrogen Chloride	345
Sulfuric Acid	38.7
Fluorides	8.0
Small Particulate Matter	43.5
Fly Ash	300

18.3 WASTE GENERATING SYSTEMS

The public is anticipating a technological solution to the environmental problems, especially those dealing with the generation and disposal of wastes. Engineers today and certainly in the future will be called upon to improve ecoefficiency[8]. Strategies for realizing improvements include:

---

[8] The World Business Council for Sustainable Development has suggested that ecoefficiency is the delivery of competitively priced goods and services that satisfy human needs and bring quality of life and the progressive reduction of ecological impacts and resource intensity throughout the life cycle to a level commensurate with the Earth's estimated carrying capacity.

1. Minimizing emissions, increasing yields, reducing the generation of nonproductive material streams, using energy more efficiently, and substituting benign materials for hazardous materials.
2. Consider wastes as raw materials for other products.

Engineers are interested in examining processes as systems. With a systems approach, one examines the complete process—not a small portion of it. The advantage of systems analysis is that techniques for reducing or eliminating waste become more apparent. Let's begin a systems analysis by considering a linear waste model that is typical of a poorly designed process generating a significant amount of waste.

18.3.1 A Linear Waste Model

A linear waste model that depicts the generation of the maximum amount of waste and the consumption of the maximum amount of power per unit of production is illustrated with a block and arrow drawing in Fig. 18.3.

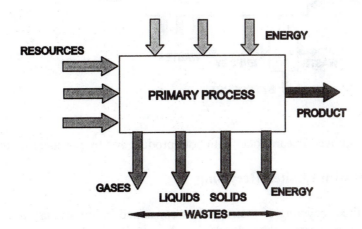

Fig. 18.3 A linear waste model that generates maximum waste without energy recovery.

The linear waste model is for a process that utilizes 100 percent virgin resources as input material. Energy in the form of electricity, steam and hot water is also provided as input to the primary production process. The resources and energy are consumed in the process as the product is produced. The product might be a lawn mower, plastic for soda bottles, or carpeting for floors.

The wastes created in the process are rejected and transported from the facility as gases, liquids, or solids. The gases are usually emitted from stacks (often tall) into the atmosphere. The small particulate matter and chemical compounds carried by the gas are distributed over a large area, which usually exceeds several hundred square miles. The liquid waste is usually carried in a pipeline that eventually dumps into a municipal sewerage line or perhaps directly into a stream or river. Have you ever noticed the large number of factories located on water front property? Have you wondered why?

The solid waste generated in the linear waste process is either incinerated or shipped directly to a disposal site as described previously.

Energy losses are also a form of waste. Steam exhausted to the atmosphere is not a pollutant since it is pure water. However, there is a significant amount of available energy in that steam, which is wasted when it is released to the atmosphere. It could be employed to heat the water used throughout the facility. If hot water is released into a stream or river, energy is again wasted. That hot water could be used to heat the building in the winter or for some other useful purpose.

An improved linear waste model is presented in Fig. 18.4. This model is similar to that shown in Fig. 18.3 except that some of the waste is utilized in a secondary process to produce a second product (a by-product). There are five benefits, which result from this improvement in the manufacturing process.

1. The resources input to the secondary process are obtained at essentially no cost.
2. The by-product has value and its distribution increases the company's profits.
3. Society benefits from a more effective use of resources and energy.
4. Less waste is generated and the impact to the environment is less severe.
5. The cost to the company for disposing of the waste is reduced.

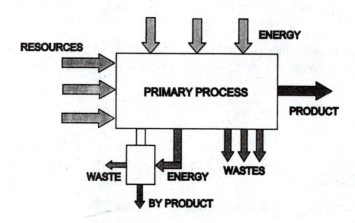

Fig. 18.4 An improved linear model with both product and by-product production.

18.3.2 A Waste Model with Limited Recycling

A waste model with limited recycling of the resources consumed in the primary process is shown in Fig. 18.5. The model demonstrates three fundamental methods that may be employed to reduce wastes and to reduce the consumption of virgin resources.

*   Utilize waste as a resource to create by-products of value to society.
*   Recycle waste internally, processing it as required, to convert it to the quality of a virgin resource.
*   Employ recycled materials obtained from external companies that recycle industrial and municipal wastes.

Let's consider some examples that demonstrate the savings accrued by employing each of these three waste reduction methods. Dupont, one of the largest chemical companies in the world, produces an enormous amount of waste. Responding to growing social pressures, Edgar S. Woolard, Dupont's CEO, initiated a broad and aggressive program in 1989 to reduce the quantity of waste generated in their production processes. One of their most successful approaches was to view chemical wastes as a production product looking for a market. They found two waste products generated in the primary production process for nylon—methylglutaronitrite and di-basic-acid. With some additional processing methylglutaronitrite is a suitable resource for the manufacture of epoxy coatings for concrete and for general-purpose epoxy adhesives. Di-basic-acid is converted into a strong solvent that is an excellent substitute for acetone and methylene chloride, which are hazardous solvents. In this case, wastes were converted into valuable by-products with ready markets, and the costs and environmental impact of disposal were eliminated.

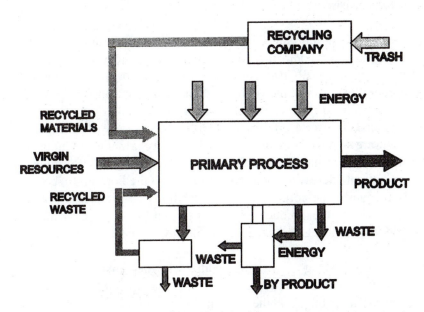

Fig. 18.5 A waste model with limited recycling.

Gillette, the razor blade company, saves money and reduces the negative effects of wastewater disposal on the local municipality and its environment by recycling internally generated wastewater. They installed a water recirculation system permitting them to recycle water used in the manufacturing process. They reduce the consumption of water from 730 million to 156 million gallons per year. The result was an annual reduction of $1.5 million in water and sewerage costs and a decrease of 574 million gallons of wastewater imposed on the municipal sewerage system. Gillette found that working to eliminate or reduce waste was not only environmentally friendly, but also provided increased profits.

Resources (materials) that feed primary or secondary processes need not be virgin materials. Materials that contain significant recycled content may often be employed. Most metals, such as aluminum, steel, lead, zinc and copper, contain a significant percentage of "scrap" without loss of performance. Both Ford Motor Company and Chrysler have changed the composition of several components installed in some of their automobiles to 100% recycled plastic and have made significant reductions in costs. The position taken by the automakers and most other manufacturers is to use recycled materials if they cost less and perform as well as virgin materials.

3M (Minnesota Mining and Manufacturing) was one of the first large corporations to address waste management by initiating their "Pollution Prevention Pays (3P)" program in 1975. Since then they have implemented hundreds of employee suggestions for reducing wastes and cutting emissions. From 1975 to 1990, they have documented saving in excess of $500 million and have reduced air emissions by 120,000 tons, solid wastes by 410,000 tons and waste water by one billion gallons [1]. The 3P program at 3M probably offers the best guide to other corporations for improving their manufacturing processes to reduce wastes while enhancing corporate profits.

It is clear to most high-level corporate managers that generating wastes is an inefficient utilization of resources, which increases costs, decreases profits and pollutes the environment. Environmental regulations, discussed in a later section of this Chapter, will force changes to improve both primary and secondary manufacturing processes. However, the recognition of the costs associated with the generation and disposal of wastes by corporate managers and employees will produce even more changes. Most corporate executives are moving from a compliance phase, where regulations were followed and met, to a leadership and persuasion phase, where managers and employees are encouraged to make changes to improve ecoefficiency.

18.3.3 Life Long Ownership

Life long ownership is a concept introduced by the town of Sanger, California and the court [1]. The town sued three large chemical companies for contaminating the town's water supply with the pesticide dibromochloropropane (DBCP). The court ruled in favor of the town and established the precedent that the corporations were responsible for their product even after the time of the sale. The town was awarded $15 million to clean up its water supply.

Corporate responsibility for a product was established much earlier in cases where products with inherent dangers were sold to the public. Product liability lawsuits are an every day occurrence that has alerted corporations to the costs associated with releasing unsafe products to the market. Since the class action lawsuit addressing the gasoline fires with the Ford Pinto, both the public and the corporations clearly recognize the corporation's responsibility to produce a product that is safe to operate for the entire life of the product. In the near future, it appears that the corporation's responsibility for producing products that are safe for the environment will be more firmly established by the courts.

The concept of life-long ownership addresses the basic environmental problem—responsibility for toxic wastes and pollution belong to the manufacturer and not the customer. The advantage of this approach is to place the responsibility and the cost of the disposal with the corporation making the product. With the cost of disposal added to the original cost of producing the product, there is a strong economic incentive for redesign of both the processes and the products to reduce life cycle costs[9].

While the concept of lifelong ownership may appear to be new and radical, some countries have already enacted laws requiring corporations to assume responsibility for disposal. In 1992 Japan passed legislation requiring durable goods manufacturers to establish resource recovery centers. In Belgium a new law requires the last owners to hand in end-of-life vehicles to licensed recovery centers or to new car dealers if they purchase a new car. The percentage of reused or recycled content in the vehicles is also mandated [5]. In Germany manufacturers and distributors are required to take back and to recycle product packaging. Customers are permitted to leave all packaging materials for the products they purchase in the retail outlet for recycling.

A few companies already recognize their environmental responsibilities and the potential profits in recycling or remanufacturing their own products. Again in Germany, the automaker BMW has established a disassembly plant to recycle most of the components in its scrapped autos. Also, the newer models are being designed to permit rapid disassembly. In Belgium, an agreement was reached for the auto dealers to take back old models for recycling [6]. In the U. S., Ford is opening a chain of used auto parts centers [7]. Ford's new president, Mr. J. Nasser, is interested in Ford being perceived as the world's most environmentally sensitive automaker. Ford has set a goal of increasing the amount of a junked car that is recycled from the current 80% to 95%.

In the U. S., Xerox—the copy machine manufacturer—has established an "Asset Recycle Management" program. Early in the development of the copy machine, Xerox recognized the potential health hazard associated with disposal of photoreceptor drums with a selenium and arsenic coating. Xerox accepted ownership of the discarded photoreceptor drums, and discovered that they could be remanufactured at a fraction of the cost of making new drums. Xerox has expanded their remanufacturing program and today they collect older models of copy machines, which are traded in when a customer purchases a newer more sophisticated model. The old machines are remanufactured by checking all of the components and replacing only those that are worn or those with a limited life expectancy. These remanufactured, but older models, are sold to customers in Central and South America or Eastern Europe where cost is more important than the latest new feature. Xerox believes in reusing serviceable components to reduce costs and their demand for virgin resources. The

---

[9] Life cycle costs include the costs of production, operation, maintenance, and disposal.

customer benefits from lower prices for new/remanufactured copiers and society benefits from the conservation of virgin resources and reduction of wastes imposed on the environment.

In some instances, two companies cooperate to reduce waste and produce a new product using waste as a raw material from one company in a product produced by another. Chaparral Steel produces steel and in the process, a waste product called slag, is generated. Texas Industries is a company producing Portland cement. In a joint research program, Chaparral Steel and Texas Industries developed a patented process that uses steel slag in a cement kiln to create high quality Portland cement [8]. The joint venture has increased profits for both companies, reduced waste disposal impact on the environment, cut energy consumption in producing cement, and reduced CO_2 emissions.

18.3.4 Waste Metrics

Corporate America understands that environmental issues have a direct impact on the bottom line; therefore affecting the way products are designed. When any factor affects the bottom line, corporate America seeks to devise a set of measurements for assessing the effectiveness of the corporation in managing this factor. Robert J. Eaton, the CEO of DaimlerChrysler, has described the characteristics that any metric for measuring ecoefficiency must exhibit [9].

- The metrics must meet the needs of all of the stakeholders—community, government, and business.
- They must facilitate innovation and growth with continuous improvement as a cornerstone.
- They must be harmonized at the local, state, national and international levels.
- They must be fully compatible with existing business systems. If they do not add value, they will not support continuous improvement and will not be used.
- They must measure the right things—what is measured is what gets managed.

Let's suppose that you have been assigned the task of reducing waste generated by a production facility by 45%. Management (your boss) asks you to prepare a quarterly report describing your activities, the costs of implementing your program, and periodic measurements of the progress. How do you propose measuring the progress made in reducing the waste generated in a production process? There are two approaches—one based on the weight of the waste and the other on costs. The latter is more important than the former, but more difficult to measure.

Let's consider a waste management index based on weight. This approach is relatively easy to implement because production control staff usually maintains careful records of unit production of primary products, by-products, and the weight of all materials employed. Solid waste generated is collected and weighed. Wastewater and other liquid emissions are measured with a flow meter. Measurements of gas emissions from vents and/or stacks yield the weight of various chemical compounds released to the atmosphere. The weight of all of these wastes are summed to give:

$$W_{waste} = W_{solids} + W_{liquids} + W_{gases} \qquad (18.1)$$

The weight of the total output from the production process is determined from:

$$W_{total} = W_{product} + W_{by\text{-}product} + W_{waste} \qquad (18.2)$$

The waste measurement index (WMI) is a ratio determined from:

$$WMI = (W_{waste}/W_{total}) \times 100 \qquad (18.3)$$

Both Eqs. (18.1) and (18.3) are useful in measuring the effectiveness of a waste reduction effort. The weight of the waste released to the environment given by Eq. (18.1) is of interest to the surrounding communities and to management. However, the weight of the waste is affected by the production rate. If the production process is expanded from eight hours per day five days a week to twelve hours per day six days a week, the weight of waste generated is probably going to increase even if your waste reduction efforts have been markedly successful. However, the waste reduction index, given in Eq. (18.3) is not affected by the production rate because it is a ratio. Using a ratio gives a much better indicator of progress in a waste reduction program when the production rate varies.

18.3.5 Cost Analysis of Waste

In a waste reduction program one of the first tasks is to identify all of the sources of waste. While identifying sources seems obvious, it is easy to overlook some sources such as cleaning solvents, packaging materials, lubricating oils and greases, scrapped equipment and tools, and other debris. The weight of these less obvious items of waste is measured and added to the weight of the waste as determined in Eq. (18.1). Finally, considering the following factors makes an estimate of the cost of the waste generated in a production process.

1. Cost of raw materials to replace the wasted resources.
2. Labor costs to remove and treat wastes.
3. Storage costs to hold wastes at the production facility prior to final disposal.
4. Shipping costs to load and transport wastes to a reclamation center or a disposal site.
5. Disposal fees at the disposal site or at an incinerator.
6. Intangible costs such as insurance fees, regulatory costs, or the cost of EPA and other local regulatory agency issued pollution permits.

A cost analysis is essential for a waste reduction program because the expense of dealing with waste is much larger than most managers appreciate. Also, the EPA imposed regulations require extensive reports describing the quantities of waste generated and the methods used in disposing of these wastes. Preparing these reports requires time and effort. Regulatory costs, which exceeded $115 billion or 2.1 % of the gross domestic product (GPD) in 1990, are estimated to increase to 2.8% of the GPD by the year 2,000 [10].

A second benefit of the cost analysis of waste is in the identification of high cost waste streams. Maximum cost savings are realized when investments are made in equipment to reduce high cost waste. Employee training can be focused on elimination of these specific wastes. Research expenditures may be justified in creating a by-product employing these wastes.

18.4 AIR QUALITY

Perhaps the environmental issue that is most widely discussed in the news media is air quality. Stories of the effect of acid rain on the foliage of the North Eastern states are common. The pollution of air over many of our major cities such as Los Angeles, San Diego, Denver, and Washington is a major concern to large segments of our population. The air quality index is reported daily in the weather section of our newspapers. This concern lead Congress to pass the 1990 Clean Air Act, which enables the U. S. Environmental Protection Agency (EPA) to establish limits on the amount of various pollutants allowed in the air anywhere in the U. S.

18.4.1 The 1990 Clean Air Act

The EPA has established two sets of limits on the amount of specified pollutants in the atmosphere—one set, called the primary standard, is to protect health. The other set, called the secondary standard, is to protect the environment. If one of the several EPA designated geographic regions in the U. S. exceeds these limits, it is classified as a non-attainment region; today about 90 million Americans live in non-attainment regions [11] indicating that much remains to be accomplished in reducing air pollution.

The 1990 Clean Air Act (CAA) vests the EPA with significant regulatory and enforcement power. The EPA requires industry, state, and local governments to periodically report on emissions to the atmosphere. CAA gives the EPA teeth and claws; they can fine violators about as easily as a police officer can write a traffic ticket. The fines are large and prison sentences[10] even for first offenders are common. CAA defines seven categories of pollution offenders—one misdemeanor and six felonies. Moreover, the EPA and the Federal Bureau of investigation (FBI) have a memorandum of understanding. The EPA can call on the services of nearly 10,000 FBI field agents in 56 field offices to investigate potentially responsible parties. Clearly, cleaning up the air is intended to be a very serious business.

The EPA has established deadlines for the states, local government and industry to reduce air pollution. States are required to develop implementation plans outlining procedures for reducing specified pollutants such as carbon monoxide, sulfur dioxide, nitrogen oxides, particulate matter, etc. If in the opinion of the EPA, the state's implementation plan is deficient, the EPA has the power to enforce its mandates in that state.

The 1990 Clean Air Act is long and complicated. In this Section we will briefly describe select aspects of the Act to demonstrate the magnitude of the effort currently underway to improve air quality and the environment in general. Today, it is estimated that industry spends more than $150 billion each year complying with regulations imposed on them by the EPA [10]. The costs of reducing air pollution are significant and as the EPA reduces the allowable limits in the near future[11], these costs will increase dramatically. The regulations and the burden of reports required by the CAA are costly and confining. In addition, industry must also deal with several other environmental laws including:

1. The Clean Water Act.
2. The Comprehensive Environmental Response, Compensation and Liability Act.
3. The Hazardous Materials Transportation Act.
4. The Resource Conservation and Recovery Act.
5. The Safe Drinking Water Act.
6. The Superfund Amendments and Reauthorization Act.
7. The Toxic Substances Control Act.
8. The Endangered Species Act.

Since each Act is the basis for a long stream of Federal and State regulations, the number of regulations limiting corporate activities has increased markedly in recent years. Regulations clearly affect business—compliance takes time, effort and significant capital investment. Indeed, in 1991 the

---

[10] Of the 65 criminal cases the EPA referred to the Justice Department in 1991, 55 convictions were obtained in the courts. The sentences resulted in a total of 62 years of imprisonment or an average of 1.8 years for each individual found guilty. This sentence is equivalent to the time given to those convicted of first time possession of heroin [11].

[11] In July of 1997, the EPA issued new regulations requiring states to dramatically reduce the amount of ozone and microscopic soot that they may allow in the air. In a recent lawsuit by a number of industrial groups, a three-judge panel of the U. S. Court of Appeals, ruled that the EPA failed to justify the pollution levels selected as minimum requirements to protect public health. The ruling is being appealed by the Justice Department.

Pharmaceutical Manufacturers Association stated "If the current trend continues, environmental regulation, not FDA[12] approval, may cause the greatest delay in new drug introduction. The time to deal with regulatory agencies is also a major factor in the design of automobiles. Eaton [9] has stated "easily half of the three-year product introduction cycle can be consumed by acquiring air emission permits." Construction or modification of facilities need for producing the new design may not be initiated until a permit is issued. Clearly, dealing with the bureaucrats not well trained in technology or familiar with the principles of continuous improvement, is an issue.

18.4.2 Auto and Truck Emissions

Much of our air pollution comes from three primary sources—auto and truck emissions, coal burning power plants, and the exhausts of large and small businesses. Let's consider the pollution generated by the internal combustion engines used to power our autos and trucks.

When we burn gasoline or diesel fuel in an internal combustion engine, the chemical reaction produces hydrocarbon emissions that pollute the atmosphere. In some cities with high-density traffic and confining hills or mountains, smog forms which is always irritating and sometimes dangerous. Reduction of hydrocarbon emissions has been addressed with considerable success using two different approaches. The first was to install catalytic converters on all new autos manufactured after 1971. These converters were very effective and reduced emissions by 95%. Unfortunately, the smog did not disappear; we added 50 million more autos to the national fleet since introducing the catalytic converters; we drive each vehicle more now than in 1971; and many of these new vehicles are pick-up trucks and sport utility vehicles[13] (SUVs) which consume copious quantities of gasoline. We mitigated the pollution problem with the converters, but it was not eliminated, and continues to grow with increased population and affluence.

A second, more costly approach is to reformulate the gasoline so that it releases a lower concentration of hydrocarbons in the combustion process. Makower [12] has estimated that the cost of reformulating gasoline to meet the emissions standards imposed in California will be in excess of $10,000 per ton. This cost will escalate in a few years when more stringent standards are adopted. Catalytic converters, with an estimated cost of $1,000 per ton, represent a much more cost-effective method for reducing hydrocarbon emissions than gasoline reformulation.

There are other novel approaches to reducing emissions from the internal combustion engine. In 1990 Unocal Corporation, in an effort to reduce pollution in Southern California, purchased many old clunkers (pre 1971 autos without catalytic converters) from their owners for a flat price of $700. For less than $6 million, they removed 8,376 high-polluting autos from the roads and placed them in recycling centers (junkyards). Unocal environmental engineers estimate that this action was equivalent to removing 150,000 new autos from the road. The costs were relatively low at about $940 per ton of hydrocarbon emissions[14] (12.8 million pounds) removed from the atmosphere.

A very simple approach to reducing hydrocarbon emissions from trucks and autos is to reduce the number of miles driven and/or the amount of gasoline and diesel fuel consumed. This can be accomplished with one of the fundamental laws in economics—the law of supply and demand. Gasoline and diesel fuel is incredibly cheap in the U. S. when compared to the costs in Europe. The difference is about 400% (roughly one dollar in the U. S. and $4 to $5 in Europe. The tax policy in Europe is to limit consumption of gasoline and the number of miles driven. In the U. S., our low taxes

---

[12] The very long time required for approval for introducing new drugs by the Federal Drug Administration (FDA) is well recognized by the population in general and in particular by those anxiously waiting for a drug known to be effective that has been available in Europe for several years.

[13] It is interesting to observe the high percentage of SUVs driven by individuals who claim to strongly support environmental issues.

[14] At the same time Unocal was working to reduce hydrocarbon emissions from its refinery. Their engineers estimated that removing the equivalent weight of emissions from the refinery would have cost $160 million.

on gasoline and diesel fuel encourage consumption, unnecessary driving, and the purchase of gas guzzling vehicles. We will discuss the effect of government policies on environmental issues later in Section 18.6.

18.4.3 Power Plant Emissions

Many of the utility companies in the U. S. burn coal in the production of electricity that is distributed to our homes and businesses. They consume an enormous quantity of coal; unfortunately, this coal is not a pure hydrocarbon. It contains small quantities of many different elements. Perhaps the most troublesome of these is sulfur. When sulfur is involved in the combustion process it reacts with oxygen to create sulfur dioxide (SO_2). Scrubbers installed in the smokestacks of the utilities to remove the sulfur from the flue gases are effective in reducing, but not eliminating, the release of SO_2 to the atmosphere.

The emissions from the smokestacks of our utilities vary from one plant to another. Some utilities burn natural gas or oil with much lower sulfur content; consequently, they do not contribute significantly to the quantity of SO_2 in the atmosphere. Other utilities burn coal with a high sulfur content that overburdens the pollution control equipment installed in their smokestacks. Clearly in 1990, compliance with the provisions of the CAA was not uniform.

To address the problem of non-uniform compliance, the CAA enabled the EPA to issue permits permitting the coal burning utilities to release 2.5 pounds of SO_2 for each million of BTU[15] of heat generated. The idea in issuing permits is to have the market drive the efforts to reduce SO_2 emissions. If a utility releases less than the allotted amount of SO_2, it may sell its permits to a company that has difficulty in meeting the limits established. This practice rewards those utilities that have achieved early compliance and penalizes those not yet in compliance.

The Tennessee Valley Authority (TVA) [16] in 1995 made one of the first purchases of a pollution permit. This agency purchased the right to emit 10,000 tons of SO_2 from the Wisconsin Power and Light. The cost was nearly $3 million or about $300 per ton. The TVA basically purchased the time needed to procure and install more efficient scrubbers, to convert to a coal with lower sulfur content, or to convert their boilers to burn oil or natural gas.

The CAA established standards to reduce SO_2 emissions by 50% by the year 2001. The costs to the utilities for new pollution control equipment to achieve this level of compliance have been estimated at $4 to $7 billion [12]. The pollution permit concept establishes a market that places a price on pollution. It provides an economic incentive for the utilities to purchase pollution control equipment to reduce their emissions of damaging and obnoxious gases.

18.5 THE KYOTO PROTOCOL CARBON EMISSIONS GOALS

The Kyoto Protocol was developed at the International Kyoto Climate Change Conference held in Japan in early December 1997. The objective of the conference was to accelerate the pace of international efforts under the United Nations Framework Convention on Climate Change. After intense negotiations, officials from 160 countries reached agreement on a legally binding Protocol under which industrialized countries would reduce their collective emissions of six greenhouse gases 5.2% below their 1990 levels by 2008-2012. For the U. S. these reductions are even more severe—a 7% reduction. The American Society for Mechanical Engineers (ASME)[17] has prepared a position

---

[15] The British Thermal Unit (BTU) is a measure of heat.

[16] The TVA is an agency of the Federal Government producing electrical power from hydroelectric dams, nuclear power plants, and coal fired plants.

[17] The 125,000-member ASME is a worldwide engineering society that deals with technical, educational, and research issues. It operates an extensive publishing operation, sponsors many technical conferences and professional development courses each year, and establishes many industrial and manufacturing standards.

paper on the technical implications of the Kyoto Protocol [13]. While the ASME does not take a position on the need to achieve the reductions specified in the Protocol, it has evaluated the feasibility of achieving these reductions in the U. S. We will briefly discuss some of the issues raised in this position paper.

In 1990 the carbon emissions in the U. S. were about 1,346 million metric tons[18] (MMT); however, under a business-as-usual scenario carbon emissions are expected to increase by about 34% to 1803 MMT by 2010. To meet the goals of the Protocol, we must reduce these anticipated carbon emissions by 551 MMT to 1252 MMT by 2010.

Is this a realistic goal knowing the propensity of Americans to consume fuel? Does technology exist to reduce the amount of carbon employed in our combustion processes?

The major sources of greenhouse gases are transportation, industry, buildings, and electric power generation. Let's consider these sources individually, and summarize the findings of the ASME technical committee regarding each of them.

18.5.1 Transportation Sector

In a life-goes-on-as-usual scenario, energy consumption for transportation is expected to increase by about 2.4% per year until 2010. This estimate is based on expectations of continued low prices for fuel and for higher disposable personal income. With more money and cheap gas, people will buy more powerful and larger vehicles in increasing numbers and vehicle-miles driven will increase by an estimated 1.5% per year. At the same time, engine improvements will increase efficiency—but only by about 0.2% per year. The business expansion is anticipated to continue and the growth of the economy will result in an increase in freight transport with an attendant increase in the use of diesel fuel by 1.4% per year. Also, air travel will grow during this period causing jet fuel usage to increase by 2.8% per year.

All segments of the transportation sector indicate a relatively large increase in the consumption of carbon. By 2010 the transportation sector is expected to generate 628 MMT of carbon emissions, which is an increase of 196 MMT above the 1990 level. These estimates consider technology, economic growth, and human behavior. It is apparent that the carbon consumption for transportation is expected to increase at a high rate—not decrease in accordance with the Kyoto Protocol.

The transportation sector presents significant opportunities for advanced technology to reduce carbon emissions. Some possible areas for improved efficiencies or lower emissions are listed below:

1. Engine improvements—the direct-injection, stratified charge gasoline engine and turbocharged direct-injection diesel engine should provide efficiency improvements of 15 to 30%.
2. Hybrid electric vehicles—with a combined electric drive, an auxiliary power unit and an energy storage device offer opportunities for reductions of carbon emissions of up to 50%.
3. Advanced fuel cells—proton-exchange-membrane fuel cells that operate at relatively low temperature (180-300 °F) would offer significant reductions in emissions. Unfortunately, fuel cell development is in the research stage and may not become commercially feasible in the near future.
4. Alternative fuels—fuels such as compressed natural gas or propane would reduce carbon emissions by 10 to 20%. More effective would be a bio-fuel such as ethanol or methanol produced from feedstock. The uncertainties with this approach are the increased cost of alcohol relative to gasoline and the availability of sufficient quantities of feedstock materials to produce the large quantities needed.

---

[18] A metric ton is a unit of mass equal to 1000 kilograms. When converted to a force on Earth, it is 2205 pounds.

5. A high-speed integrated light rail system—operating at 200 to 300 MPH— magnetically levitated or with steel wheel-rail cars. This system would substitute electricity for gasoline or jet fuel while relieving both air traffic and highway congestion.
6. An effective information system—providing up-to-date information on traffic congestion and alternative routes would save time and reduce fuel consumption.

18.5.2 Industrial Sector

Small increases in carbon emissions are expected in the industrial sector in the coming decade. The trend of energy usage in this sector is toward electrification and a shift to energy produced by electric utilities. Some larger industrial installations will continue to use coal to generate electrical power and to provide heat for manufacturing processes. Increasing use of cogeneration, where the steam from turbine driven generators is utilized as heat for production processes, will reduce carbon consumption. The challenge in cogeneration is not technology, but the necessity for coordination between industry, business, and municipal governments. Another major area where reductions of carbon emissions are possible is by improving production processes and eliminating wastes as discussed previously in Section 18.3.

18.5.3 Building Sector

The carbon consumption in the building sector is for heating, air conditioning, lighting and other building services. The building sector accounts for about 1/3 of the total carbon emissions. With the business-as-usual scenario, the energy usage is expected to continue to grow with carbon emissions increasing to 612 MMT by 2010. This represents an increase of 33% from the 1990 levels—not a 7% reduction. Some of the methods to reduce carbon consumption in the building sector are listed below:

1. Advanced insulation technology—utilizing gas filled and evacuated panels with much higher R-values for construction.
2. Smart climate control systems—capable of adjusting temperatures automatically to accommodate actual usage of each room in a building 24 hours a day seven days a week.
3. Smart occupancy based light systems—capable of adjusting light intensity with voice-activated controls.
4. Advanced lighting systems—include more efficient light sources, efficient dimmable power supplies, and highly reflective and efficient optical designs.
5. Cogeneration—small natural gas fired turbines to provide both the hot water and electricity required for individual facilities with improved efficiencies.

18.5.4 Electric Power Sector

The electric power sector emitted 477 MMT of carbon in 1990 and is expected to emit 663 MMT by the year 2010, which represents growth in demand for electricity of 1.25% per year. The fuels used to generate the power in 1995 were coal (54.3%), nuclear (21.8%), hydro (10.4%), natural gas (10.4%), and oil (2.1%). The prospect for reducing carbon emissions in the electric power sector is not promising. A significant fraction of the power produced is from nuclear plants, and many of these plants are nearing the end of their licensed life. If these plants are not relicensed, it is unlikely that electrical supply will satisfy the growing demand. Moreover, as the nuclear plants are phased out, it will be necessary to replace them with new plants fueled with either coal or natural gas. The consequence is even higher carbon emissions than predicted.

More than half of the electricity generated in the U. S. is from burning coal. We could convert these coal-burning units to natural gas; however, the demand for natural gas will increase its price significantly and seriously deplete our limited reserves. A more rational approach includes inspection and relicensing all of the nuclear plants deemed safe to operate, development of advanced gas and coal cycles, and development of advanced nuclear plants. It appears that it will be necessary for society to decide among the following alternatives.

- Agree to accept severe reductions in consumption of both gasoline and electricity.
- Agree to the construction and operation of additional nuclear power plants[19].
- Agree not to conform to the limits set by the Kyoto Protocol.

Much has been written about the use of renewable energy technologies, but the hope of generating significant amounts of power from renewable sources is dim. Hydropower represents the best potential but the environmental permits required for adding turbines to existing dams will be difficult to obtain. Indeed, dams at potential hydro sites are scheduled for removal because they are harmful to fish. The application of solar energy suffers from its limited availability and the relatively low temperatures that can be generated without focusing devices. The potential for biomass in generation of electricity is limited by the availability of farmland needed for energy crop production. Wind energy is limited by available sites and by availability of cost-effective, energy-storage methods.

18.5.5 Discussion

The limits imposed on carbon consumption by the Kyoto Protocol will be difficult if not impossible to achieve unless significant changes are made in government policy and the attitudes of our population relative to conservation. Achievement of the goals will require several strategies to be implemented in an extremely short time[20]. Since actions by our population and Federal, State and Local governments are not swift, there is little reason for optimism.

The transportation and electric power sectors afford the best opportunities for improvement. Some of the strategies for reduction of carbon emissions include:

- A carbon tax to reduce consumption.
- Relicensing nuclear plants to extend their safe life.
- Replace coal-burning plants with state-of-the-art natural gas combined cycle plants.
- Expansion of hydropower-generating capacity.
- In the long term, construction of new safe nuclear power plants.
- Development of electric vehicles with fuel cell power sources.
- Development of an effective information system that significantly reduces traffic congestion.
- Development of an integrated air transport and high-speed rail system.

To achieve sustainability, environmental impacts will have to be reduced and a balance between growth and the environment established. There are three approaches to accomplish this balance [14]:

1. Stabilize population.
2. Decrease wealth and affluence.
3. Apply existing technology to the fullest extent and develop new methods for producing goods and services with minimal impact on the environment and the resource reserves.

---

[19] Implicit in this agreement is the acceptance of one or more disposal sites for nuclear wastes.
[20] It should be recognized that the time to construct a new power plant today is about ten years with much of that time consumed in obtaining permits and dealing with environmental constraints.

Population stabilization does not appear to be feasible because of different political and cultural values among the populations of different countries. Decreasing wealth and hence consumption by design is also an unlikely event. People will not voluntarily reduce their standard of living. Only a crisis of major magnitude will result in reduced wealth and consumption without political upheaval. Indeed, the wisdom of this approach for achieving environmental objectives should be questioned. It is a well-known fact that birth rates decrease when standards of living and educational levels are raised.

By a process of elimination, we find technology the most suitable approach for achieving the balance between economic growth and the environment. As engineers, we must develop sustainable processes and systems that create, deliver and manage high quality goods and services.

18.6 GOVERNMENT POLICIES

The policies of the governments—federal, state and local—markedly affect the way we live and act. During World War II, in a state of emergency, the Federal government suspended production of almost all durable equipment including automobiles, rationed food, severely restricted gasoline consumption, and drafted into the armed services nearly every physically fit man between the ages of 18 and 35. Clearly, the governments have the power to markedly change the way we live and the amount of resources we consume. On the other hand, we have the power to vote the rascals out of office.

In times of peace, the governments affect the way we live using more subtle methods. Tax policy is one of the most important and effective government tools. While taxes are levied to raise money to conduct the affairs of governments, these taxes also encourage or discourage certain activities by the population. For example, deducting interest cost for home mortgages encourages home ownership. Similarly taxing income derived from interest on bonds, certificates of deposit, etc. discourages savings.

The tax system at all levels of government is difficult to understand. By taxing— income, business profits, capital gains, interest from savings, and sales—governmental tax policy discourages many good activities, such as new investment and new business creation, which lead to a robust economy and prosperity. Moreover, the annual filing of the income tax forms for the Internal Revenue Service (IRS) has become a costly headache of major proportions. It is estimated that $0.65 is expended on compliance for every $1.00 collected by the IRS [15]. Perhaps the federal government should lead the way in eliminating waste by markedly reducing the costly and wasteful process employed in the collection of taxes.

18.6.1 A New Approach to Taxation

Paul Hawken [1] has outlined a new approach for governments to follow in writing tax laws. Instead of taxing the good things like income, capital gains, interest on savings, etc, the government should tax the bad things like utilization of virgin resources, emissions, disposal of wastes, and consumption of products and services. Of course, people and the governments need time to adjust to a new system; hence, Hawken proposes that the shift from one system to the other take place over a twenty-five year period. Also the shift would be revenue neutral. For every dollar collected with a green fee or green tax, taxes on the good things like income would be reduced by a dollar.

Expenditures by the Federal government for the fiscal year 2,000 are estimated at $1.7 trillion. While this is a lot of money, it is only 20% of the Gross Domestic Product (GDP) that is estimated at nearly $8.5 trillion. Transforming the tax system over a twenty-five year period would imply changes of less than one percent of the GDP per year. This rate of change is very slow particularly when compared to drastic actions taken by the federal government during the crisis of World War II.

The advantages of taxing actions that deplete our resources and degrade our environment are numerous. Green taxes would be included in the price a customer pays for consuming products and services. The taxes provide motivation for companies to improve their processes and the design of their product lines to reduce these costs. The consumer will recognize the added cost of the green tax and have the opportunity to modify their buying habits to avoid these charges. This approach gives the customer and corporation strong incentives for minimizing green taxes. In doing so, society will save virgin resources, reduce pollution, and lower carbon consumption.

18.6.2 Gasoline Tax

A barrel (42 gallons) of light sweet crude oil is priced on the New York market at about $30.00[21] per barrel or 71 cents per gallon. Shipping, refining and distribution costs including profit add about another 25 to 35 cents. Current state and federal taxes combined average about 34 cents per gallon taxes; thus, unleaded regular gasoline is sold at about $1.40 to $1.70 per gallon in most regions of the U. S. In other countries, the tax on gasoline is much higher—$2 to $3 per gallon. Americans love cheap gasoline; they buy large and heavy vehicles that consume large quantities of fuel, and they drive to the exercise facility. The problem is the high rate of consumption of finite virgin resources of petroleum and the emissions of smog producing hydrocarbons that result from this tax policy.

Currently, about 125 billion gallons of gasoline is consumed each year in the U. S. Suppose taxes on this gasoline were increased from the average of 34 cents to $2.00. The increased tax collections would amount to $207 billion per year if driving habits did not change. Increased gasoline taxes should markedly decrease consumption, conserving valuable reserves of petroleum and significantly decreasing pollution and carbon emissions.

Howls of protest were heard when Ross Perot, a presidential candidate in 1992, suggested increasing the gasoline tax ten cents a gallon each year for five years to raise the revenue needed to eliminate the deficit that existed then in the federal budget. People complained that it was not fair to those living in rural areas because they had to drive longer distances than those living in urban regions do. This may be true, but those living in rural areas have the opportunity to increase the prices of the products from their farms and ranches to better reflect the energy consumed in their production. Trucking firms complained that they would have to increase freight rates. Increasing freight rates is appropriate because the cost of a virgin resource (petroleum) would be more accurately assessed. Indeed, higher freight rates by the trucking industry would lead to increasing use of both rail and water transportation which are much more energy efficient methods of transport. It would also reduce the number of trucks on the interstate highways and reduce highway fatalities.

Paul Hawken [1] argues persuasively that "by imposition of incremental and eventually large green fees, businesses are positively encouraged not merely to meet regulations, but to embrace them, to exceed them, because the lower the green fees, the lower their costs".

18.6.3 Carbon Tax

We have already discussed the Kyoto Protocol, which constituted a worldwide effort to address the environmental problem of excess emissions of CO_2. The U. S. is responsible for a large share of this problem. Today about 25 billion tons of CO_2 is released worldwide to the atmosphere each year. We are responsible for almost 25% of this total while supporting only 4% of the world's population.

A careful study of uses of carbon in the U. S. by the ASME cast doubt on the ability of the U. S. to meet the 7% reduction on carbon emissions proposed by the protocol. Perhaps a better approach

---

[21] The price of crude oil on the world market varies from day to day. In the past two years, the price of crude oil varied from about $10 to nearly $35 per barrel. The cost of crude oil has increased sharply since OPEC members began to curtail production and OPEC members abided by their quotas. Retail prices in some regions of the country do not always reflect true costs, and prices above the national average are common.

to reducing CO_2 emissions would be to implement a tax on carbon consumption. In the U. S. we waste significant amounts of energy. Industries in Japan and Europe operate with much more efficiency using only about half the energy per unit output than comparable industries in the U. S. With significantly higher costs for energy in Japan and Europe there is much more incentive to strive for more efficient processes and practices.

Technologies for significant reductions in energy consumption exist, but often are not fully implemented. Lighting is an excellent example of a newly developed technology[22] waiting for full implementation. The Electric Power Research Institute (EPRI) estimates that if highly efficient lighting were fully implemented nationwide, electricity required for lighting would be reduced by 50% [10]. Aggregate demand for electricity would be reduced by 10%, emissions of both CO_2 and SO_2 would be reduced and fuel would be conserved. The problem is that only a small portion of our population has taken advantage of these new lamps. Green taxes imposed on consumption of electricity would motivate both industry and ordinary households to convert to more efficient lighting.

Reducing energy consumption often requires a capital investment. With our very low cost of energy, which does not adequately assess external environmental costs, management and homeowners often decide to continue wasteful energy practices. With green fees or taxes, the external environmental costs would be reflected in higher costs of energy and individuals and companies would be motivated to invest in capital improvements to reduce energy consumption.

As green taxes and fees are increased, it is important to reduce taxes on income, profits, and capital gains. The changes to the tax system should be revenue neutral. Legal tax avoidance provides incentives promoting investment in capital equipment to reduce energy consumption.

18.7 ENGINEERING RESPONSIBILITIES

Environmental issues are currently a serious concern to the engineering profession and will become even more significant in the coming decades. As the finite limits of resources becomes better understood by all concerned and the effects of carbon emissions on global warming become more apparent, engineering efforts to improve efficiencies and to design more environmentally friendly products and processes will intensify.

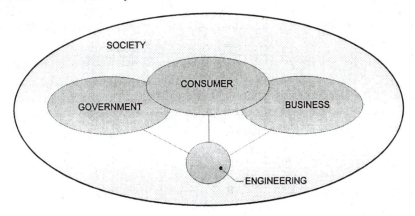

Fig. 18.6 Three constituencies served by engineers.

In their efforts to design environmentally better systems, engineers serve three different constituencies—industry, governments and society as illustrated in Fig. 18.6. Unfortunately the signals received from each of the constituencies are mixed. Society clearly wants clean air and water, yet they show no desire to conserve. They buy bigger and bigger autos and drive more and more

---

[22] A 20-watt electronic compact fluorescent bulb provides light comparable to a 75-watt incandescent bulb and has a life of 10,000 hours.

miles. They spend every dime they earn. The government establishes the EPA and passes laws placing limits on pollutants released to the atmosphere. Yet the government pursues tax policies that encourage consumption and discourage conservation. It is like the Surgeon General warning of the dangers of smoking tobacco while the Agriculture Department continues to subsidize farmers growing tobacco. The attitude and policies followed by industry is not uniform. Some companies like 3M have vigorously pursued environmentally friendly ways of doing business for the past thirty years. Other companies are still in the denial or avoidance stage; they do the minimum required by government-imposed regulations to avoid penalties.

As a new engineer beginning the first assignment in your career it is important to carefully assess the position of the company relative to environmental issues. Perhaps they recognize the importance of eliminating wastes, making the production processes more efficient, designing lightweight energy efficient products with a long life. On the other hand, they may largely ignore environmental issues and focus on making profits without much concern for the environment. It is also important to assess the attitude of your manager. Without the boss on your side, giving his or her approval, you will not make significant progress on any ideas you have for changing the company's practices.

The actions you take in the first month or two on the job are very important because you establish your reputation. Do not try to become perfectly green. If you become known as a tree hugger, any attempts to initiate programs for eliminating wastes or making processes more efficient will not be given the attention they deserve. Observation is in order for the first few months to learn about the processes used and the company's attitudes and practices. When you understand the implications of waste generated in the processes and apparent inefficiencies, you can make suggestions for reducing wastes or improving the efficiency of a process. These suggestions should be accompanied with an analysis of the costs to indicate the saving accomplished by reducing wastes. If the suggestion involves investment in capital equipment, the cost analysis should incorporate the interest expenses associated with this equipment. There are significant economic returns associated with the elimination of wastes. The costs of disposal of wastes have escalated in recent years and new assessments of old waste problems may indicate more profitable solutions that are environmentally friendly.

Perhaps your first opportunity as an engineer will be in designing a new product or process or the modification of one or the other. In either case, the design process should include careful consideration of the environment. While we design for easy manufacturing, assembly and maintenance, it is important to add the environment to this list. Good environmental design is inherently good for business. If the use of virgin resources is minimized, the costs of producing a product are reduced. If the wastes are recycled there will be additional cost benefits. If the production process is made more energy efficient, costs are again reduced. If paper or plastics used in packaging the product are minimized, it costs less. There is a consistency between good environmental design, lower cost, and increased profits. Well-managed corporations recognize this consistency.

18.8 SUMMARY

We have described the fundamental environmental issue—sustainability. We cannot predict when shortages in virgin resources will develop, but we are certain that many of our resources are finite. Today the generation and disposal of wastes is one of our most severe environmental problems. It is serious for two reasons—we waste valuable virgin resources and we pollute the environment.

Waste has been categorized into several different types and current methods for either storing or disposing of these wastes have been briefly described. The particularly acute problems associated with handling both hazardous and radioactive wastes are discussed. Many methods for handling wastes are known, but the best solution is to eliminate waste. Several waste models have been introduced beginning with the most wasteful (the linear model), and evolving to the concept of life long ownership. Laws, court actions, public attitude, and the recognition of corporate responsibility will eventually force significant reductions in waste generation and the utilization of virgin resources. Methods for measuring progress of waste reduction programs based on both cost and weight of the waste generated have been outlined.

Congress has, in the past two decades, enacted several different environmental laws that constrain the activities of corporations operating in the U. S. We have briefly described some aspects of the 1990 Clean Air Act (CAA) because it is the most widely discussed of the many different environmental acts. The EPA is empowered to enforce the CAA. Offenders are usually charged with felonies, and large fines and significant prison sentences may result after a conviction. Corporations often consider the environmental acts a severe constraint; however, there is a growing tendency for the more progressive companies to recognize public attitudes, the intent of the law, and the opportunities for cost reductions by eliminating wastes. We discussed methods used in the auto and truck industries for reducing hydrocarbon emissions. We have also described the concept of issuing permits for SO_2 emissions for electric utilities, which uses market forces to promote the installation of pollution control equipment by assessing a cost for producing a ton of pollution.

The Kyoto Protocol is an international effort aimed at reducing the amount of CO_2 released in the atmosphere. The concept is to relieve the greenhouse effect that is believed to produce global warming. The Protocol limits the carbon emissions from each country in the world. This basically amounts to restrictions placed on the industrialized countries since they are responsible for most of the carbon emissions. The results of an ASME position paper describing the impact of the Protocol on energy utilization for transportation, industry, buildings and electrical power are presented. Strategies for reducing carbon emissions in the future are listed.

Paul Hawken's ideas about tax policies by the federal government have been described. Conversion of a tax structure from one that imposes penalties on the good things we do to one that taxes the bad things is presented. A green tax, imposed on the consumption of gasoline and electrical energy, is discussed. The economic incentives to reduce wastes and pollution and to conserve virgin resources with a new green tax policy are described.

Finally, we describe the dilemma that you will face upon entering the engineering profession. You will serve three different constituencies and the message each provides is not consistent. We recommend that you proceed carefully and avoid being classified by management and peers as an environmental extremist. With time and a favorable corporate policy regarding the environment, you will recognize opportunities for reducing wastes, designing lightweight efficient products, and conserving energy. Suggestions are presented for pursuing engineering activities associated with improving the environment and conserving resources.

REFERENCES

1. Hawken, P., The Ecology of Commerce: A Declaration of Sustainability, HarperBusiness, 1993.
2. Malthus, T. R., An Essay on the Principal of Population, 1798.
3. Weaver, G. R., The Mainspring of Human Progress, 13th Edition, Foundation for Economic Education, Irvington-on-Hudson, New York, NY, 1984
4. Frosch, R. A., "Editorial," The Bridge, Vol. 29, No. 1, Spring, 1999.
5. Makower, J., Beyond the Bottom Line: Putting Social Responsibility to Work for Your Business and the World, Simon & Schuster, New York, NY, 1994.
6. Anon, Business and the Environment, Vol. IX, No. 11, November 1998.
7. Akre, B. S., Ford Opening Auto Parts Recycling Centers, Knoxville, News-Sentinel, May 1, 1999.
8. Forward, G. and A. Mangan, "By-Product Synergy," The Bridge, Vol. 29, No. 1, Spring, 1999.
9. Eaton, R. J. "Getting the Most Out of Environmental Ethics," The Bridge, Vol. 29, No. 1, Spring, 1999.

10. Denton, D. K., Enviro-Management: How Smart Companies Turn Environmental Costs into Profits, Prentice Hall, Englewood Cliffs, NJ, 1994.

11. Anon, Cleaning up air pollution: the programs of the 1990 Clean Air Act, EPA web site, http://www.epa.gov/oar/oaps.html.

12. Makower, J., The e Factor: The Bottom Line Approach to Environmental Responsible Business, Tilden Press, Times Books, Random House, New York, NY, 1993.

13. Anon, "Technology Implications for the U. S. of the Kyoto Protocol Emission Goals," Global Climate Change Task Force of the Council on Engineering and Council on Public Affairs, ASME, New York, NY, December, 1998.

14. Richards, D. J., "Harnessing Ingenuity for Sustainable Outcomes," *The Bridge*, Vol. 29, No. 1, Spring, 1999.

15. Payne, J. L., "Costly Returns," Institute for Contemporary Studies, San Francisco CA, 1993.

EXERCISES

18.1 Prepare a graph showing the world population from 1,750 to 2,000. Fit an exponential equation ($P = Ae^{Bt}$) to this curve of population versus time. Finally, predict the world population in the years 2020, 2050 and 2100. Comment on sustainability of the population in the future. Note P is the population, t time, and A and B constants in the equation shown in parenthesis.

18.2 Prepare a list of items you consume per week and estimate the weight of these items. Compare the total weight with the average of 136 pounds per week consumed by a typical American.

18.3 Where is the landfill that is used by the municipality in which you live? How many miles is the municipal solid waste shipped before being disposed of in this landfill? Is it possible to see or smell the landfill? What are the tipping fees at this landfill? What are the charges per month for the disposal service that takes care of the waste you and your family generates?

18.4 Does the municipal government in which you reside collect yard waste? Do they chip tree limbs and branches to make mulch that is available to the citizens of the municipality?

18.5 Does the municipal government in your hometown require that you sort and separate the wastes from your home? If so determine the final disposition of the newsprint, glass containers, steel and aluminum cans, and plastics.

18.6 Identify the three corporations in your home state that are the major sources of toxic wastes. Hint, the EPA issues a Toxic Release Inventory (TRI) annually that may be accessed on their web site—http://www.epa.gov.

18.7 Where is the location of the closest site for the temporary storage of radioactive wastes relative to the university that you are attending?

18.8 Where is the location of the closest site for permanent storage of radioactive waste relative to your hometown?

18.9 Where is the location of the closest incinerator that burns:

- Municipal wastes;
- Hazardous wastes;
- Low-level radioactive wastes?

18.10 The data in Table 18.1 indicates that incineration of 1,000 tons of municipal solid waste results in an output in excess of 2,000 tons. Please explain the doubling of the waste product in the incineration process.

18.11   Prepare an engineering brief describing a linear waste model. Include in the brief an outline of a plan to reduce the wastes generated in the manufacturing process.

18.12   Cite the benefits of a waste model of a process that utilizes wastes as the primary resource for a by-product.

18.13   If you were the vice president of engineering, what policy would you adopt for the use of recycled materials in your product?

18.14   Do you believe engineers should redesign products so that recycled materials with lower performance characteristics than virgin materials would be suitable?

18.15   Search the literature and write a 200-word paper describing a successful program to reduce waste that was implemented by a corporation.

18.16   Write an engineering brief describing life long ownership of a product by the corporation, which produces that product.

18.17   Write a paper describing the implications of life long ownership of a product and the possible benefits of life long ownership to society and the environment.

18.18   Write an engineering brief describing a new automobile that would be manufactured by the Forever Durable Corporation.   Understand that this corporation has a cradle to grave philosophy and assumes ownership of the automobile over its entire life.

18.19   You have been in charge of a waste reduction program for two years.   Data collected from the process control staff over this period are summarized in the table shown below.

PERIOD	W_{solids} (tons)	$W_{liquids}$ (tons)	W_{gases} (tons)	$W_{product}$ (tons)	$W_{by-product}$ (tons)
1ST QTR	500	2,000	10	1,000	0
2ND QTR	600	2,400	12	1,200	0
3RD QTR	550	2,400	12	1,200	50
4TH QTR	600	2,800	11	1,400	100
5TH QTR	600	1,700	13	1,700	150
6TH QTR	800	2,000	15	2,000	200
7TH QTR	600	1,500	11	1,500	150
8TH QTR	725	1,800	13	1,800	175

Determine the waste management index for each of the eight quarters.   Also determine the total waste generated each quarter.   Write an engineering brief assessing the success or lack thereof of your waste reduction program.

18.20   Write a short paper describing the effectiveness of the Clean Air Act of 1990.   In the paper give reasons for your opinions.

18.21   In your hometown or city, do you have any problems with air quality?   What is the air quality index on a good day and on a bad day?   Where do you find the air quality index?

18.22   Is the formulation of gasoline sold in your region of the country changed from time to time to improve air quality?   If so does the price increase or decrease when the formulation is changed to improve air quality?   Does the performance of your auto as measured by the miles per gallon improve or degrade with the new formulation?

18.23   Consider one of the recent environmental laws enacted by Congress, and write a paper describing its effect on industry located in your hometown.   A list of these laws is given in Section 18.4.1.

18.24   How many gallons of gasoline do you consume per year when driving your automobile? Would you be willing to reduce this consumption?   If so by what percentage?

18.25   Would you be in favor of gasoline rationing to improve the air quality in select cities?

18.26   Would you be in favor of gasoline rationing to save a valuable finite resource?

18.27   Review the usage of electricity at the College you are attending. Prepare a list of suggestions for reducing the consumption of electricity. The new practices that you suggest should not interfere with the educational process.

18.28   What is the name of the utility company in your hometown? What fuel is used to fire the boilers in this utility? Are they considered as a significant polluter in your locality? Is it necessary for them to purchase SO_2 permits?

18.29   Write a paper describing both the advantages and the disadvantages of the law, which enables the EPA to issue pollution permits.

18.30   Describe the Kyoto carbon emission goals. State your assessment of the ability of the U. S. to meet these goals.

18.31   Prepare a list of improvements in the transportation sector that will result in a reduction of carbon emissions.

18.32   Prepare a list of improvements in the building sector that will result in a reduction of carbon emissions.

18.33   Prepare a list of improvements in the electric power sector that will result in a reduction of carbon emissions.

18.34   Do you concur with Paul Hawken's philosophy that tax policies of the state and federal governments are misdirected? Please state the arguments for your conclusion.

18.35   Do you believe green taxes and fees in place of income taxes will result in the conservation of resources? Please state the arguments for your conclusion.

18.36   Upon graduation with a B. S. in engineering you begin to work for a typical corporation at a salary of $40,000 per year. Determine your federal and state income taxes assuming that you are single.

18.37   Upon graduation with a B. S. in engineering you begin to work for a typical corporation at a salary of $40,000 per year. Determine your federal and state income taxes—assume that you are married with no children and your spouse's income is equal to yours. Have you ever heard of the marriage penalty? Determine this penalty for the assumed situation.

18.38   Would you prefer to pay income taxes or green taxes? Do you believe the politicians will give you a choice?

18.39   Write a paper describing your position relative to environmental issues when you begin your career in industry.

18.40   In a recent article[23] Arthur Robinson and Noah Robinson, chemists at the Oregon Institute of Science and Medicine, wrote that global warming, which is the primary reason for the Kyoto treaty, is a long-term trend that began about 300 years ago. In the article they show a graph of temperatures over the past 3,000 years that indicates our current temperature is about one degree Fahrenheit below the average temperature over this period. They also show that in medieval times (1,000 AD), the temperature was about 3 degrees higher than the current temperatures. These conditions produced what is considered an optimum climate. Prepare a paper stating your position with respect to the Kyoto treaty that considers the information in this article.

18.41   Discuss with your team methods that you plan to employ in gaining access to the article referenced in Exercise 18.40.

---

[23] See article titled "Global Warming is 300-Year-Old News," The Wall Street Journal, January 18, 2000.

APPENDIX A

GUIDES AND FORMS FOR DESIGN TEAMS

Team Assessment Form

TEAM MEMBERS' NAMES						
SCORES	1 -10	1 -10	1 -10	1 -10	1 -10	1 -10
LEADING ROLES						
1. Initiate						
2. Provide						
3. Seek						
4. Clarify						
5. Test						
6. Summarize						
SUPPORTING ROLES						
1. Monitor						
2. Encourage						
3. Compromise						
4. Sense						
5. Harmonize						
6. Test						
HINDERING ROLES						
1. Uncooperative						
2. Degrading						
3. Dominating						
4. Avoiding						
5. Withdrawing						
6. Distracting						
TOTAL SCORE						

TEAM MEMBERS' AGREEMENT

- The information discussed by our team members will remain confidential.
- We will acknowledge problems and deal with them.
- We will be supportive rather than judgmental.
- We will respect differences.
- We will provide responses directly and openly in a timely fashion.
- We will provide information that is specific and focused on the task and not on personalities.
- We will be open, but will respect the right of privacy.
- We will not discount the ideas of others.
- We are each responsible for the success of the team experience.
- The team has all of the resources needed to solve the problems that arise during the development process.
- Every member of the team will contribute to its success.
- We will try to know and understand our fellow team members so as to identify ways for enhancing everyone's professional development.
- We recognize the importance of the team meeting to the success of the group, and accordingly we agree to:
- Always attend the scheduled team meetings.
- When attendance is impossible, notify the team leader and as many other members as possible in advance.
- Use meeting time wisely
- Start on time.
- Limit our breaks and return from them on time.
- Keep focused on our goals.
- Avoid side issues, personality conflicts and hidden agenda.
- Share responsibility for briefing members when they miss a meeting.
- Avoid making phone calls or engaging in side conversations that interrupt the team.
- _____.
- _____.
- _____.
- _____.
- _____.
-

SIGNATURES:

_____   _____

_____   _____

_____   _____

Date_____

Agenda
Weekly Meeting
Date _____ Time _____

Team Member Responsible

1. Weekly status report   _____

2. Review of outstanding action items

   - Action item #_____   _____
   - Action item #_____   _____
   - Action item #_____   _____
   - Action item #_____   _____
   - Action item #_____   _____

3. Report on progress

   - Subsystem _____   _____
   - Subsystem _____   _____
   - Subsystem _____   _____
   - Subsystem _____   _____
   - Subsystem _____   _____

4. Identify new problems

   - Problem #_____   _____
   - Problem #_____   _____
   - Problem #_____   _____
   - Problem #_____   _____
   - Problem #_____   _____

5. Assignment of action items

   - Action item #_____   _____
   - Action item #_____   _____
   - Action item #_____   _____
   - Action item #_____   _____
   - Action item #_____   _____

6. New business
7. Summary
8. Adjourn

Action Item Record

Team Meeting of _____

Decisions made	Follow-up required	Responsible member	Date complete
1.	1.	1.	1.
2.	2.	2.	2.
3.	3.	3.	3.
4.	4.	4.	4.

Action at next meeting	Preparation required	Responsible member	Date complete
1.	1.	1.	1.
2.	2.	2.	2.
3.	3.	3.	3.
4.	4.	4.	4.

Team Meeting of _____

Decisions made	Follow-up required	Responsible member	Date complete
1.	1.	1.	1.
2.	2.	2.	2.
3.	3.	3.	3.
4.	4.	4.	4.

Action at next meeting	Preparation required	Responsible member	Date complete
1.	1.	1.	1.
2.	2.	2.	2.
3.	3.	3.	3.
4.	4.	4.	4.

Team Meeting of _____

Decisions made	Follow-up required	Responsible member	Date complete
1.	1.	1.	1.
2.	2.	2.	2.
3.	3.	3.	3.
4.	4.	4.	4.

Action at next meeting	Preparation required	Responsible member	Date complete
1.	1.	1.	1.
2.	2.	2.	2.
3.	3.	3.	3.
4.	4.	4.	4.

CONCEPT SELECTION
PUGH CHART

CRITERIA	CONCEPTS			
	B	C	D	A
				DATUM

S - CONCEPT EQUAL TO DATUM

+ CONCEPT SUPERIOR TO DATUM

− CONCEPT INFERIOR TO DATUM